META-ANALYSIS
FOR EXPLANATION

META-ANALYSIS FOR EXPLANATION

A Casebook

Thomas D. Cook
Harris Cooper
David S. Cordray
Heidi Hartmann
Larry V. Hedges
Richard J. Light
Thomas A. Louis
Frederick Mosteller

RUSSELL SAGE FOUNDATION   NEW YORK

The Russell Sage Foundation

The Russell Sage Foundation, one of the oldest of America's general purpose foundations, was established in 1907 by Mrs. Margaret Olivia Sage for "the improvement of social and living conditions in the United States." The Foundation seeks to fulfill this mandate by fostering the development and dissemination of knowledge about the political, social, and economic problems of America.

The Board of Trustees is responsible for oversight and the general policies of the Foundation, while administrative direction of the program and staff is vested in the President, assisted by the officers and staff. The President bears final responsibility for the decision to publish a manuscript as a Russell Sage Foundation book. In reaching a judgment on the competence, accuracy, and objectivity of each study, the President is advised by the staff and selected expert readers. The conclusions and interpretations in Russell Sage Foundation publications are those of the authors and not of the Foundation, its Trustees, or its staff. Publication by the Foundation, therefore, does not imply endorsement of the contents of the study.

Library of Congress Cataloging-in-Publication Data

Cook, Thomas, 1941–
    Meta-analysis for explanation : a casebook / Thomas Cook.
      p.   cm.
    Includes bibliographical references and index.
    ISBN 0-87154-220-X
    ISBN 0-87154-228-5 (pbk)
    1. Social sciences—Research—Evaluation. 2. Meta-analysis.
    I. Title.
H62.C5856   1992   91-17999
300'.72—dc20

First Papercover Edition 1994

RUSSELL SAGE FOUNDATION
112 East 64th Street, New York, New York 10021

10 9 8 7 6 5 4 3 2

Contents

Foreword

The practical value of social science depends upon its ability to deliver usable knowledge about the causes of social problems and the effectiveness of policies and programs designed to alleviate them. The immense diversity of social life, however, and the great welter of factors underlying any social phenomenon make it difficult, if not impossible, to derive conclusive knowledge from any single study, no matter how well designed or intelligently analyzed. The causal process that appears so essential at one time or place may prove less important in another. The program that works well with one group under certain conditions may be less effective with another group when the circumstances are a bit different.

These basic facts of social life render the success of social science crucially dependent upon its ability to accumulate results across the many studies of a given social process or program. The accumulation of results and the gradual convergence on information of higher quality is one hallmark of progress in any science, but it is particularly key in social science, where there may be no single, uniform answer to a given question, but rather a family of answers, related by principles that emerge only over the course of much research.

Traditionally, this process of distilling reliable generalizations from the history of research on a given problem has been considered something of an art. Experience, good judgment, and a sound understanding of methodological principles should enable a seasoned scientist to make useful sense of a related set of studies, but with no guarantee that a similarly experienced analyst would reach a similar conclusion. This potential for disagreement among the experts—often realized in spades for certain socially important issues—has undoubtedly weakened social science as a source of social policy. If social scientists cannot agree, then how can policymakers be guided by their results?

The relatively new and still developing practice of "meta-analysis" holds out a partial solution to this dilemma. By developing a clear set of methodological guidelines for reviewing prior research and using statistical principles to summarize the results of previous studies, meta-analysis offers the possibility of making the process of reviewing a research literature more a science than an art. The aim of meta-analysis is to discipline research synthesis by the same methodological standards that apply to primary research. This goal implies that research reviews should be just as replicable as any other piece of scientific work. Two scientists applying the principles of meta-analysis to a given research literature should arrive at similar conclusions. And if not, their differences should be traceable to explicit analytic choices that can be independently assessed. Disagreement among the experts should become more a matter of method than opinion.

Meta-analytic practice has grown steadily since its introduction into social science in the mid-1970s, and important statistical research greatly refined and strengthened meta-analytic techniques in the 1980s. Nevertheless, we remain a long way from achieving the ideals of scientific research reviewing. Much of the technical power of meta-analytic methodology has yet to be put to regular use in social science; nor has the potential policy impact of these methods been fully realized. A good deal of what passes for meta-analysis is simple vote-counting across studies, and even when effect sizes are computed, much current meta-analysis fails to go beyond simple assessment of main effects. The power of meta-analytic techniques to evaluate the conditions under which effects occur and to explore the mediating processes that may underlie those effects is rarely exploited.

In response to this state of affairs, the Russell Sage Foundation initiated a program of support in 1987 designed to improve the state of the art in research synthesis by encouraging more effective use of statistical techniques for summarizing research in social science. This volume is the second of several book projects that aim to advance this objective.

The first of these volumes, *The Future of Meta-Analysis*, published by the Foundation in 1990, originated in a conference set up by the Committee on National Statistics of the National Research Council to assess the role of meta-analytic methods in current practice, their applicability to policy-relevant research, and their prospects for improved use in the future. Among its many concerns, this volume explicitly addressed the issue of replicability by reviewing an exercise in which multiple meta-analysts were commissioned to synthesize the same research literature (with encouragingly convergent results).

The current volume is less conventional, both in its organization and its provenance, and as its title suggests, it explicitly confronts the difficult question of how to use meta-analytic techniques to address issues of explanation. One of the central activities pursued under the Russell Sage program in research synthesis has been a series of case studies illustrating innovative uses of meta-analysis. To develop these cases, the Foundation invited a small number of practitioners of meta-analysis to propose research synthesis projects of their choosing in which the questions under study involved particularly challenging use of meta-analytic techniques. The Foundation assembled a distinguished advisory committee to select the most promising of these proposals and to assist in developing these case studies by consulting on a continuing basis with the meta-analysts as they pursued their projects. We wagered that this consulting process would provide the expert advice needed to make the cases under study truly illustrative of the best current practice in meta-analysis, and we suspected that our committee of experts would also learn a good deal in the process about the practical problems of making the techniques of meta-analysis work "on the ground."

This book provides the results of this two-sided learning experiment. The core chapters of the volume present a selection of four cases, each chosen to illustrate the problems of using meta-analysis to go beyond the simple description and estimation of effects and to begin to address the problems of explaining why, or under what conditions, a given effect can be observed. These efforts to use "meta-analysis for explanation" confront challenging technical issues and sometimes break new methodological ground. To give these methodological and strategic issues a full airing, we asked our meta-analysts to provide a running commentary on their own decision process as they made the analytic choices that guided their work. The resulting chapters provide a much more revealing description of the process of doing meta-analysis (and doing it well) than can ordinarily be found in journals.

Surrounding the case studies are four chapters written by the members of the advisory committee. These chapters begin with a brief introduction to meta-analysis and the problems involved in using the approach for explanation. Following the four cases, the committee members consider the general problems of conducting state-of-the-art meta-analysis and illustrate these problems by referring to particular issues that arise in the four case studies. The book then concludes with an interesting discussion of what the committee members learned—about using meta-analysis for explanation, about the strengths and weaknesses of meta-analytic methods when used for this purpose, and about the potential relevance for policy of this new brand of research synthesis.

We believe that the cases and the expert commentary provided in this unique volume constitute more than just a tour de force. Indeed, we hope that this exercise will be as useful to others as it has been for ourselves. By supplying well-worked examples of what meta-analysis can accomplish and by discussing the specific issues that must be addressed in conducting these sophisticated applications of meta-analytic techniques, we have tried to provide a useful entry point for students and an effective learning tool for anyone interested in realizing the full potential of this powerful methodology in social science.

ERIC WANNER
President, Russell Sage Foundation

Annotated Bibliography of Meta-Analytic Books and Journal Issues

Richard Light
Frederick Mosteller

Chalmers, I.; M. Enkin; and M. J. N. C. Keirse, eds.
1989 *Effective Care in Pregnancy and Childbirth.* New York: Oxford University Press.

In a landmark two-volume work in medicine, the editors collected all randomized clinical trials in pregnancy and childbirth and organized the field according to topic. They persuaded many teams of physicians to analyze the appropriate trials in a standard meta-analytic manner so that recommendations for practice were based on documented evidence—perhaps 1,000 meta-analyses in all. In four appendixes, they list 171 topics where the evidence seems firm for the recommendation and 63 where the treatment seems promising but more data are needed; 146 where information is needed but not available; and 61 instances where practices sometimes used in the past should definitely be abandoned. This work gives the details of the meta-analyses that form the basis of the evidence. In a less research-oriented paperback (Enkin, M.; M. J. N. C. Keirse; and I. Chalmers, eds. *A Guide to Effective Care in Pregnancy and Childbirth.* New York: Oxford University Press, 1989), they provide recommendations for practitioners and others without the specifics of the meta-analyses.

Cooper, Harris M.
1989 *Integrating Research: A Guide for Literature Reviews.* Beverly Hills, CA: Sage.

This book is particularly strong in presenting the detailed steps that any meta-analyst must take when organizing an analysis. Careful descriptions explain how to decide which subgroup of a large population of studies to include; how to gather a full set of studies; how to search existing literature; and how to assess research quality. The book focuses, especially in Chapters 4 and 5, on what might go wrong in a meta-analysis and how to avoid pitfalls and traps.

Glass, Gene V.; B. McGaw; and M. L. Smith
   1981   *Meta-Analysis in Social Research.* Beverly Hills, CA: Sage.

This was the first book to give concrete suggestions for doing a meta-analysis. The authors describe in detail how they and their colleagues developed some early conceptions of meta-analysis. They present many formulas for converting certain outcome measures into other, equivalent measures. They delve deeply into one example—a meta-analysis of outcomes from psychotherapy, using this case to illustrate various calculations and to explain how to interpret the resulting findings.

Hedges, Larry V., and Ingram Olkin
   1985   *Statistical Methods for Meta-Analysis.* Orlando, FL: Academic Press, 1985.

This book gives the most mathematically sophisticated treatment of meta-analysis currently available. It emphasizes the importance of a meta-analyst's understanding of the underlying model assumed for a meta-analysis. It also explains how to examine a group of outcomes to see whether their variability is roughly what we would expect due to chance, or whether the variability among outcomes exceeds chance expectations. It exploits more fully than do other works the concept of effect size to bring more studies into a meta-analysis.

Hunter, John E., and Frank L. Schmidt
   1990   *Methods of Meta-Analysis: Correcting Error and Bias in Research Findings.* Newbury Park, CA: Sage.

This work treats problems produced by the designs of the original studies and offers various methods of statistical analysis needed for meta-analysis. It reviews the steps needed in the search, analysis, interpretation, and reporting, with some attention as well to computing. Illustrations are chosen from industrial organizational psychology and the organizational behavior literature.

Light, Richard J., ed.
   1983   *Evaluation Studies Review Annual, Vol. 8.* Beverly Hills, CA: Sage.

This edited collection contains many published meta-analyses from such fields as education, social science, medicine, and policy. It provides the reader with a quick introduction through a choice of examples, some now famous, together with separate discussions of selected meta-analytic issues by well-known scientists.

Light, Richard J., and David B. Pillemer
   1984   *Summing Up: The Science of Reviewing Research.* Cambridge, MA: Harvard University Press.

This book is written at an introductory level. The authors emphasize simple techniques for beginning a meta-analysis; they stress the fact that a precise formulation of a question for meta-analysis must drive any decision about what procedures to use. This book suggests that qualitative information can often strengthen a meta-analysis, and it gives several simple graphical techniques that are easy to implement. The authors also discuss several meta-analyses that have affected social policy.

Rosenthal, Robert
    1984   *Meta-analytic Procedures for Social Research.* Beverly Hills, CA: Sage.

A full overview of actual procedures used in meta-analysis is provided in this book. Many examples illustrate the calculations. Chapter 4 gives a detailed framework for using meta-analysis to compare different results among different studies as well as to combine different results across different studies. The author describes a detailed treatment of eight methods for combining significance tests from a group of studies when the meta-analyst wants to maximize the statistical power of an effort by including all studies.

Wachter, K. W., and M. L. Straf, eds.
    1990   *The Future of Meta-Analysis.* New York: Russell Sage Foundation.

This book summarizes views presented at a conference whose participants were stimulated by a collection of meta-analyses on the effectiveness of desegregation for improving education in schools. Some innovative presentations, some controversies, and some areas for the future of meta-analytic research and for more general scientific development emerged from these considerations.

Wolf, Frederick.
    1986   *Meta-Analysis.* Beverly Hills, CA: Sage.

A brief overview of various steps that each meta-analyst must think through to carry out the work is provided in this book. The exposition is nontechnical and is presented at an introductory level. Several numerical examples illustrate each step in a meta-analysis.

International Journal of Technology Assessment in Health Care
    1989   5 (4).
    1990   6 (1).

In a collection of articles entitled "Alternative Methods for Assessing Technology," organized by David Eddy, these two journal issues treat many aspects of data synthesis especially oriented toward health and medicine.

Statistics in Medicine. Special Issue 6 (3). April-May, 1987. New York: Wiley.
    1987   This special issue of the journal gives a detailed summary of the Proceedings of the Workshop on Methodological Issues in Overviews of

Randomized Clinical Trials, held in May 1986 at the National Institutes of Health, Bethesda, Maryland. Several of the early chapters describe why meta-analysis can be helpful in health and medicine, especially when a meta-analysis has access to many randomized, controlled field trials. Special techniques are presented in detail, as are examples from aspirin as a treatment for coronary heart disease; overviews of cardiovascular trials; and the role of meta-analysis in cancer therapeutics. Many figures and graphs illustrate how meta-analysis is actually used.

1

The Meta-Analytic Perspective

Purposes of Social Science

Social science seeks to produce descriptions and explanations of phenomena that apply across a wide range of persons and settings. The hope is that such knowledge will improve both substantive theory and public policy. Policymakers particularly value descriptive knowledge if it can be generalized to many population groups and settings because they need to identify successful practices that can be transferred to a wide range of settings. Such knowledge transfer is enhanced if policymakers also know why relationships come about. Knowing which processes bring a desired end makes it easier to design new programs or modify existing ones in ways that set these processes in motion. Indeed, with full explanatory knowledge program officials are free to design procedures that are uniquely suited to their locality, provided only that these procedures set in motion the processes known to be effective.

Scholars particularly value explanatory knowledge if it identifies the processes through which causal forces shape the social world. Although they value knowledge that describes basic relationships—for example, whether juvenile delinquency programs reduce recidivism— they are likely to assign even greater value to explaining how or why such relationships come about. Explanation promises greater prediction and control. More particularly, it enriches substantive theory, by making it more complete, aesthetically pleasing, and possibly providing more clues as to the new theories that need developing. Thus, both Cronbach (1980, 1982) and Weiss (1980) advocate explanation as the most useful type of knowledge for both basic science and public policy.

Single Studies

We rarely expect comprehensive or generalizable knowledge to result from a single effort at data collection, whether it be from a survey, an ethnography, a laboratory experiment, or a field experiment. Nearly all experiments, for example, take place with a restricted population of persons from a small number of cities, factories, hospitals, or whatever. Furthermore, in individual studies investigators are seldom able to implement more than one variant of a planned treatment. As a result, the theoretical construct that the treatment is meant to represent is inevitably confounded with whatever conceptual irrelevancies are contained in the chosen treatment version. How can one juvenile delinquency program stand for a particular class of programs, let alone for juvenile delinquency programs writ large? A similar point can be made about assessing causal effects. To measure "aggression" requires comprehensive assessment of both verbal and physical aggression (both with and without the intent to harm), and each of these types of aggression should be assessed using cognitive, behavioral, and perhaps even physiological instruments. The resources typically available for single studies rarely permit such breadth of measurement. Instead, researchers tend to select measures that either reflect their individual preferences or the dominant substantive theories. Finally, it is obvious that a single study takes place at a single time. By itself, this offers no guarantee that the same results would occur at another date.

Single studies not only fail to provide knowledge that is widely generalized, they also fail to produce comprehensive explanatory knowledge. Glymour (1987) has noted the infinitely large number of models that, *in theory*, might explain any given phenomenon. He has also noted the much smaller (but often still quite large) number of models that can often *plausibly* explain a phenomenon given what is already known about it. Few single studies attempt to assess the viability of multiple contending explanations of a relationship and, of those that do, even fewer involve high-quality measures of all the constructs in all the models. A single study can explore a single explanatory theory, or preferably a circumscribed set of theories. However, it is likely that single studies will result in uncontroversial and comprehensive explanatory findings about the processes that explain a stable descriptive relationship. There is little disagreement among epistemologists or most practicing scientists, particularly social scientists, that explanation is a more difficult task than description. In sum, individual inquiries are limited in the generalizability of the knowledge they produce about concepts, populations, settings and times, and single studies frequently illuminate only one part of a larger explanatory puzzle. Consequently, most research of this type fails to convince scholars interested in either general description or *full* explanations. Yet most of the cause-probing empirical research being conducted today consists of single studies that aim to describe or explain a phenomenon. Is it possible to combine information from many studies so as to produce more general knowledge, including knowledge about causal explanation? That is the topic of this book.

Traditional Qualitative Literature Reviews

Scholars often use literature reviews to establish generalized knowledge claims and to specify some of the conditions on which relationships might depend. Useful literature reviews (1) make the study topic very clear and (2) include only substantively relevant studies. In their totality it is better if these studies represent (3) a wide range of populations of persons, settings, and times; (4) a heterogeneous collection of examplars of both the treatment and effect; and (5) a broad range of potential explanatory variables. In addition, the review should (6) be sensitive to possible biases in individual studies lest the total bias across all the studies be more in one direction than the other. There is no point in naively synthesizing many studies with the same bias or even many studies with different biases that, in the aggregate, are stronger in one direction than another.

Yet traditional literature reviews often fail to meet all six criteria. If a heterogeneous group of studies has been identified and collected, it may be difficult to determine the results of each study. Many traditional nonquantitative reviews depend on statistical significance tests to decide whether a study has uncovered a particular relationship. But whether an effect is "significant" or not depends on the statistical power to reject the null hypothesis. Sample size is a major determinant of power, but sample size is not a substantively relevant factor. What is the logic, then, for the same size of effect being "significant" with one sample size but not another?

Traditional reviews often use a "vote count" method to accumulate the results across a collection of relevant studies. That is, the reviewer counts how many results are statistically significant in one direction, how many are neutral (i.e., "no effect"), and how many are statistically significant in the other direction. The category receiving the most votes is then declared the winner because it represents the modal or most typical result. Lost in this approach is the possibility that a treatment might have different consequences under different conditions.

Also lost is any notion that some studies are more valuable than others and deserve special weighting. This might be because they are particularly substantively relevant, because they have better methodology for drawing causal inferences, or because their sample sizes and quality of measurement permit more precise estimates. Whatever the reason, it cannot be naively presumed that all studies entering a review merit the same importance.

Although a few traditional reviews deal with the above issues, the majority do not. Moreover, a common problem with traditional qualitative review methods is that they are not easy to carry out once the number of studies becomes large and attention has to be paid to effect sizes, differential weighting, and the possibility that other variables modify the basic bivariate causal relationship under study. Analytic convenience is not a strong point of even the most thoughtful qualitative reviews, but it is a strong point of quantitative methods of synthesis. Given these limitations of traditional reviews, other procedures for synthesizing prior research results are needed.

Meta-Analysis

Although many methods have been advanced for synthesizing a particular literature and are described in Light and Pillemer (1984), meta-analysis constitutes the best-known and probably most flexible alternative available today (Smith and Glass 1977; Hedges and Olkin 1985). In addition, it promises to solve the problems associated with both single studies and traditional narrative reviews.

Meta-analysis offers a set of quantitative techniques that permit synthesizing results of many types of research, including opinion surveys, correlational studies, experimental and quasi-experimental studies, and regression analyses probing causal models. In meta-analysis the inves-

> All studies combine evidence over units of analysis. In a meta-analysis, the units are published reports (or studies).

tigator gathers together all the studies relevant to an issue and then constructs at least one indicator of the relationship under investigation from each of the studies. These study-level indicators are then used (much as observations from individual respondents are used in individual surveys, correlational studies, or experiments) to compute means, standard deviations, and more complex statistics. For most intents and purposes, study-level data can be analyzed like any other data, permitting a wide variety of quantitative methods for answering a wide range of questions.

A Brief History

Meta-analysis makes research synthesis an explicit scientific activity. Notions that research reviews could be conceptualized as scientific research did not develop within the social sciences until the 1970s. Feldman (1971) wrote that "systematically reviewing and integrating . . . the literature of a field may be considered a type of research in its own right—one using a characteristic set of research techniques and methods" (p. 86). He suggested that the work of others should be viewed as the reviewer's raw data. While Feldman did not formalize the process, he did identify four parts to research integration: sampling studies, developing a systematic scheme for indexing and coding material, integrating studies, and writing the report.

Taveggia (1974) argued that much of the supposed inconsistency in social science literatures was due to traditional reviewing procedures that failed to consider the probabilistic nature of research results. He also suggested that one of the major stumbling blocks that scientific research reviewers would face was the nonindependence of outcomes within and between studies.

Jackson (1980) presented a model of research reviewing that divided the process into four tasks, using the same terminology commonly used to describe primary research. Cooper (1982) pushed the isomorphism a step further by proposing a five-stage process of research reviewing. Included in his model were descriptions of the functions, procedures, and threats to scientific validity engendered by each step in the reviewing process.

Until the mid 1970s applications of rigorous research review were rare. Perhaps the central event stimulating interest in reviewing procedures was the appearance of a review on the effectiveness of psychotherapy by Smith and Glass (1977), which introduced the term "meta-analysis" to stand for the statistical combination of results of

Early Examples of Meta-Analysis

Although the term "meta-analysis" was first used by Gene V. Glass in 1976, the problem of combining the results of quantitative research studies has a much longer history. Stigler (1986) documents the development of statistical methods for combining astronomical data collected under different conditions. The "conditions" frequently corresponded to data collected at different observatories or in different studies. Thus, the problem of combining such data is what we today might call meta-analysis. In fact, Stigler argues that Legendre invented the principle of least squares in 1805 to solve this essentially meta-analytic problem.

Meta-analyses that involve some form of averaging estimates from different studies have been part of the statistical literature since the beginning of this century. For example, Karl Pearson (1904) computed the average of five tetrachoric correlations from separate sets of data to summarize the relationship between immunity from infection and inoculation for enteric fever.

Later examples of meta-analysis appear in literature in the social sciences. In 1932 Raymond T. Birge published a paper in the *Physical Review* devoted to the methodology of combining the results of a series of replicated experiments, calculating the standard error of the combined estimate, and testing the consistency of the results of the experiments. Analyses based on Birge's methods are still used today by physicists to combine estimates from several experiments.

Not much later than Birge's work in physics, papers on combining the results of agricultural experiments began to appear in the statistical literature. For example, Cochran (1937) and Yates and Cochran (1938) described the problems of combining the results of agricultural experiments that would be familiar to contemporary meta-analysts.

References:

Birge, R. T.
  1932 The calculation of errors by the method of least squares. *Physical Review* 40:207–227.

Cochran, W. G.
  1937 Problems arising in analysis of a series of similar experiments. *Journal of the Royal Statistical Society (Supplement)* 4:102–118.

Glass, G. V.
  1976 Primary, secondary and meta-analysis of research. *Educational Researcher* 5:3–8.

Pearson, K.
  1904 Report on certain enteric fever inoculations. *British Medical Journal* 2:1243–1246.

Stigler, S. M.
  1986 *The History of Statistics: The Measurement of Uncertainty Before 1900.* Cambridge, MA: Harvard University Press.

Yates, F., and W. G. Cochran
  1938 The analysis of groups of experiments. *Journal of Agricultural Science* 28:556–580.

independent studies. While examples of meta-analysis can be found prior to Smith and Glass (see Beecher 1953; Rosenthal 1984), this paper certainly brought the procedures to the attention of a broad audience of social scientists.

In the mid 1980s several refinements in meta-analytic techniques enhanced their acceptability. These included (1) the development of statistical theory and analytic procedures for performing meta-analysis (Hedges and Olkin 1985) and (2) the integration of quantitative inferential techniques with narrative and descriptive ones, meant to ensure that the summary numbers of a meta-analysis did not lose their substantive meaning (Light and Pillemer 1984). Moreover, the methodology of meta-analysis has developed into a growth field in its own right. It has spawned many books and articles on such matters as coding, document retrieval, and statistical analyses to adjust for study differences in sample size and to deal with fixed- and random-effects models.

Stages of Meta-Analysis

Table 1.1 (Cooper 1982) presents a five-stage model of the integrative review as a research process. At each stage, it lists the question asked, primary function, procedural variations, and associated threats to validity. Here, we present a brief conceptual introduction to the stages. More detailed discussions of the issues associated with each stage, as illustrated by the cases in this volume, will be given in Chapters 7 and 8.

Problem Formulation

Scientific endeavors start with the formulation of the research problem. Primary research or research review can focus on a single relation or it can deal with a series of interrelated issues, asking "whether" or "why" a relation exists.

Further, in all types of social research the scientist must grapple with the correspondence between the theoretical definition of variables and their operational definition. Primary researchers must specify the defi-

A high-quality meta-analysis requires the same attention to design, conduct, analysis, and reporting as does an experimental study, and the same issues need to be considered.

Table 1.1   The Integrative Review as a Research Project

Stage Characteristics	Stage of Research	
	Problem Formulation	Data Collection
Research Question Asked	What evidence should be included in the review?	What procedures should be used to find relevant evidence?
Primary Function in Review	Constructing definitions that distinguish relevant from irrelevant studies.	Determining which sources of potentially relevant studies to examine.
Procedural Differences That Create Variation in Review Conclusions	1. Differences in included operational definitions. 2. Differences in operational detail.	Differences in the research contained in sources of information.
Sources of Potential Invalidity in Review Conclusions	1. Narrow concepts might make review conclusions less definitive and robust. 2. Superficial operational detail might obscure interacting variables.	1. Accessed studies might be qualitatively different from the target population of studies. 2. People sampled in accessible studies might be different from target population of people.

Source: Cooper, 1982. Reprinted by permission.

nitional boundaries of their concepts before the research begins. For example, what is meant by the term "juvenile delinquency" or "psychoeducational care"? Primary researchers also must choose a way of realizing a variable that corresponds to the concept. For example, what constitutes a marital therapy? Which test will be given to measure achievement? Primary data collection cannot begin until variables have been given an empirical reality. Reviewers also must begin their work with a clear understanding of the definitional boundaries of their task. The research reviewer, however, might redefine these boundaries as

	Stage of Research	
Data Evaluation	Analysis and Interpretation	Public Presentation
What retrieved evidence should be included in the review?	What procedures should be used to make inferences about the literature as a whole?	What information should be included in the review report?
Applying criteria to separate "valid" from "invalid" studies.	Synthesizing valid retrieved studies.	Applying editorial criteria to separate important from unimportant information.
1. Differences in quality criteria. 2. Differences in the influence of non-quality criteria.	Differences in rules of inference.	Differences in guidelines for editorial judgment.
1. Non-quality factors might cause improper weighting of study information. 2. Omissions in study reports might make conclusions unreliable.	1. Rules for distinguishing patterns from noise might be inappropriate. 2. Review-based evidence might be used to infer causality.	1. Omissions of review procedures might make conclusions irreproducible. 2. Omission of review findings and study procedures might make conclusions obsolete.

the review progresses. The reviewer also has the relative luxury of evaluating the correspondence of operations to concepts as operations appear in the literature. For example, does a study in which nurses gave patients booklets on heart surgery constitute an instance of psychoeducational care? Is this the same idea that underlies another study that provided two hours of discussion about surgery involving the patient, a social worker, and the patient's family?

The reviewer must decide what is in and what is out and provide a credible rationale for these decisions.

Data Collection

The data collection stage of the literature review can be viewed as analogous to survey sampling in primary research. In both cases there exists a target population about which the researcher wishes to draw inferences. Different procedures for obtaining data will determine the likelihood of whether any individual member of the population is sampled. In contrast to a primary surveyor, who usually does not intend to obtain responses from an entire population, a literature reviewer probably would do so if it were possible. Through a literature search, the reviewer collects data. This step is also the most distinctive aspect of the reviewing task. Cooper (1987) identified 15 different techniques used by reviewers for retrieving research literature. Each procedure results in a biased sample of studies. For example, online reference database searches underrepresent the most recent research and unpublished research. Using the reference lists of journal articles overrepresents the operations (and results) compatible with the research paradigm tying the particular journal to its journal network. The searching techniques should complement one another so as to ameliorate systematic bias in the methods and results of the retrieved studies.

Data Retrieval and Evaluation

If a literature search is analogous to a survey, then the data retrieval procedure in a review is analogous to an interview schedule. Data retrieval in interviews begins by deciding what questions to ask of each respondent and what form the questions will take. In research review, the same decisions are made but the data are extracted from study reports. In both forms of research, the questions asked are typically guided by theoretical concerns and by issues in measurement and method. Thus, the first two criteria for judging the adequacy of a coding frame in a literature review is its fidelity to the problem area's theoretical underpinnings and its sensitivity for disclosing potential methodological artifacts in the literature. A third criterion concerns the reliability of codings. Survey researchers make sure that interviewers

> A meta-analysis is an observational study and must be conducted with considerable care. The meta-analyst must investigate and attempt to control potential threats to validity that arise in nonexperimental research.

share clear definitions of responses. Likewise, review coders must be shown to be equivalent interpreters of the content of reports.

Data evaluation involves judging the quality of the individual data points. In research review, the reviewer makes a complex set of judgments about the validity of individual studies and their results. These judgments often include assessments of how treatment and control groups were constituted, the psychometric properties of measuring instruments, and the suitability of statistical analyses. The chosen criteria must be stated explicitly in advance of their application, and they must be applied to studies without prejudice.

Analysis and Interpretation

During data analysis and interpretation, the researcher synthesizes the separate data points into a unified statement about the research problem. Whether the data are responses of individuals within primary studies or are the results of studies themselves, some rules must be employed to distinguish signal from noise.

Until meta-analytic techniques came into use, primary researchers were required to adopt a set of statistical assumptions, choose inference tests appropriate to the research design, and report the outcomes of test applications, while research reviewers were not required to make any statistical accounting for their inferences. Most often, the decision rules behind summary statements (the in-sums and the-research-appears-to-supports that pepper all reviews) were obscure, even to the reviewers themselves.

Since meta-analytics, both forms of research are on a more equal footing. Proponents of meta-analysis believe that the same rules for interpretation required of the first users of the data ought to be re-

---

The accumulated findings of dozens or even hundreds of studies should be regarded as complex data points, no more comprehensible without full statistical analysis than hundreds of data points in a single study could be so casually understood. (Glass 1977, p. 352)

Source:

Glass, G. V.
    1977   Integrating findings: The meta-analysis of research. *Review of Research in Education* 5:351–379.

quired of all users thereafter. Meta-analysis is a simple extension of the rules for primary data analysis.

Still, choosing to use quantitative procedures to synthesize studies does not ensure scientific validity. Meta-analytics can be applied incorrectly. The necessary assumptions about data can be inappropriate. Calculations can be in error. And of course, reviewers can misinterpret the meaning of a meta-analytic statistic. Any of these mistakes may threaten the validity of a review. At the least, with quantitative reviews the suppositions underlying inferences are made explicit, and therefore are open to public scrutiny and test.

Public Presentation

Producing a document that describes a project is a task that confronts all researchers. Primary researchers have been required to follow a rigid format, in terms of both form and the type of information that must be disclosed. Research reviewers have had no formal guidelines. This proved no great drawback until the review came to be viewed as a type of research.

Most commentary on scientific writing rests on the assumptions that the central goals of a report are to (1) permit readers to assess the rigor of the research procedures and (2) provide enough detail so that other interested researchers might replicate a study's methods. Reviewers who view their work as research, therefore, have come to mimic the form of primary data reports. The appearance of methods sections in research integrations has become commonplace. These sections include information on how the literature was searched, what criteria were used to judge the relevance of studies, description of prototypical primary research methods, characteristics of primary studies that were extracted on coding sheets, how independent hypothesis tests were identified, and what statistical procedures and conventions were employed. Separate results sections detailing the outcomes of statistical tests are also not uncommon. Obviously, the accuracy and clarity of these descriptions will bear on the utility of the review.

The model of research synthesis portrayed in Table 1.1 is neutral with regard to the purpose of the review (Cooper 1988). Reviews that take on the goal of assessing the validity of descriptions or causal generalizations might lead to different procedural decisions (and problems) than a research review with the goal of assessing evidence about an explanatory model. The difference might be especially keen with regard to how the problem is formulated, how research is evaluated, and how evidence is synthesized and interpreted.

Meta-Analyses Teach Us About Small Effects

By looking at many meta-analyses of both social and medical treatments and programs, it is possible to develop some general observations about new treatments, and how well they are likely to work. One strong finding from various meta-analyses is that most new treatments have, at best, small to modest effects. Gilbert, Light, and Mosteller (1975) found this for their summary of three dozen social and medical programs. Few of these innovations showed large, positive effects—indeed, only a modest number showed any significant positive effects. Light (1983) presents the results of two dozen meta-analyses that are exemplary and finds a similar result. He concludes that most innovations have at best small positive effects, that such small positive findings should be treasured rather than ignored, and that "the importance of this finding is that managers of programs should understand they shouldn't expect large, positive findings to emerge routinely from new programs."

References:

Gilbert, J. P.; R. J. Light; and F. Mosteller
   1975   Assessing social innovations: An empirical base for policy. In C. A. Bennett and A. A. Lumsdaine, eds., *Evaluation and Experiment.* New York: Academic Press.

Light, R. J., ed.
   1983   *Evaluation Studies Review Annual.* Beverly Hills, CA: Sage.

Meta-Analysis Today

Today, the appliation of meta-analytic techniques is growing at a rapid rate and promises to continue to do so for some time. According to Guzzo, Jackson, and Katzell (1987), the number of documents indexed by PsychINFO using the term "meta-analysis" has increased steadily over the past decade and reached nearly 100 for the year 1985.

The achievements of meta-analysis have been considerable for a method with such a short history. Some practical questions that formerly fomented wide disagreement now seem to have been resolved by the method. Gone, for instance, are the days when a conference on individual psychotherapy would devote many hours to discussing whether it was effective in general. Since the work of Smith and Glass (1977), and its follow-up by Randman and Dawes (1982), among others, the debate is stilled. Debates continue, of course, but they are dif-

ferent ones. Meta-analyses have also generated a new source for opti-
mism among social scientists of all types, because so many positive
findings have been reliably identified. In the field of program evaluation, for instance, many effects seem to be smaller than program ad-
vocates expected, but most are larger than the Cassandras extracted
from their narrative reviews. No longer is it possible to entertain the
pessimistic, simplistic, and energy-sapping hypothesis that "nothing
works."

Most meta-analyses have concentrated on assessing whether a given
type of intervention has a particular type of effect. The better meta-
analyses have also gone on to explore some of the method factors,
some of the populations and settings, and some of the treatment var-
iants that influence the size of the effect. But exploration of such contingency variables is rarely systematic and even less often involves assessing the intervening mechanisms through which a treatment causes
its effects. This volume seeks to present exemplary instances of meta-
analyses that seek both to answer the descriptive questions to which
meta-analysis is traditionally directed in the social and health sciences
and to explore how explanatory questions might be answered more
systematically so as to enhance the scholarly and policy yield of reviews.

Chapter 2 discusses general issues that surround the use of meta-
analysis to further causal explanations. It begins with an analysis of the
nature and use of explanation in science. It then focuses on meta-analysis
in the social policy domain, though many of the same issues apply to
the development of basic theory.

Chapters 3 through 6 present four examples of meta-analysis. Three
of the examples deal with issues of causal generalization, probing
whether a manipulable causal agent has a particular effect across a broad
range of treatment variations, outcome constructs, persons, settings,
and times.

In Chapter 3, Devine asks whether patient education enhances the
recovery from surgery and reduces the length of hospital stays. If education causes patients to leave the hospital sooner than they would
otherwise and achieves this without adverse consequences, then education might be one means of reducing the costs of medical care in the
United States.

In Chapter 4, Lipsey explores whether juvenile justice programs prevent recidivism. This is an important question with obvious policy implications, because of the high volume of criminal offenses perpetrated
by juveniles, who need help in moving into adult roles as productive

workers, parents, and citizens. The issue also has considerable importance for the scholarly community interested in criminology. Heretofore, widespread disagreement has existed about the efficacy of programs in reducing recidivism among juveniles.

In Chapter 5, Shadish examines how family and marital therapies affect mental health. For the policy world, this question is important because marital break-up has severe economic consequences for women and children. Finding ways to conduct brief and effective therapy with family units that might otherwise go asunder is important. It is also valuable to identify ways to reduce the conflict between children and parents within families. Moreover, a great deal of money is spent on such therapy. As in the Lipsey case, the scholarly literature disagrees as to the general effectiveness of family and marital therapies, and it is also not clear whether some types of therapy are more effective with different kinds of family groups.

In Chapter 6, Becker begins with the fact that math and science achievement differs for boys and girls. She then seeks to explain why they differ in science achievement. To a degree that is unusual in meta-analytic practice, this explanation is the major focus of Becker's work.

As the cases are presented, readers should note how the authors specified their research questions, how they handled issues of study selection for the review, and how they developed a coding scheme. Readers should also note how adjustments were made for study-level differences in methodological irrelevancies, including sample sizes; how tests were made for the homogeneity of effect sizes so as to learn about important moderating and mediating variables; how analyses were conducted to examine the robustness of relationships through probing variables that might have statistically interacted with the treatment and how analyses were performed to control for the correlated error that can arise when the same study produces multiple estimates of a relationship. Finally, readers should note how the authors reported their findings and discussed their significance for further theoretical work and for policy.

Chapter 7 examines some generic problems that arise in the conduct of meta-analysis, whether the effort focuses on explanation or has more traditional goals. Finally, Chapter 8 returns to explanatory meta-analysis. It uses the four cases to highlight specific problems and promises for the use of meta-analysis in the search for scientific explanation.

In highlighting the explanatory aspects of the meta-analysis in this volume, we seek to move meta-analytic practice in more of an explanatory direction than has been the case heretofore, not to replace the dominant descriptive emphasis, but to complement it.

2

Explanation in Meta-Analysis

Definition of Explanation

Explanation entails understanding how or why something happens as opposed to describing what has happened. We can describe changes in academic achievement levels, but to explain why achievement rises or falls over time we must somehow take into account those antecedents that produce or generate the observed variation. Or we might, by means of randomized experiments and valid measures, determine whether a particular type of patient education is associated with longer or shorter hospital stays for particular population groups. But to do so would not necessarily imply anything about how or why one type of education is effective for some groups and not others.

Explanation always presupposes a clearly designated and reliably measured phenomenon, whether it is variability in an outcome or a causal relationship between a treatment and an outcome. If, however, a measure of achievement contains much bias, or if a causal claim about an educational program is measured erroneously because of sample selection bias, then the analyst runs the danger of trying to explain something that does not exist.

Explanation is a complex process that epistemologists seek to understand. For the purpose of this volume, a brief excursion into some of this thought is helpful. Writers of philosophical treatises differ considerably among themselves about the nature of explanation and the procedures required to justify explanatory knowledge claims. (A brief, accessible overview for nonphilosophers is provided by White 1990.) For most epistemologists, explanation entails identifying the total set of contingencies on which an event or relationship depends (e.g., Bhaskar 1975; Mackie 1974). If the contingencies are discovered or even closely approximated, several desirable consequences follow. One is better prediction of events; another is greater control over these events; a third is better theories of the events in question; and a fourth is more suggestive theories about events like those predicted yet different from them. Good explanations promote new theories as well as fill in details about existing ones.

Collingwood (1940) has made some important distinctions about how explanation differs when the phenomena under analysis are historical events, manipulable events, or "scientific" processes. He notes that historical explanation is always context-specific, depends on assumptions about individuals' motivations that are difficult to test, and usually provides little information that may help control human events in the future. It is difficult to know with much certainty why Napoleon invaded Russia or why Rome fell. We shall not explore this type of explanation any further, because the research reported in this volume is not historical in Collingwood's sense. However, his two other types of explanation—of manipulated events and scientific processes—are relevant and deserve special scrutiny here.

Explanation Through Manipulable Agents

Collingwood (1940) notes that lay understanding of explanation is based on knowledge about the manipulable agents that bring about desired consequences and avoid noxious ones. Manipulability is also central to the rationale for experimentation in the tradition that Fisher (1935) pioneered, in which one variable is controlled and varied by the experimenter to see how it influences another dependent variable.

Information about manipulable variables and their effects is likely to be incomplete, however. An explanation can rarely specify the total set of conditions on which a causal relationship depends (Mackie 1974). Most experiments are limited by time and resources. Therefore, they will fail to specify completely the components of the manipulated variable responsible for the outcome, the components of the outcome affected by the manipulation, and the pathways or mechanisms through which the manipulation influences the effect. For example, academic achievement is determined by many factors, including home learning patterns, siblings, quality of teachers, a child's intelligence, the curri-

Several Explanations May Be Consistent with the Same Data

It is of course wise to remember that several explanations may be consistent with a set of research studies. Even when essentially all of the variation in a set of study results is explained by a plausible explanatory variable, other variables, suggesting other explanations, may be just as effective in explaining between-studies variation in results. Thus, empirical methods based on tests of goodness of fit are not sufficient to determine the validity of an explanation in research synthesis.

An interesting example is provided by the work of Becker (1986), who analyzed studies of gender differences in conformity that used the fictitious norm group paradigm. In a previous meta-analysis Eagly and Carli (1981) identified the sample of ten studies and suggested an explanation for the variability in effect sizes. They proposed that sex-typed communication was responsible for the variation in conformity: Studies conducted predominantly by male experimenters communicated a subtle message to female subjects to conform.

Becker's analysis confirmed that there was statistically significant variability among the study results (effect sizes); the homogeneity statistic was significant at the .01 level. Dividing the studies into groups according to the percentage of male authors (25, 50, or 100 percent), Becker found that the percentage of male authors was significantly related to the effects size ($p < .001$). Moreover, the variation among the effect sizes remaining after controlling for the percentage of male authors was no more than would be expected by chance if the model fit the data perfectly ($p > .15$). This seems to offer strong support for Eagly and Carli's hypothesis.

However, Becker also examined several other plausible explanatory variables, including the number of items on the conformity measure (the number of items on which the subject had the opportunity to conform). Fitting a weighted regression model to the effect sizes with the logarithm of the number of test items as the predictor variable, Becker obtained just as good an empirical fit to the data. The test statistic for goodness of fit of the regression model was $p > .25$. Thus, a methodological variable produced just as strong an empirical relation with effect size as did the substantive variable. This is not surprising since the percentage of male authors and the number of items are highly correlated ($r = .60$). Given this high a correlation, no statistical procedure could reasonably be expected to sort out the effects of the substantive and the methodological variable.

Reference:

Becker, B. J.
  1986 Influence again: An examination of reviews and studies of gender differences in social influence. In J. S. Hyde and M. C. Linn, eds., *The Psychology of Gender Progress Through Meta-Analysis*. Baltimore: Johns Hopkins University Press.

The Role of Theory

The fact that research results in the physical sciences often fail to meet the criterion of statistical consistency has important implications for social and behavioral science. New physical theories are *not* sought on every occasion in which there is a modest failure of experimental consistency. Instead, reasons for the inconsistency are likely to be sought in the methodology of the research studies. *At least tentative confidence in theory stabilizes the situation so that a rather extended series of inconsistent results would be required to force a major reconceptualization.* In the social sciences, theory does not often play this stabilizing role. (Italics added)

Source:

Hedges, Larry V.
   1987   "How hard is hard science, how soft is soft science? The empirical cumulativeness of research." *American Psychologist* 443–445.

cula the student is exposed to, peers, and so on. Many of these factors cannot be manipulated. Those that can be manipulated far outnumber our ability to do so. It is not realistic to expect an experiment to provide full explanation, either of a phenomenon in general or of a change in a phenomenon caused by a manipulated treatment.

Experiments are often advocated as the method of choice in science. This is surprising at first blush since science strives to accomplish the very theoretical explanation that experiments rarely achieve. How can it be that experiments are incomplete and yet so esteemed? In discussing this, Popper (1959, 1972) restricts himself to experiments that explicitly test hypotheses derived from substantive theory.[1] The utility of Popper's hypothetico-deductive approach to explanation via experimentation depends on four principal assumptions. The first is that the hypothesis under test has been validly deduced from the superordinate theory. The second is that the independent and dependent variables chosen for study faithfully incorporate the constructs specified in the guiding substantive theory. The third is that the theory guiding the selection of treatments and measures is explicit about the mechanisms

---

[1] He also deals with cases where a particular theory makes precise quantitative predictions and the available measuring instruments are so precise that the theory can be tested even without an experiment. But we concentrate here on experiments.

that produce or generate a causal relationship. And the fourth is that no other mechanisms can be invoked that might alternatively explain the relationship. To this end, Popper particularly emphasizes experiments that promise to distinguish between two or more theories which make different predictions about a study's outcome, often because each theory specifies unique mechanisms through which the cause comes to influence the effect. Only when all the above assumptions are met, does "black box" experimentation promote causal explanation.

It is unfortunate that Popper's assumptions apply even less in the social sciences than the natural sciences. This is because social theories are rarely as specific as the method requires and because social measures are never theory-neutral (Kuhn 1970). Social psychology is a subfield of the social sciences where experiments take place routinely and where theory-testing of the type Popper discusses should be viable if it is viable at all. So, let us take one example from there to illustrate the complexities. In a well-known experiment (Festinger and Carlsmith 1959) subjects were paid for advocating an attitude position counter to their own. The payment was varied, being either $1 or $25. The experimenters reasoned that the $1 payment would generate "cognitive dissonance"—the tension state that follows when on cognition (a private belief) implies the obverse of another (the public advocacy of a discrepant belief). Hypothesizing that this tension had to be reduced, Festinger and Carlsmith predicted that subjects would change their private belief more if they were paid $1 than $25. The data "confirmed" this prediction. However, Festinger and Carlsmith did not measure directly their explanatory construct "cognitive dissonance," only inferring it from the results. Critics have subsequently argued that these results were not due to an internal tension state, but to individuals thinking that the only rationale for publicly advocating what they do not believe *in the absence of extrinsic financial reasons for the advocacy* is that they privately believe the belief position they see themselves advocating (Bem 1972). The descriptive link between payment and attitude change is not at issue here; but the causal mechanism is.

By themselves, experiments are rarely useful for causal explanation, though their explanatory yield can be enhanced by direct and careful measurement of hypothesized explanatory processes and by selecting treatments and outcomes whose relationship no explanatory mechanisms can explain other than the one the experimenter sets out to test. But to increase the explanatory yield of an experiment through these strategies is no guarantee of full explanation!

The meta-analyses in this volume do not involve experiments deliberately designed to test theoretical propositions about explanatory processes. Instead, the independent variables were typically selected because they seemed powerful enough to affect outcomes that are socially important but have proven resistant to past attempts to modify them. The length of hospital stay, prison recidivism, and unsatisfactory family relations are all of this kind. Likelihood of impact was the experimenter's first criterion for selecting treatments and social importance was the first criterion for selecting outcomes. Fidelity to substantive theory was a subsidiary priority, and rarely was the theory in question a well-specified one. The language of hypothesis-testing was regularly used; but the hypotheses were seldom about causal mediating processes and were rarely so unique as to permit only one explanation of why a treatment and outcome were related to each other.

Despite these limitations, partial explanation can still be achieved from experiments, and it is often useful in enhancing prediction, control, and theory development. In their theories of research design, both Campbell (1957; Campbell and Stanley 1966) and Cronbach (1982) contend that person and setting factors are especially likely to moderate causal relationships and help explain why a treatment has the effects it does. (Campbell adds time to this list.) Both authors assume that social affairs are multiply determined in complex ways and that the diversity typically found among people, settings, and historical climates creates a unique context for each study. This study-specific context then somehow transforms the "meaning" of treatments that, on the surface, appear identical, setting in motion unique causal processes with various of the populations, settings, and times studied. From such ontological assumptions about real-world causal complexity comes the expectation that studies with a heterogeneous array of persons, settings, and times will result in many statistical interactions rather than a simple main effect (Cronbach and Snow 1981). Knowing about such interactions informs us about the specific types of conditions under which a treatment effect is large or small, is positive, negative or perhaps even null.

Identifying moderator variables does more than just facilitate prediction and sometimes control. The more complex a pattern of results, the more likely it is that it will provide distinctive clues about why two variables are related differently under different conditions. To give a hypothetical example relevant to a case in this volume, if patient education were to promote the speedier release of surgical patients in private hospitals but not in public hospitals, this would help hospital administrators better predict when patient education is likely to be effective. But it might also induce theorists to develop causal-explanatory hypotheses about the conditions under which patient education is more effective. They might, for instance, conjecture that it is more effective when physicians have considerable discretion over how long the patient should stay in the hospital, the subsidiary assumption being that physicians in private hospitals have more discretion over discharge decisions compared with their colleagues in public hospitals.

Some contingency variables affect the magnitude of a causal relationship but not its sign, implying that the treatment usually has, say, a positive effect but that this effect is sometimes larger or smaller. Other contingency variables may affect the sign of a relationship, indicating that it is sometimes positive and sometimes negative. In theoretical work, reliable differences in effect sizes between population groups or social settings can often help differentiate between theories even when all the relationships have the same causal sign. In much policy research, on the other hand, actors do not have enough discretion or political support to implement one class of treatment with one type of person or in one setting and another class of treatment with different types of persons or settings. Policy and program officials in central planning offices usually struggle to influence the grand mean of all the projects under their administration in towns, cities and counties throughout the United States; they rarely have the time, energy, knowledge, control, or freedom to engage in the local fine-tuning that seems desirable if an effect is larger in one set of circumstances than another (Cook, Leviton and Shadish 1985). However, members of the policy-shaping community are more likely to pay close attention to population, setting, or time factors that affect the *sign* of a causal relationship. This implies the possibility of unintended negative side effects that are often politically (and humanely) undesirable.

Meta-analyses, including the examples in this book, are rife with tests of how much the relationship between an independent variable and a dependent variable is moderated by a broad range of person, setting, and time variables. Hence, we need to examine how well the various data-analytic techniques used in this volume help specify causal contingencies. Devine, for example, prefers to stratify the data by a large number of population and setting attributes, *taken singly*, in order to probe if the direction of effect is constant. Lipsey and Shadish, on the other hand, prefer a multivariate approach in which they simultaneously assess how much variability in effect sizes is accounted for by the particular population, setting, and time characteristics they examine. Whatever the method used, the goal is the same and is not without explanatory implications—how to know whether the sign or magnitude of a causal relationship varies with attributes of the person, settings, and times included in a database.

Explanation Through "Scientific" Processes

Collingwood's third type of explanation is what he calls "scientific." Scientific explanation entails identifying the total set of circumstances that inevitably produce a given event or relationship. Theories of the structure and function of DNA aspire to this, as does natural selection theory and quantum mechanics. Such explanations are likely to be reductionistic, to approach full prediction of a particular event or relationship, and they usually provide powerful clues about what to manipulate in order to bring about a particular end.

To give a hypothetical example, if we learned that "Sesame Street" improved preschoolers' achievement because the nonhuman characters elicited attention and the learning materials were clearly and repeatedly presented, we could use attention-getting devices, clear materials, and multiple presentations in many different learning contexts. The crucial component is reproducing the generative process, not the same causal agents. The potential transferability that knowledge of causal principles offers probably explains why identifying such principles has been the Holy Grail of basic science since the Enlightenment.

Few explanatory analyses in the social sciences involve causal forces as generalizable, well-substantiated, predictive in their consequences, and flexible in their cross-situational transfer as some in the physical and natural sciences. Most social experimenters have to struggle to come up with the treatment and outcome components that might be responsible for producing a causal effect much less a substantive theory that specifies the chain of influence from a treatment to an outcome. Even if they make such predictions, their tests are rarely very strong. Identifying the factors that reliably produce important phenomena is the scientific ideal; but it is very difficult to achieve in practice, perhaps especially so in the social sciences.

Nonetheless, many methods exist in the quantitative social sciences to promote scientific explanation of the type under discussion. Such methods come under names like path analysis, structural equation modeling, or causal process analysis. All these suffer from a high likelihood of bias resulting from misspecifying the causal model and also (in varying degrees) from error in the measurement of variables (Glymour, Sprites, and Scheines 1987). But since explanation through the identification of mediating principles is so important for theoretical understanding and cross-situational transfer, critics who reject quantification but want to understand causal processes have to look for other methodologies. Some now suggest the greater use of qualitative methods to probe mediating processes, principally the methods used in ethnography, journalism, or history (e.g., Cronbach 1982). The call to abandon quantitative analysis of causal processes will fall on many deaf ears in the social sciences and nearly all the medical sciences. Nonetheless, it serves to illustrate problems in all current methods for promoting the theoretical understanding of causal mechanisms.

Explanation Through Prediction of Outcome Variability

A fourth model of explanation, not dealt with in detail by Collingwood, equates explanation with predicting all of the variability in the object to be explained (e.g., Bridgeman 1940). This model is widely used in the social sciences, particularly by those who use multivariate regression methods to model causal relationships. The strong assumption is that a relationship is perfectly modeled (i.e., explained) when the multiple correlation coefficient (R) equals one (1.0); the weaker assumption is that the relationship is better modeled (explained) the higher the R coefficient.

Several practices stand out that are worth noting in this prediction-dominated theory of explanation. The need to be certain about the object under study dictates that outcome variability due to methodological and substantive irrelevancies (e.g., those in Cook and Campbell's list of internal and construct validity threats) must be removed before analyzing the "explanatory" variables of substantive interest. Only after such threats have been ruled out as causes is it logical to claim that any more substantive predictors might account for variability in the outcome.

Various substantive attributes of the treatment loom large in trying to "explain" whatever variability remains after irrelevancies have been dealt with. Among these, the fidelity of the implemented treatment to the original plan or theory is important. Treatment dosage issues are also important, for they specify the thresholds required for obtaining an effect and the thresholds beyond which no further increment in effect can be detected. Also important is treatment class. In single comparative studies, different treatments can be directly compared with each other. The same is true in comparisons between studies, once it can be assumed that all irrelevant differences between studies have been taken into account. Being able to generalize about the differential effectiveness of treatment classes can often provide clues to explain why some studies result in relatively larger effects.

The treatment attribute that comes closest to explanation in Collingwood's third sense is the analysis of treatment components. Here the researcher wishes to identify the variability in effect size associated with different treatment components taken singly or in combination. In nearly all social experiments the treatments are multivariate hodgepodges rather than unitary constructs. Oftentimes, broad treatments are planned out of the fear that narrow treatments might not affect historically recalcitrant outcomes (see Cook, Anson, and Walchli in press). No level of analysis can be specified at which the analytic breakdown of treatment components has to stop; but with nearly all social experiments useful breakdowns can be accomplished.

The "variance accounted for" model of explanation is extremely flexible and can accommodate the person, setting, and time variables. But, for the multivariate prediction model to be overtly explanatory, substantive theory has to guide both the selection of variables and their order of entry into the analysis.

Important Tasks of Explanation

This volume should be read with eight attributes of explanation in mind. The most common explanatory tasks are identifying (1) mediating processes that causally link one construct to another; (2) the causal components of treatments that influence an outcome, and (3) the components of the effect on which the cause has impact. Less common tasks are analyzing (4) the person, setting, and time variables that might moderate a relationship; (5) the impact of differences in classes of treatment; (6) the theoretical integrity of treatments and outcomes; and (7) the consequences of different dosage levels. None of these analyses is meaningful if the underlying phenomenon to be explained is spurious. Thus, it is important to (8) construct an argument that the phenomenon-to-be-explained is not due to an artifact.

Given the poor state of theory in each of the substantive areas addressed in this volume, we should not expect that the case illustrations of meta-analysis will use manipulation of variables to explain main effects. Instead, they will focus on person, setting, and time contingencies that might moderate a relationship. This strategy goes some small way toward increasing explanatory yield.

Also, the conception of explanation as prediction leads them to probe how effect sizes are related to many different but often interdependent sources of variability, including methodological irrelevancies, treatment attributes that might help specify the nature of the causal agent and its relationship to outcomes, as well as person, setting, and treatment attributes.

Explanation of Experimental Variation in Physical Chemistry

We have suggested that one notion of explanation of between-experiment variation in syntheses of social research is to specify the variables that are associated with this variation. It perhaps is useful to note that the same principle of explanation is employed in research syntheses (so-called critical evaluations) of research results in certain areas of the physical sciences. An interesting example is provided by the critical evaluations of the spectral reflectance of aluminum, conducted by Touloukian and DeWitt (1972).

Spectral reflectance is a fundamental physical property of a material, corresponding crudely to the "color" of the material. Spectral reflectance is the proportion of the incident light at a given wavelength that is reflected off a material. Because the spectral reflectance varies with the wavelength, it is best conceived as a function of the wavelength of the incident light. That is, for any given wavelength of an incident light beam, a certain proportion of that light is reflected and the shape of the function for a given material corresponds with its familiar color.

Touloukian and DeWitt first compiled the results of experiments reported in 22 papers that had measured the spectral reflectance of aluminum by plotting the empirical spectral reflectance functions from each of the studies (see figure A). Each of the curves in the figure represents the results of several measures of spectral reflectance at different wavelengths. These results do not seem to indicate a single simple relationship. Indeed the results seem rather chaotic. Touloukian and DeWitt explained the variation in results by grouping experiments according to the way in which the sample of aluminum used in the experiment was prepared (e.g., evaporated films, polished surfaces, etc.). They found that these post hoc groupings of experiments according to method of surface preparation explained much of the variation in results and made it possible to present synthesized results (called recommended values) for each of their types of surface preparation (see figure B).

References:

Touloukian, Y. S. and DeWitt, D. P.
  1972 *Thermophysical Properties of Matter-The TPRC Data Series*, vol 7. New York: IFI/Plenum Data Corp. © Purdue Research Foundation. Figures reprinted with permission.

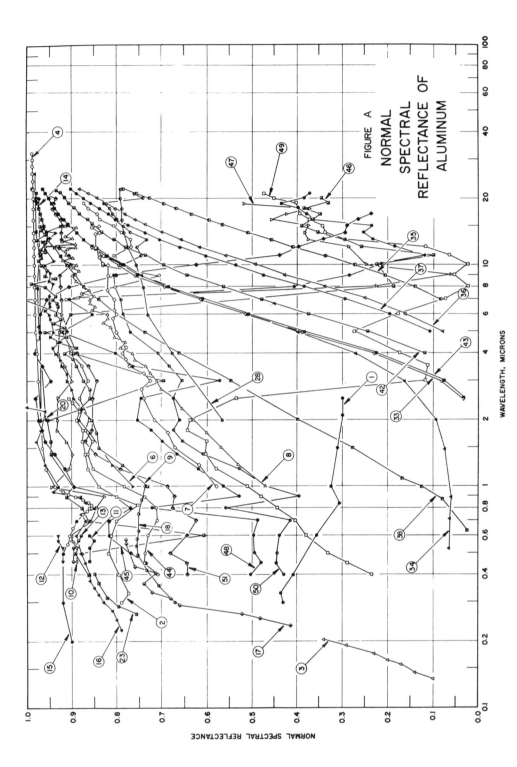

FIGURE A

NORMAL SPECTRAL REFLECTANCE OF ALUMINUM

NORMAL SPECTRAL REFLECTANCE

WAVELENGTH, MICRONS

28

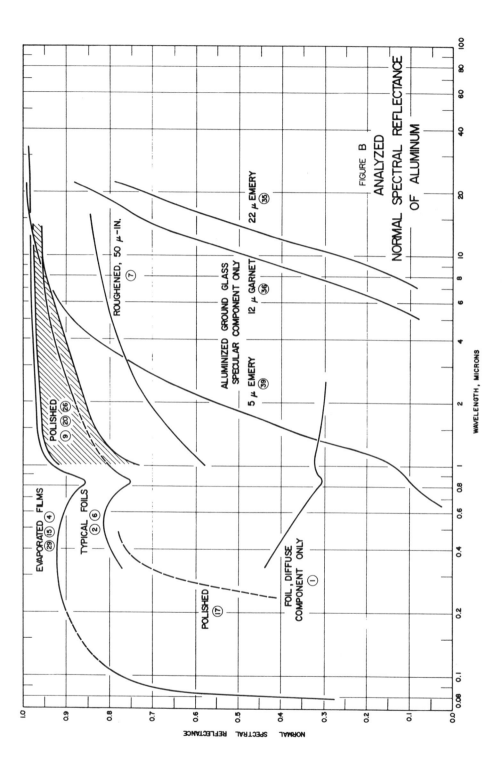

EVAPORATED FILMS ㉙ ⑮ ④

TYPICAL FOILS ② ⑥

POLISHED ⑨ ⑳ ㉖

POLISHED ⑰

ROUGHENED, 50 μ-IN. ⑦

ALUMINIZED GROUND GLASS
SPECULAR COMPONENT ONLY

5 μ EMERY ㊴

12 μ GARNET ㊱

22 μ EMERY ㉟

FOIL, DIFFUSE
COMPONENT ONLY ①

NORMAL SPECTRAL REFLECTANCE

WAVELENGTH, MICRONS

FIGURE B
ANALYZED
NORMAL SPECTRAL REFLECTANCE
OF ALUMINUM

29

> As in all scientific studies, intimate knowledge of the subject at hand is a necessary (but not sufficient!) ingredient of a successful meta-analysis.

Finally, these meta-analysts use the most commonly held understanding of explanation in terms of modeling the antecedent conditions that force, generate, mediate, or necessitate an effect. In nonexperimental contexts, such modeling usually requires identifying a specific constellation of interrelated antecedents that are responsible for the obtained variability. In experimental contexts the modeling relies on identifying those consequences of a treatment that are also causal mediators of a later outcome.

This volume takes these three related approaches to explanation and asks how well each helps meta-analysts fulfill the eight tasks of explanation.

Approaches to Explanation Illustrated in This Volume

While the case illustrations in this volume all attempt to use meta-analysis for explanatory purposes, they do not all subscribe to the same model of explanation or combine the models they use in the same ways.

Devine emphasizes a model of explanation that attempts to determine how robust a causal relationship is across setting, time, and person contingencies. Her chapter provides a compelling example of robust replication, demonstrating that positive effects of psychoeducational care for surgical patients are found across many variations in time, study design, patient characteristics, and hospital types. Devine also finds considerable robustness of effect across such treatment components as the content, timing, and mode of delivery and across a variety of outcome constructs, such as length of stay, pain medication taken, and the anxiety that patients report experiencing. In her analyses, Devine examines these separate effects on between-studies variation, using a criterion of robustness of causal sign rather than of average effect size. Close examination of her results shows considerable variability in average effect sizes between many of the subgroups she created when partitioning studies by patient, physician, hospital, or temporal attributes. But there was no such variability in causal direction.

Although the demonstration of robustness of findings effectively serves one conception of explanation, Devine also attempts to probe the causal mechanisms that produce the treatment effects. She articulates several distinct theories about treatment mechanisms and proposes a division of treatment components into groups that are theoretically meaningful. But her aspirations for examining mediating variables are not fully realized because the required data were rarely, if ever, collected or reported in the primary studies.

Lipsey emphasizes a different model of explanation. He is primarily concerned with predicting variation in study results from study characteristics that might plausibly be related to outcomes. His collection of studies is even larger, more diverse, and in many ways more complicated that Devine's. He uses a variety of characteristics of the clients, settings, and treatments that he judges to be substantively irrelevant and controls for these in a single linear model analysis, removing sources of method variance before examining substantive (treatment-related) sources of variation. His analysis includes the partitioning of treatments into classes which are related to philosophies of treatment, and he examines treatment dosage and intensity. The outcomes he examines are limited to recidivism, but there is considerable variety in its operationalization according to time of follow-up, types of delinquency, and sources of the data. In sum, Lipsey relies most heavily on a "variance accounted for" model of explanation. The model is closely linked to clearly articulated statements about which study factors are and are not substantively relevant. Controlling for these factors then allows him to focus on substantively important treatment attributes that he can presume to be free of method irrelevancies.

Unlike Devine, Lipsey does not attempt to probe mediating processes explicitly, leaving this to future papers. However, he is concerned with the generalizability of findings across settings, persons, and times. He examines the degree to which setting and person characteristics predict study results after the variability due to irrelevancies has been removed. This allows him to probe the extent to which such factors limit the generalizability of findings and allows his model coefficients to be used in predicting treatment effectiveness in any given setting with any particular group of offenders.

Of all the authors in this volume, Shadish is the most eclectic in his pursuit of explanation. He uses all three of the models presented here. For example, he stresses prediction or accounting-for-variation-in-effects as one of his principal analytic strategies. Shadish examines between-studies variation via a single linear model, in much the same way as Lipsey. Among the factors he examines are type of treatment (kind of therapy), its intensity or fidelity (e.g., via the indicator of university-based treatments), and its duration and intensity. Although he does not explore variations in outcome constructs, he does explore a range of variations in the measurement of his one outcome. Using the same technique, he also examines the contribution of setting and person characteristics to prediction.

Although Shadish does not explicitly decompose treatments into theoretically meaningful components, in some analyses he concentrates on understanding causal-mediating processes. In this connection, a novel and controversial aspect of his work is the use of a structural equation model for study-level characteristics. He examines the direct effect of a psychotherapist's behavioral orientation and also indirect effects mediated through the specificity of the dependent variable. He also examines whether latent variables might be used instead of observed measures to help understand methodological effects.

Becker has a different purpose from the other contributors. Whereas Devine, Lipsey, and Shadish examine whether and under what conditions a relationship exists, Becker assumes the existence of a causal relationship between gender and achievement in science and seeks only to understand why it exists. She explores whether the antecedents of science achievement differ for girls and boys. Since her study is the only exclusively explanatory meta-analysis in this group, her questions and methods are somewhat different and more specific to explanation. In concentrating on the structural relationships among antecedents that might mediate the relationship of gender to science achievement, Becker utilizes a single model but pursues it more intensely than do the other contributors to this volume.

She begins by posing a complex model with many mediating variables. As with Devine, this proves difficult to test in practice, given the paucity of measures of mediating variables available in the existing studies. Hence, she develops a simpler substantive model more amenable to empirical investigation, examing how between-studies variations in subjects, settings, and the particular construct measured influence estimates of the model coefficients. She takes seriously the problem for her model of the dependence among the multiple inputs from each study and also explores the problems created by missing data. Finally she addresses the problem of empirically testing the specification of the model for causal mediating processes.

Each of the meta-analyses included in this volume illustrates the explanatory potential of meta-analytic methods, while also presenting exemplary meta-analytic techniques, given today's state of the art. We hope that these examples will be so compelling and so clearly presented that other practitioners will be able to use them to improve their own meta-analytic research. By embedding the methodology in actual cases, the volume provides a context for improved understanding of meta-analytic methods that would not be possible if we had relied only on an abstract presentation of the same material.

.

3

Effects of Psychoeducational Care with Adult Surgical Patients: A Theory-Probing Meta-Analysis of Intervention Studies

Elizabeth C. Devine

Background and Theoretical Base

There is a large research base of controlled, clinical trials on the effects of patient education and/or psychosocial support (hereafter called psychoeducational care) administered to adult surgical patients. Initial meta-analyses of this research (Devine and Cook 1983, 1986) have demonstrated three important findings.

First, patients receiving additional psychoeducational care recovered more quickly, experienced less postsurgical pain, had less psychological distress, and were more satisfied with the care they received than patients receiving the psychoeducational care usually provided in the setting. For these four outcomes, average d values, based on a sample of studies, were .50 ($n = 73$), .39 ($n = 31$), .40 ($n = 36$), and 1.49 ($n = 5$), respectively (Devine and Cook 1986).

Second, four threats to internal and construct validity of these findings were examined and found to be implausible alternative explanations for the results of the meta-analysis. Beneficial effects were quite evident

1. in studies that had not been published and hence were not threatened by the bias resulting from the tendency of authors, editors, and reviewers to publish only studies with statistically reliable effects (Greenwald 1975; Rosenthal 1978);

Note: Studies used in this analysis are indicated by a †.

2. in studies with high internal validity, that is, studies with random assignment and low attrition;
3. with measures having little measurement subjectivity—for example, measures in which raters were unaware of the treatment condition of subjects; and
4. in studies with a placebo-type control group.

Finally, in this research the beneficial effects of psychoeducational care were found in both males and females, in adults of different ages hospitalized for many different types of surgery, in different types of hospitals, and across the time span of 1960 to 1982. The inference drawn from these results was that beneficial effects of psychoeducational care with adult surgical patients were quite robust and generalizable.

It is important to note that the psychoeducational care provided as the experimental treatment in this research was inexpensive to administer. Usually it lasted less than an hour and involved only one treatment provider (most often a registered nurse). In some studies patients were treated in groups, or treatment was provided primarily through printed or audio-visual materials. The relatively brief and inexpensive nature of treatment coupled with the fact that some of the outcomes obtained have a direct impact on lowering health care costs (e.g., decreased length of hospital stay and fewer medical complications) suggest that these results are important for health care policy. With this in mind a series of follow-up studies have been conducted to address deficiencies in past research and to refine analysis of existing research.

The first follow-up study was used to determine whether beneficial effects would continue to be obtained in the health care environment of the mid 1980s and when staff nurses rather than researchers administered the treatment.

In this study, which involved 354 surgical patients, it was found that staff nurses could increase the amount of psychoeducational care they provided (O'Connor et al. 1989). In terms of patient outcomes, patients in the experimental treatment group had shorter hospital stays, fewer of them used antinausea or antianxiety medications, and fewer of them used hypnotic medications. Statistically significant differences were not found on the use of pain medications (Devine et al. 1988). It is clear from this clinical study that the findings of prior meta-analyses on the effects of psychoeducational care are of more than just historical interest. The treatment continues to have relevance for the current health care environment.

The two major purposes of this chapter are (1) to reexamine the overall efficacy of psychoeducational care provided to adult surgical patients through meta-analysis of an expanded and updated sample of studies and (2) to probe theoretical explanations of the effects in order to further develop a treatment theory.

Two somewhat distinct theoretical orientations have been used in research on the effects of psychoeducational care on surgical patients. The first is concerned primarily with promoting physiologic recovery through moderating the adverse effects of general anesthesia, shallow respirations, and immobility after surgery. The second and more prevalent orientation is focused on alleviating psychological distress and enhancing coping. These orientations overlap, in that physiologic recovery is viewed as an end in itself by some researchers and as one indicator of successful coping by others. Because of this, interventions designed to promote physiologic recovery (e.g., coughing and deep breathing) can fit in either theoretical framework.

The primary model addressing the adverse physiologic effects of surgery will be called the nonstress, physiologic model. Developed largely by anesthesiologists (Bendixen et al. 1965; Dripps and Demming 1946), this model specifies that performance of certain activities—for example, deep breathing, coughing, bed exercises (hereafter referred to as stir-up exercises), as well as early ambulation—will reduce respiratory and vascular complications after surgery.

Among the four models addressing psychological distress and coping, the seminal work of Janis (1958) is most frequently cited. Based largely on interview and survey data, the theory that Janis developed has come to be called emotional drive theory. He found that patients with either too much or too little anticipatory fear had negative emotional disturbances after surgery. He proposed that providing preparatory information and specific assurances before surgery would stimulate the "work of worry" (Marmor 1958). Through a kind of emotional inoculation, individuals would develop more realistic views of the upcoming situation and could use coping strategies at their disposal to promote postoperative adjustment.

The parallel response model (Leventhal 1970) differs substantially from emotional drive theory. This theory and its sequel, the model of self-regulation under stress (Leventhal and Johnson 1983), posit that fear behaviors and behaviors for coping with danger can be independent. According to Leventhal and Johnson, fear arises from the perception of threat and concern about the adequacy of coping resources (Lazarus 1966, 1968). In the absence of intervention, the model predicts a linear rather than a curvilinear relationship between preoperative fear and negative postoperative emotional reactions. Coping, on the other hand, stems from the perception of danger and the availability of resources and skills to decrease the danger.

Other models and theories have been used less explicitly in the intervention-based research with surgical patients, but have some applicability to psychoeducational care provided to surgical patients. These include Lazarus's work on coping (Lazarus 1966, 1968) and control theories (Averill 1973; Miller 1979; Thompson 1981). While some aspects of Lazarus's theory of coping are integrated into the parallel response model discussed above, other aspects are not. For example, one type of psychoeducational care involves teaching patients techniques for cognitive reappraisal of anxiety-provoking events, calming self-talk, and/ or selective attention. These coping strategies are interesting for two reasons. First, they are diametrically opposed to Janis's proposition about the importance of stimulating the "work of worry." Second, they are coping strategies to be used early in what Lazarus calls primary appraisal (when the degree of threat and the adequacy of coping resources are being assessed). If used, they may reduce the likelihood that situations will be perceived as threatening.

The importance of control is mentioned with some regularity in this research. The supposition, based largely on laboratory research, is that individuals are less likely to appraise a potentially adverse situation as stressful if they have or perceive some degree of control over the situation (Corah and Boffa 1970). Providing subjects with information about expected experiences and/or teaching them exercises to perform could increase their sense of control.

Bandura's social learning theory (1977) does not address psychological distress and coping directly. Yet, it is directly applicable to treatments in which patients have a mastery experience (e.g., they demonstrate their ability to do the exercises correctly). According to social learning theory, teaching followed by a mastery experience should be more effective than teaching without it.

Methods

Sample and Selection Criteria

One hundred and eighty-seven studies were included in the sample; 105 of these studies were obtained and coded for the earlier meta-analyses (Devine and Cook 1983, 1986). To update and expand the sample of studies available for the current meta-analysis, the following major approaches were used. A computerized search was made of *Dissertation Abstracts*, *Psychological Abstracts*, and *Medlar* using such keywords as surgical patients, patient education, and evaluation/outcome study. To identify potentially relevant unpublished studies two approaches were used. Lists of theses and dissertations were requested from 138

National League for Nursing accredited graduate programs in nursing; studies that appeared relevant from their title or abstract were requested through interlibrary loan or purchased from University Microfilms. In addition, lists of all dissertations conducted by nurses were obtained from University Microfilms and studies that appeared relevant were purchased. Finally, using the ancestry method, reference lists of relevant studies and review papers were examined to identify other potentially relevant studies. There was an 85 percent response rate to the survey of graduate programs in nursing. The success rate of obtaining identified unpublished studies (mostly master's theses and projects) was 89 percent. Interlibrary loan was the primary source of these studies, although occasionally studies were obtained directly from researchers or were purchased from University Microfilms.

Studies included in the meta-analysis were those in which

1. psychoeducational care was provided to patients in the experimental treatment group;
2. experimental and control treatments differed in psychoeducational content;
3. subjects were adults, hospitalized for surgery;
4. treatment and control subjects were obtained from the same setting;
5. at least four subjects were included in each treatment group;
6. outcome measures of recovery, pain, and/or psychological distress were included.

Studies sampling subjects scheduled for therapeutic abortions or diagnostic procedures (e.g., cardiac catheterization) were excluded from the sample. Also excluded were studies of the effects of medications (e.g., antiblood-clotting drugs), devices (e.g., counter-pressure stockings or intermittent positive pressure breathing machines), or the institution of discharge planning services.

Treatments

Interventions included in this research were quite varied in content. Analysis of treatments, as described in each research report, revealed three reasonably distinct domains of content: health-care-relevant information, exercises to perform (skills teaching), and psychosocial support. Many interventions included elements from two or more domains of content. Even in treatments including the same domains of content, there often was noteworthy variability in the number of elements of content from each domain included in the intervention.

Health-care-relevant information often included details about preparing the patient for surgery and timing of the various procedures and activities, as well as the functions and roles of various health-care providers involved in this preparation. Many treatments included information about self-care actions to be performed (e.g., requesting pain medications when needed) and dietary restrictions. In addition, the normalcy, intensity, and duration of postoperative pain and other typical discomforts frequently were described. In a few studies, a wide range of sensations occurring in the preoperative and early postoperative time periods were described or other health-care-relevant information was provided. Usually, many types of health-care information were included in an intervention, although some treatment and control groups differed in only one type of information (e.g., information about usual postoperative pain).

Skills teaching frequently included coughing, bed, and breathing exercises. Other treatments included a variety of relaxation exercises, hypnosis, cognitive reappraisal of events, or surgery-specific exercises (e.g., arm exercises for women who had breast surgery). Except for the stir-up exercises (coughing, deep breathing, and bed exercises), which almost invariably were taught in the same intervention, it was typical for treatments to include instructions in only one skill.

Psychosocial support included identifying concerns of individual patients and attempting to alleviate those concerns, providing appropriate reassurances, fostering the patient's problem-solving skills, encouraging the patient to ask questions throughout the hospitalization, and providing a supportive treatment provider on more than one occasion. Frequently treatments included two or more of these elements.

Separating treatment content into the domains of information, skills teaching, and psychosocial support is done mainly for heuristic purposes since some overlap exists among the three domains. For example, both information and skills teaching may be reassuring to the patient or be delivered in a psychosocially supportive manner. In some interventions, coded as containing only information, exercises were mentioned although not formally taught. This may have increased the patients' receptivity to the skills teaching they received from hospital staff as part of the usual care provided to all patients. Finally, it is likely that most psychosocially supportive interventions contained relevant information about the hospitalization experience. In fact, given the patient-directed nature of some psychosocial interventions it is clear that no information, or even skills teaching, was precluded from the treatment if the provider judged it to be the best way to alleviate an individual patient's concerns.

Method of treatment delivery and timing of treatment administration were somewhat less variable. Most treatments were delivered by a treatment provider (usually a nurse) to each patient individually; in some instances patients were taught in groups or the treatment was provided primarily through printed or audio-visual materials. Most interventions involved only one treatment, ranging in length from 7 minutes to 90 minutes, with a median duration of 30 minutes. Treatment was most frequently administered the night before surgery. In a few instances, treatments were provided prior to hospitalization or both before and after surgery.

Measures

The primary outcome constructs of interest in this meta-analysis were recovery, postsurgical pain, and psychological distress. Prevalent measures of recovery included length of hospital stay, incidence of respiratory and other medical complications, and postsurgical respiratory function. Prevalent measures of postsurgical pain included use of analgesics and various ratings by patients (e.g., McGill pain questionnaire, amount of painful sensations, and amount of distress from pain). Prevalent measures of psychological distress included state anxiety or mood, as well as use of sedatives, antiemetics, and hypnotics. Theory-relevant outcomes were coded as well (e.g., the extent to which experiences were congruent with expectations or the extent to which subjects performed the exercises that were taught). Other outcomes such as blood loss, blood pressure, or number of items remembered from the recovery room were disregarded.

Selected characteristics of studies, subjects, treatments, and settings were coded using a slightly modified version of the coding form developed for the earlier meta-analyses (reported in Devine 1984; see Table 3.1).

The initial coding form was tested for reliability by having ten randomly selected studies coded by two nurses with research training. Satisfactory reliability was achieved since intercoder agreement was 92 percent (Devine and Cook 1983, 1986). All new studies obtained for the current meta-analysis were double-coded. After coders were thoroughly trained, they coded studies independently, compared responses, and discussed coding differences to achieve consensus. If consensus was not readily achieved a third person (usually the principal investigator) was consulted. Only 10.6 percent of the items needed to be discussed to achieve consensus. No instances of irreconcilable differences were encountered. To correct for chance agreement, Kappa

Table 3.1   Major Coded Characteristics of Studies, Subjects, Treatments, Settings, and Outcomes

Studies
    Publication form
    Date of issuance (e.g., publication date)
    Professional preparation of first author
    Manner of assignment to treatment condition
    Type of control group
Subjects
    Average age
    Gender
    Type of surgery
Treatments
    Content
    Timing
    Duration and frequency
    Mode of treatment delivery
Settings
    Type of hospital
    Country
Outcomes
    Measurement subjectivity[a]
    Sample size
    Effect size

[a] Adapted from Smith, Glass, and Miller's scale of reactivity (1980) and reported in Devine and Cook (1986).

(Cohen 1960) was calculated by dividing the difference between the observed percentage agreement and the by-chance expected percentage agreement by one minus the by-chance expected percentage agreement. For this calculation, 50 percent was used as the by-chance expected agreement. The obtained Kappa was satisfactory (.79). Since many items were multichotomous rather than dichotomous, using 50 percent as the expected level of agreement provides a conservative estimate for Kappa.

Procedures

To reduce the likelihood of experimenter expectancies (Rosenthal 1973, 1974) introducing bias into the data, the following precautions were used. In the initial meta-analyses a conscious effort was made to code the characteristics of a study, its treatment, its subjects, and its setting without making reference to outcomes. Similarly, effect size values were calculated without making reference to the other aspects of the study. Additional precautions were taken with studies added for the current meta-analysis. The tasks of (1) screening studies for inclusion, (2) coding the characteristics of studies, (3) coding the characteristics of the treatment, and (4) calculating effect size values were performed by different individuals. Coders were instructed to examine only the relevant parts of the research report. This was done to decrease the likelihood that knowledge of other aspects of the study would adversely influence coding decision.

The primary magnitude of effect statistic used in this meta-analysis was effect size (d). It is based on Cohen's statistic δ (1969) and represents the standardized mean difference between treatment and control groups measured in standard deviation units. Effect size was estimated by dividing the between-groups difference in mean scores by the pooled within-group standard deviation[1] or was derived from selected statistics (e.g., t values) or from proportions according to formulas and tables provided by Glass, McGaw, and Smith (1981). The basic formula for effect size is:

$$d = \frac{M_c - M_e}{S}$$

where M_c = mean of the control group, M_e = mean of the experimental group, and S = pooled within-group standard deviation. The observed effect size value was assigned a positive sign when results indicated that beneficial effects for patients were obtained (e.g., postsurgical pain was less or recovery was speedier for subjects in the experimental treatment group). A negative sign was used when the reverse was true. Then, effect size values then were corrected for the bias due to the likelihood of studies with small sample sizes to overestimate the population effect size (Hedges 1981). When there were insufficient data from which to calculate d values, the direction of effect was coded whenever this was available.

For descriptive purposes, standard deviations of d values are reported in several places. However, another measure of variability, the square root of the variance component estimate ($SVCE$), is used as well throughout this chapter. (The $SVCE$ is an estimate of the variance across

---

[1] Estimates of effect size using the standard deviation of the control group instead of the pooled within-group standard deviation also were calculated. The direction of findings and conclusions were the same when this estimate of effect size was used.

studies of the δ values that give rise to the d values observed.) Thus the *SVCE* is a better measure of the true population variability because its calculation involves removing the contribution of sampling error of d about δ from the variance of d.

Pre-treatment and post-treatment scores on the same dependent variable were reported for 45 outcomes. These were primarily measures of respiratory function or psychological distress. For those outcomes, observed effect size values were adjusted for pre-treatment differences between the groups by subtracting d estimated from pre-treatment data from d estimated from post-test data. There was no systematic pre-treatment difference across these 45 outcomes. Pre-treatment d values were unimodally and symmetrically distributed around a median of $-.06$. Nonetheless even in studies with random assignment to treatment condition, individual pre-treatment differences were sometimes larger than one standard deviation unit. It was not possible to adjust all observed d values for pre-treatment group differences, largely because there were no pre-treatment measures for most of the outcomes in this meta-analysis. (Pre-treatment measurements are impossible for many of the outcomes—e.g., length of postsurgical hospital stay, incidence of medical complications, postsurgical pain, incidences of the use of sedatives or hypnotics.) However, it was decided to adjust observed d values for pre-treatment group differences whenever these data were available.

To reduce redundancy, the simple unweighted averages of some observed d values were calculated before being coded for the meta-analysis. This was done primarily when there were multiple subscales of a single measure (e.g., the McGill pain questionnaire), when the same measure was used on multiple occasions (e.g., analgesic usage on each of the first four days after surgery), and when multiple tests of respiratory function were used (forced expiratory volume and vital capacity).

To avoid doubly weighting studies with two control groups, comparisons with usual-care control groups were used in all analyses except those in which a placebo-type control group was needed. When contrast with placebo-type control treatments was desired, studies in which this was the only type of control treatment were aggregated with those in which it was the second control group.

UNITS OF STATISTICAL ANALYSIS. For many dependent variables three units of analysis were possible: *outcomes, comparisons,* and *studies.* Analyses based on the sample of outcomes included all effect size values calculated for the dependent variable under analysis. For each outcome construct, analyses based on the sample of comparisons included only one effect size value for each experimental treatment group. In studies with multiple measures of a particular outcome construct (e.g., pain), a single estimate of effect was created for each treatment-control comparison by averaging all the effect size values for that outcome construct. In contrast, for each outcome construct, analyses based on the sample of studies included only one effect size value for each study. For studies with only one treatment-control comparison, there was no change from the procedures used to obtain the sample of comparisons. However, when studies had multiple experimental treatment conditions, each of which contrasted with the same control group, the following procedures were followed. If a prediction was made about which of the experimental treatments should have the largest effect, the effect size value from the sample of comparisons for that treatment was selected to represent the study. If no prediction was made and the experimental design was factorial, the effect size value from the sample of comparisons for the treatment that included the most factors was selected to represent the study. In all other instances, a single effect size value for each study was obtained by calculating the unweighted average of the effect size values from the sample of comparisons for all experimental treatment groups in the study.

REJECTING QUESTIONABLE DATA.   Of 547 outcomes, 58 outcomes from 29 studies were judged to be of questionable validity. These included the following:

1. studies in which the manner of assignment to treatment condition was not reported or had a high chance of introducing bias; examples include assigning subjects to treatment condition according to whether or not they attended preadmission education or according to researcher convenience ($n = 37$ outcomes from 12 studies);
2. outcomes for which the pre-test difference was greater or equal to 1.0 standard deviation unit ($n = 5$);
3. outcomes for which v (ratio between the standard deviations in treatment and control groups) was either 4.0 or greater or .25 or less ($n = 8$);
4. studies or outcomes that were questionable based on narrative descriptions by the author; examples include using total length of stay as an outcome when most subjects in the control group were admitted the night before surgery and most subjects in the experimental group were admitted the morning of surgery, studies in

which treatment diffusion had a high probability of occurring, and studies with discrepancies in the results reported (n = 8 outcomes).

Throughout this chapter, the "restricted" sample of studies, comparisons, or outcomes refers to samples that excluded the aforementioned data. In all other instances these data are included. All analyses in the models testing section of this chapter are conducted with restricted samples of studies or comparisons.

Results

Sample Characteristics

The sample included 187 studies (see Appendix 3.A). Sample size and at least one d value were available from 169 studies. In 16 studies only direction of effect was discernible. In 2 studies sample size was not reported. Study, subject, and setting characteristics of the 169 studies are summarized in Table 3.2, which also includes average d values calculated on a global measure of patient well-being. The d value for patient well-being was calculated for each study by averaging the d values from the sample of studies for the constructs recovery, pain, and psychological distress.

STUDY CHARACTERISTICS.   Just over one-third of the studies were published in a journal or a book. The majority of studies were master's or doctoral theses that, to this author's knowledge, have not been published elsewhere. In 82 percent of the studies nurses were first author; in 13 percent psychiatrists, psychologists, or pastoral counselors were first author. In the rest, either other professionals (e.g., anesthesiologists, physical therapists, or educators) were first author or professional affiliation could not be determined. Studies ranged in date of issuance from 1961 to 1988.

Random assignment to treatment condition was used in 69 percent of studies. For the purpose of this project, types of nonrandom assignment were grouped into high, medium, and low quality. High-quality nonrandom assignments included studies that used pre-test–post-test design with separate cohorts from the same hospital (24 percent of studies). Medium-quality nonrandom assignment included studies based on sequential assignment of a convenience sample with nonrandom start or matching (2 percent of studies). Low-quality nonrandom as-

Table 3.2   Average d Value for Patient Well-Being and Distribution of Studies by Selected Characteristics of Studies

Characteristic	Mean	Number	Percentage
Publication Form			
Journal	.55	49	29.0
Book	.39	6	3.6
Doctoral Dissertation	.28	16	9.5
Master's Thesis or Project	.32	98	58.0
Professional Affiliation of First Author			
Nurse	.38	138	81.7
Psychiatrist, Psychologist, or			
Counselor	.43	22	13.0
Other	.35	9	5.3
Publication Data			
1961–1968	.30	13	7.7
1969–1972	.43	21	12.4
1973–1976	.42	35	20.7
1977–1980	.47	43	25.4
1981–1984	.28	33	19.5
1985–1988	.34	24	14.2
Manner of Assignment to Treatment			
Condition			
Random Assignment	.40	117	69.2
High-Quality Nonrandom	.39	40	23.7
Medium-Quality Nonrandom	.55	4	2.4
Low-Quality Nonrandom	−.27	3	1.8
Not Reported	.34	5	3.0
Type of Control Group[a]			
Usual Care for Setting	.38	127	73.0
Usual Care plus Placebo-type			
Treatment from Researcher	.40	47	27.0

[a] Five studies had both kinds of control groups.

signment included studies that assigned patients based on their availability to attend the education session and studies in which the manner of assignment to treatment condition was not reported (5 percent of the studies).

There were 174 control groups in the 169 studies. Patients received usual levels of psychoeducational care for the setting ("usual care") in 73 percent of the control groups. In the rest, patients received usual care plus a placebo-type treatment from the researcher.

Table 3.3   Average *d* Values for Patient Well-Being and Distribution of
Studies by Selected Characteristics of Subjects and Settings

Characteristic	Mean	Number	Percentage
Type of Surgery			
Abdominal	.35	69	40.1
Thoracic	.42	24	14.3
Orthopedic	.46	11	6.5
Gynecological or Urologic	.60	8	4.8
Other Minor	.45	16	9.5
Day Surgery	.11	3	1.8
Other Major	.35	13	7.7
Major plus Minor	.38	24	14.3
		168	99.0
Gender of Subjects			
1–49% Females	.40	41	26.6
50–99% Females	.35	66	42.9
All Females	.44	34	22.1
All Males	.30	13	8.4
		154	100.0
Average Age of Subjects			
29–40 Years	.30	27	19.4
41–50 Years	.40	68	48.9
51–76 Years	.38	44	31.7
		139	100.0
Type of Hospital			
Teaching	.42	49	44.5
General	.38	47	42.7
Veterans or Military	.27	11	10.0
HMO Affiliated	.36	3	2.7
		110	99.9
Location of Hospital			
United States	.39	157	92.9
England	.43	7	4.1
Canada	.16	5	3.0
		169	100.0

SUBJECT CHARACTERISTICS.   Subjects were adults hospitalized for surgery. Across studies, a broad range of major and minor types of surgeries were included. Abdominal surgery (e.g., gall bladder, bowel, or
gastric surgery) and thoracic surgery (e.g., heart or lung surgery) were
the most prevalent major surgeries represented (Table 3.3).

Most studies included both males and females. Average ages of subjects ranged from 29 to 76 years. In almost half of the studies, the average age of subjects was between 41 and 50 years old (Table 3.3).

SETTING CHARACTERISTICS.   The vast majority of studies in this sample (93 percent) were conducted in hospitals located in the United States. The rest of the studies were conducted in Canada (3 percent) and England (4 percent). Most hospitals were teaching hospitals (45 percent) or general hospitals (43 percent) (Table 3.3).

Treatment Effects

The direction of treatment effect was obtained for 737 outcomes from 187 studies. This includes 289 measures of recovery, 239 measures of pain, and 209 measures of psychological distress. A sample of studies shows that average effect size values for these three outcome constructs range from .31 to .43, and the percentage of outcomes indicating beneficial effects range from 79 to 84 percent. In all three instances these values are different from 50 percent, which is the percentage of positive values one would expect if there were no treatment effect ($p < .001$, z test for difference in sample proportions). It is worth noting that the obtained average d values and the percentage of positive outcomes based on both the sample of comparisons and the sample of outcomes are remarkably similar for those obtained from the sample of studies (Table 3.4).

Because length of hospital stay is already in a common metric (days of hospital stay),[2] three measures of effect are used: days difference (DD), percentage difference (PD), and effect size (d).[3] The formulae for DD and PD are as follows:

$$DD = M_c - M_e$$

$$PD = \frac{M_c - M_e}{M_c} \times 100$$

where M_c = mean of the control group and M_e = mean of the experimental group.

---

[2]In one study (†Foreman 1982) length of stay was measured in hours rather than days.
[3]Because LOS is often skewed rather than normally distributed, effect size values were calculated two ways: from the means and standard deviations as presented (the traditional method) and from approximately $\log_e$ transformed values. The results were essentially the same and no conclusions changed. Only the traditionally calculated effect size values are reported in this chapter.

Table 3.4   **Results on Selected Dependent Variables by Studies, Comparisons, and Outcomes**

Measure	Mean	Number	Standard Deviation	Percentage of Outcomes in Positive Direction	Number
Recovery					
Studies	.43	109	.46	83.7	123
Comparison	.44	151	.48	83.9	168
Outcomes	.43	241	.52	79.6	289
Pain					
Studies	.38	82	.45	81.4	102
Comparisons	.36	106	.48	78.5	135
Outcomes	.40	157	.54	79.9	239
Psychological Distress					
Studies	.31	76	.51	78.5	93
Comparison	.31	96	.48	81.9	127
Outcomes	.35	149	.53	76.8	209

Length of stay (LOS) was measured in 118 contrasts between a treatment and control group (sample of comparisons). A shorter average LOS for subjects in the experimental treatment group is found in 76 percent of these contrasts. The sample of studies shows that 79 percent of 76 studies demonstrate beneficial effects on LOS (significantly different from 50 percent, $p < .001$); LOS was decreased an average of 1.5 days, 11.5 percent, or .39 standard deviation units (Table 3.5).

The present meta-analysis with its larger sample of studies replicates findings of the earlier meta-analyses (Devine and Cook 1983, 1986). Even though treatment effects are small to moderate in size, clinically and financially relevant effects are included, increasing the importance of the findings.

Threats to Validity

The four threats to internal and construct validity identified earlier were examined with this expanded sample of studies. Results are essentially the same as those reported in detail elsewhere (Devine and Cook 1983, 1986). The prevalent outcomes of length of stay, medical complications, respiratory function, pain, and psychological distress were examined using the sample of studies. For none of these outcomes were treatment effects absent or greatly diminished for the presumably less biased

Table 3.5   Average Treatment Effects on Length of Hospital Stay Based on Effect Size, Percentage Difference, and Days Difference

Effect Statistics	Mean	Number	Standard Deviation
Sample of Studies			
Effect Size	.39	65	.48
Percentage Difference	11.5	73	14.3
Days Difference	1.5	74	1.5
Sample of Comparisons[a]			
Effect Size	.40	102	.48
Percentage Difference	10.7	111	14.2
Days Difference	1.0	112	1.3

[a] Since length of hospital stay was measured only once for each experimental treatment group, no sample of outcomes is possible.

studies or measures (studies that had not been published, studies with high internal validity, studies with a placebo control group, or measures with little subjectivity). For example, average effects by publication form, manner of assignment to treatment condition, and type of control group for the global measure of patient well-being are reported in Table 3.2. Since measurement subjectivity varies by outcome rather than by study the sample of outcomes was used. Across all measures of recovery, pain, and psychological distress average d values for measures with very low, low, medium, high, and very high measurement subjectivity are .44, .46, .43, .39, and .36, respectively.

Weighted regression procedures (Hedges 1982) were used to estimate the relationship between threats to validity and size of effect. Length of stay, respiratory function, postsurgical pain, and psychological distress were examined using the sample of studies. Effects on medical complications were not examined since many d values were obtained through probit transformation and hence have different sampling distributions than those calculated from means and standard deviations. Internal validity, publication form, and type of control group were used as predictor variables. In addition, measurement subjectivity was used as a predictor of postsurgical pain and psychological distress. Measurement subjectivity was omitted for respiratory function because it had no variability and for length of stay because the relevant information so often was missing.

There were no statistically significant univariate or multivariate relationships between the threats to validity examined and estimates of effect on length of stay, respiratory function, postsurgical pain, and

Table 3.6   Average Effect Size (d) on Respiratory Function by Internal
Validity and Publication Form Based on a Sample of Studies

Characteristic	Mean	Number	SVCE[a]
Higher Internal Validity[b] and Unpublished	.47	12	.36
Higher Internal Validity and Published	.26	3	.32
Lower Internal Validity and Unpublished	.24	7	0
Lower Internal Validity and Published	.17	3	0

[a]Square root of the variance component estimate.
[b]Studies rated as higher in internal validity had random assignment to treatment condition, less than 15 percent overall attrition, and less than 10 percent differential attrition between groups.

psychological distress. The multivariate models examined included all threats to validity identified above as relevant for the specific outcome.

One significant interaction effect is noted. There is an interaction between internal validity and publication form on the effects found with respiratory function ($R^2 = .11$; $Y = .32 - .15$ internal validity $- .14$ publication form $+ .41$ internal validity x publication form). Internal validity and publication form are dummy coded with the less biased subset being coded as 1 and the more biased subset being coded as 0. Cell means reveal that average effects are highest in the least biased studies (Table 3.6).

The foregoing results strengthen the argument that the effects found in this meta-analysis are not an artifact of including studies and measures that have some threats to validity.

Testing for Homogeneity

The statistical significance of between-studies variations in effect sizes on prevalent outcomes was examined using the homogeneity test described by Hedges and Olkin (1985). In addition, a random-effects model variance component was estimated to quantify the extent of between-studies variability. Sample sizes ranged from 10 to 515 with a very skewed distribution (mean = 45; median = 30). A variance weighting procedure was used (Hedges and Olkin 1985, chap. 9) and the largest study (†Archuleta, Plummer, and Hopkins 1977) was omitted in order to avoid giving excessive weight to studies with very large samples.

Table 3.7   Homogeneity Testing of Selected Outcomes Based on the Restricted Sample of Studies

Outcome	Variance Weighted d	N	Q	p
Recovery[a]				
Length of Stay	.32	53	84.5	<.005
Respiratory Function	.33	21	22.4	.33
Resuming Normal Activities				
After Surgery	.56	7	12.6	.05
Time in ICU	−.03	7	6.6	.37
Pain				
Pain Medications	.27	70	128.3	<.005
Pain Measured by				
Questionnaires	.49	26	49.5	<.005
Psychological Distress				
Anxiety Shortly After				
Treatment	.47	12	9.8	.55
Anxiety After Surgery	.24	35	61.4	<.005
Use of Sedatives	.21	5	1.9	.75
Mood	.27	23	34.4	.05

[a]Many of the d values for the outcome "medical complications" were calculated from proportions through probit transformation. Homogeneity tests were not done on medical complications.

Results based on the restricted sample of studies are presented in Table 3.7. Results based on the unrestricted sample of studies and those based on the restricted and unrestricted samples of comparisons were essentially the same. Effects on length of hospital stay, resuming normal activities after surgery, postsurgical pain, anxiety after surgery, and mood were found to be heterogeneous. Except for length of stay, results were essentially the same when analyses were restricted to studies with random assignment to treatment condition. (Effects on length of stay were homogeneous among studies with random assignment.) Sources of variability in effects on length of hospital stay, postsurgical pain, anxiety after surgery, and mood were examined and are reported elsewhere (Devine 1990). Resuming normal activities after surgery was not measured in a sufficient number of studies to warrant further analysis.

Testing the Models

EMOTIONAL DRIVE THEORY. Correlation research not included in the meta-analysis offers little, if any, support for Janis's proposed curvilinear relationship between preoperative fear and postoperative emotional disturbances. Instead, most researchers have found a positive linear relationship between preoperative anxiety and postoperative negative reactions (Cohen and Lazarus 1973); preoperative fear and postoperative emotionality (Johnson, Leventhal, and Dabbs 1971); preoperative fear and prolonged postoperative recovery (Sime 1976); and preoperative fear and postoperative depression, anger, or complaints (Wolfer and Davis 1970).

Using intervention studies, this researcher sought evidence to test Janis's emotional drive theory. If beneficial effects of psychoeducational care are obtained through stimulating a moderate degree of fear so that patients can prepare themselves for surgery and the hospitalization experience, one would expect one or more of the following:

1. There would be an increase in fear or anxiety shortly after the treatment.
2. When pre-treatment and post-treatment anxiety scores are compared, variability should decrease more in the experimental treatment group than in the control group. This change would be the result of less anxious experimental subjects being prompted to do the "work of worry" and very anxious experimental subjects being calmed enough to do it.
3. There would be minimal or no effects from interventions unlikely to stimulate the "work of worry" such as instructions in the cognitive reappraisal of threatening events.

Is there a rise in fear or anxiety shortly after the treatment? Vernon and Bigelow (1974) specifically attempted to test Janis's theory. While they found that hernia repair patients who received accurate information about the hospitalization experience were more likely than control subjects to develop problem-oriented ideas and specific reassurances, there was no evidence that these effects were related to "anticipatory fear" or the "work of worry." Shortly after the treatment, differences between the groups in fear were small and not statistically significant. While d values could not be calculated, by three of four measures subjects in the information-treatment group had *less* fear than control subjects rather than more.

In nine studies anxiety was measured before the treatment and shortly after the treatment. In three other studies there were no pre-treatment measures, but anxiety was measured shortly after the treatment. In both of these subsets, anxiety after the treatment usually is lower in the experimental treatment group than in the control group (88.9 percent positive effects in the former, $\bar{d} = .32$, $SVCE = 0$, $n = 9$; and 100 percent positive effects in the latter, $\bar{d} = .93$, $SVCE = .05$, $n = 3$). These results are inconsistent with what one would expect to see if "work of worry" is the mechanism by which psychoeducational care has its effect. Unfortunately since these results are based on group averages, they might obscure an individual level phenomenon. To get some sense of the pattern of variability in anxiety scores, standard deviations of pre-test and post-test scores were examined next.

Is there a greater pre-test to post-test decrease in the variability of anxiety scores in the treatment group than in the control group? Pretreatment and post-treatment standard deviations were available from five studies (Spielberger's state anxiety was measured in six treatment groups and five control groups; pulse rate was measured in one treatment and one control group). For both treatment and control groups, standard deviations usually are lower in the post-treatment measure than in the pre-treatment measure. Contrary to expectation, standard deviations decrease more frequently in the control groups than in the treatment groups (83.3 and 71.4 percent, respectively). To assess the amount of difference in standard deviations, the ratio between pre-test and post-test standard deviations was calculated for each treatment and control groups (SD_{pre}/SD_{post}). Average ratios are similar for treatment and control groups (1.22 and 1.32, respectively). Contrary to expectation, the decrease in variability is slightly larger in the control group than in the experimental group. These data offer no evidence for a greater decrease in variability on anxiety scores for treatment groups than for control groups.

Finally, are beneficial effects absent or greatly diminished when the "work of worry" is not likely to be stimulated? Cognitive reappraisal–type skills training was the only intervention in four experimental treatment groups from three studies. (Note: Two of the three studies had placebo-type control groups.) All effects from those treatments on medical complications, pain, anxiety, and mood are positive. Average d values based on the restricted sample of studies are as follows: medical complication .49 ($n = 1$), pain .59 ($n = 3$; $SVCE = .51$), anxiety before surgery .69 ($n = 1$), anxiety after surgery .83 ($n = 1$), and mood .67 ($n = 1$).

One cannot completely discount threats to the foregoing analyses. For example, they are based largely on group data, which may distort an individual-level phenomenon. Also, subjects in the experimental group may have disguised their increased anxiety or fear. Taken together with the fact that the correlation research cited earlier failed to replicate the proposed curvilinear relationship, there is little if any empirical evidence to support either the underlying premise of the theory that *both* too much and too little fear is deleterious to postoperative adjustment or that treatment effects are obtained by prompting the "work of worry."

SELF-REGULATION UNDER STRESS THEORY.   According to Leventhal and Johnson (1983), the parallel response model (the precursor of self-regulation theory) posits different treatment modalities for promoting fear reduction and promoting coping. Information about what is normal (especially sensation information) is proposed to facilitate optimal emotional response to surgery while skills training is proposed to promote effective coping. Preparatory information about what is normal supposedly helps an individual form a schema about upcoming events. This "road map" of concrete and unambiguous elements of a situation serves as a standard of comparison and guides one's interpretations of the situation. Presumably then, one is less likely either to feel "on guard" all the time or to make misattributions about a situation. Teaching skills or coping strategies is supposed to increase the individual's repertoire of coping skills and thus facilitate coping.

Leventhal and Johnson originally proposed that coping instructions would have a stronger effect on coping behaviors (e.g., use of pain medications, ambulation after surgery, and length of hospital stay) than information about what is normal. However, in their own research they have not found this to be so with surgical patients (†Johnson et al. 1978a, 1978b). For example, they found that information about what is normal had a strong effect on indicators of coping with surgery and that this effect could not be explained by reductions in fear.

A second prediction from the model, according to Leventhal, is that information about what is normal (especially sensation information) affects the emotional response component of pain and thus should be expected to have a larger effect on the distress caused by pain than on the sensations associated with pain (Leventhal and Johnson 1983). This prediction is addressed below.

A model of the effects of information and skills teaching, based largely on the self-regulation under stress theory, is presented in Figure 3.1. To facilitate testing, parts of the model are designated A through G.

The first step in probing this model was to determine whether information about what was normal and/or teaching patients a skill was

Figure 3.1   Effect of Information and Skills Teaching Model

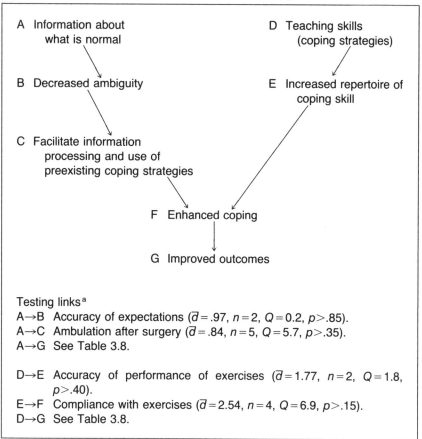

Testing links[a]
A→B   Accuracy of expectations ($\overline{d} = .97$, $n = 2$, $Q = 0.2$, $p > .85$).
A→C   Ambulation after surgery ($\overline{d} = .84$, $n = 5$, $Q = 5.7$, $p > .35$).
A→G   See Table 3.8.

D→E   Accuracy of performance of exercises ($\overline{d} = 1.77$, $n = 2$, $Q = 1.8$, $p > .40$).
E→F   Compliance with exercises ($\overline{d} = 2.54$, $n = 4$, $Q = 6.9$, $p > .15$).
D→G   See Table 3.8.

[a]No data were available on the links not reported.

associated with improved outcomes (A→G and D→G links). Average effects for frequently measured outcomes by type of content provided in the intervention are reported in Table 3.8. To allow concurrent probing of the nonstress, physiologic model, treatments with only skills teaching are presented twice—once for treatments including only stir-up exercises and once for treatments including all skills teaching.

Beneficial effects on indicators of recovery, pain, and psychological distress are found in treatments in which information and/or skills teaching is provided (Table 3.8). It is worth noting that effects on length of stay and pain are comparable between treatments providing infor-

Table 3.8 Average Effect Size Values for Selected Treatments: Restricted Sample of Studies

Outcomes	Information Only		Skills Teaching: Stir-Up Exercises Only		Skills Teaching: All Skills[a]		Information plus Skills Teaching: All Skills		Psychosocial Only	
	Mean[b]	N	Mean	N	Mean	N	Mean	N	Mean	N
Length of Stay	.53	4	.36	7	.32*	16	.35	20	.33	8
Medical Complications	—	—	.86[c]	13	.75[c]	7	.87[c]	16	—	—
Respiratory Function	—	—	.28	11	.34	14	.49	7	—	—
Pain	.56	8	.44**	6	.46**	23	.42	20	.18	8
Anxiety Shortly After Treatment	—	—	—	—	—	—	.29	4	—	—
Anxiety After Surgery	—	—	—	—	.24*	4	.16*	9	—	—
Mood	—	—	—	—	.30*	6	.25	11	—	—

Notes: A modified restricted sample of studies was used. Only one effect size value from any study was included in any mean. However, to make use of all relevant experimental treatments, if a study had an information-only treatment and a skills teaching-only treatment each of these would be included in the appropriate subgroup mean. Homogeneity of d values rejected: * $p < .05$ ** $p < .01$.
[a] Includes stir-up exercises as well as other types of skills teaching.
[b] Only averages based on four or more d values are presented.
[c] To increase sample size, effect size values calculated using probit transformation were combined with those calculated from means and standard deviations. Homogeneity testing was not done.

mation only and skills teaching only. This replicates the findings of †Johnson et al. (1978a, 1978b) and suggests that health-care-relevant information may not only reduce emotional distress, but also promote coping.

Does providing information about what is normal promote accurate schema formation and a concomitant decrease in ambiguity (the A→B link)? In this research, schema formation and ambiguity were never measured directly. The closest relevant measure was the degree of correspondence between expected occurrences and experiences. Unfortunately, this was assessed in only four comparisons from three studies. In all instances, experimental subjects reported a higher degree of correspondence between expectations and experiences than control subjects. Average d and homogeneity statistics based on a sample of studies are reported in Figure 3.1. These results are consistent with what one would expect to find if schema formation did occur.

Does providing information about what is normal or decreasing ambiguity facilitate information processing and the use of preexisting coping strategies (A→C or B→C links)? Ambulation after surgery was measured in 20 comparisons from seven studies. In 65 percent of these comparisons experimental subjects ambulated more after surgery than control subjects. Average d and homogeneity statistics based on a sample of studies are reported in Figure 3.1. These results are consistent with what one would expect to find if experimental subjects used preexisting coping strategies more effectively than control subjects. Unfortunately in only one instance did the treatment include only information. Most of the treatments included information and skills teaching. It is, however, worth noting that ambulation was not taught as a skill in any of these studies, although in some studies patients were informed of the importance of moving about after surgery as one mechanism to promote recovery. From these data it is not possible to examine whether information alone helps promote the use of preexisting coping strategies. Results from the one treatment in which only information was provided are not encouraging ($d = -1.01$). Nonetheless, information plus instructions in other skills is associated with increased use of at least one preexisting coping strategy.

Does teaching a skill or a coping strategy increase an individual's repertoire of coping skills (D→E link)? In two studies accuracy of postoperative performance of stir-up exercises was measured. In both instances positive effects were obtained. Average d and homogeneity statistics based on a sample of studies are reported in Figure 3.1. These results are consistent with what one would expect to find if subjects increased their repertoire of coping skills. Other than the stir-up exercises, performance of exercises after surgery (or related measures) were never quantified following instructions in skills.

Does teaching skills or increasing a patient's repertoire of coping skills facilitate use of coping strategies (D→F or E→F links)? In three studies compliance with stir-up exercises was measured. In all three instances large positive effects were obtained. Average d and homogeneity statistics based on a sample of studies are reported in Figure 3.1. The effect of skills teaching alone on ambulation could not be examined well because ambulation was measured only in one study in which skills teaching alone was provided ($d = -.11$). The beneficial effect of information plus skills teaching on ambulation after surgery has been discussed in the foregoing.

Are data consistent with Leventhal's proposition that information about what to expect, and especially sensation information, will have a larger effect on the distress associated with postoperative discomforts than on the sensations associated with these discomforts? The magnitude of effect on *both* sensations and distress, based on either percentage difference (*PD*) or d, was obtained for 19 comparisons from 11 studies. In most studies, patients reported on pain associated with the surgical incision. In all studies, similar visual analog scales were used for both sensations and distress. Typically, subjects marked their response on a ten-centimeter line that had descriptors associated with each of the extremes.

Information about what to expect was provided in four of these studies ($n = 12$ comparisons). In all four studies (and in 11 of 12 comparisons) beneficial effects were obtained on both distress and sensations. However, effects are larger on distress than on sensations in only 50 percent of these studies and 58 percent of the comparisons. When analyses are restricted to the treatments in which information about sensations was included in the information about what to expect ($n = 3$ studies; $n = 8$ comparisons), effects are larger on distress than on sensations in 75 percent of the studies and 75 percent of the comparisons.

These results are partially consistent with Leventhal's proposition on the effects of information about what to expect. Information about sensations had larger effects on the distress associated with discomforts than on the sensations associated with discomforts, but generic information about what to expect did not. However, four points should be noted. (1) These results are based on only a few studies. (2) Because of a lack of standard deviations, they are based primarily on the direction of effect. (3) In the three studies including sensation information, the magnitude of effect on both sensations and distress was more remarkable than the difference in effects. The effects on sensations and distress were as follows: d = .64 and .70 (†Noone 1985); PD = 14.3 and 25.7 (†Johnson et al. 1978b, experiment 1: gall bladder surgery patients; PD = 22.0 and 23.7 †Johnson et al. 1978b, experiment 2: hernia patients). (4) Two of the three studies including sensation information were conducted by Johnson, a long-time associate of Leventhal and coformulator of the theory under consideration. The foregoing suggests the need for further research on the effects of information about what to expect before firm conclusions can be drawn. It also should be noted that across the seven studies in which only skills teaching (predominately relaxation exercises) was provided to the experimental treatment group, effects are larger on distress than on sensations in 71 percent of the studies ($\bar{d}$ = .83 for distress and .37 for sensations; $SVCE$ = .93 and .39, respectively). (Note: Each study included only one experimental treatment group and hence the sample of studies is the same as the sample of comparisons.) Since these treatments did not include information, additional mechanisms of action for the differential effects of treatments on distress and sensations need to be studied.

NONSTRESS, PHYSIOLOGIC MODEL.   As a partial test of the nonstress, physiologic model, the effects of stir-up exercises alone were examined (Table 3.8). These results confirm the usefulness of stir-up exercises to reduce medical complications and to promote other aspects of recovery as well. These results are consistent with the nonstress, physiologic model.

SOCIAL LEARNING THEORY.   Not all treatments including stir-up exercises were the same. In some treatments, patients demonstrated to the treatment provider their ability to do the exercises correctly. According to social learning theory, skills teaching with a mastery experience should be more effective than teaching without it. Studies with a mastery experience had somewhat larger effects on length of stay than studies without it ($\bar{d}$ = .36, $SVCE$ = .28, n = 21; and d = .15, $SVCE$ = .14, n = 10, respectively). Medical complications and respiratory function were measured in too few of the studies without a mastery experience to warrant analysis. These results are consistent with social learning theory and suggest the need for further research in this area.

CONTROL THEORY.   Applicability of control theory could not be assessed since perceived degree control was not measured in any of the studies in this meta-analysis. Research is needed that examines the perceived control of surgical patients before and after psychoeducational care.

Conclusions

This review included 187 studies published between 1961 and 1988 on the effects of psychoeducational care on the recovery, pain, and/or psychological distress of adult surgical patients. This updated and expanded sample of studies included effect size values from 49 percent more studies measuring recovery (including 63 percent more studies measuring length of stay), 63 percent more studies measuring pain, and 111 percent more studies measuring psychological distress than the earlier review (Devine and Cook 1986). In the current review, effects of medium-small magnitude were found on recovery, pain, and psychological distress ($\bar{d}$ = .43, n = 109 studies; $\bar{d}$ = .38, n = 82 studies; and $\bar{d}$ = .31, n = 76 studies, respectively). Beneficial effects also were found on length of hospital stay, which is a subset of the recovery outcome construct ($\bar{d}$ = .39, n = 65 studies). These effects are slightly smaller than those reported earlier (Devine and Cook 1986), but continue to be reliably different from zero.

Four threats to internal and construct validity were examined with this expanded and updated sample of studies. The threats continue to be implausible alternative explanations for the results of the meta-analysis. Both univariate and multivariate analyses show that beneficial effects are not artifacts attributable to publication bias, to weak methodologic quality or to measurement subjectivity, or to a Hawthorne-type effect.

Beneficial effects were found for both male and female patients, for both older and younger adults, for patients scheduled for a wide range of surgeries, and in many different types of hospitals (Table 3.3). Also, beneficial effects were quite evident in the most recent studies (those issued between 1985 and 1988 ($\bar{d}$ = .34, n = 24) (Table 3.2). Effects were noticeably smaller in Canada ($\bar{d}$ = .16, n = 5) than in the United States or England ($\bar{d}$ = .39, n = 157; $\bar{d}$ = .43, n = 7, respectively); (Table 3.3).

Statistical significance of between-studies variations in effect size for prevalent individual dependent variables was examined across all studies and across studies with specific types of treatments (Tables 3.7 and 3.8). Results were not totally consistent, but many nonsignificant or homogenous effects were found. Given the variability of treatments, settings, and patients included in the studies reviewed, the prevalence of homogeneous effects is remarkable. This is particularly remarkable when one examines effects by type of treatment and finds that many different treatments had a substantial beneficial effect on length of stay, medical complications, respiratory function, and postsurgical pain.

The overall efficacy of psychoeducational care provided to adult surgical patients has been reconfirmed with this larger sample of studies. Two findings are particularly noteworthy: Beneficial effects persist even in the most recent studies, and they are not restricted to a narrow range of types of treatments. These findings are important from both clinical and policy perspectives. They suggest that at least in some settings there is room for improvement in the psychoeducational care provided. The modest increase in resources needed to provide a comprehensive version of psychoeducational care (e.g., about one hour of staff nurse time per hospitalization and printed or audio-visual materials) could have a sizable payoff in terms of improved patient welfare and recovery.

Unfortunately, attempts to probe theoretical explanations of effects were less successful. Most studies in the meta-analysis did not provide data suitable for model testing. There were two main problems. First, almost invariably, studies were more focused on evaluating the effects of a treatment than on testing theory. Only rarely was it apparent that either treatments were designed or dependent variables were selected to test theory. In addition, only rarely was the concordance between study results and the theoretical or conceptual framework for the study discussed. In even fewer studies were theory-relevant, micromediating variables measured.

Second, most treatments were fairly comprehensive in nature. From either clinical or evaluation perspectives, multidimensional treatments are advantageous. Having a robust treatment helps ensure that the treatment will be strong enough to enable the researcher to detect effects, if they exist. Multidimensional treatments also are better models of typical clinical practice. However, from either a theory-testing or theory-building perspective multidimensional treatments are quite problematic. Unless multidimentional treatments are included in a factorial-type design, it is usually impossible to determine the effects of individual treatment components much less test theories related to specific components.

Despite these problems, to a limited extent theory was able to be probed with the existing research. Janis's emotional drive theory is the most widely cited theoretical or conceptual framework used in this research. However, in admittedly less than perfect assessments, neither of the two critical aspects of this theory—the curvilinear relationship between preoperative fear and postoperative disturbances and evidence for the "work of worry"—were found. Given this lack of support, Janis's emotional drive theory should not be used uncritically to guide future evaluation research in this area. It is curious that so many researchers created effective treatments using this theoretical framework. Additional theory-testing research may help identify what, if any, aspects of this theory have merit.

There was some support for self-regulation under stress theory, the nonstress physiologic model, and social learning theory. However, since so few studies contributed data to these analyses, and most analyses were based on between-studies contrasts, more research is needed. Specifically, theory-testing research is needed which includes appropriate within-study contrasts of critical treatment components and theory-relevant outcome measures.

Several reporting weaknesses in individual studies were problematic for the meta-analysis. The main reporting weakness was that the content of usual care was never described. This was a problem because experimental treatments were invariably contrasted with usual care for the setting or usual care plus a researcher-delivered, placebo-type treatment. Since psychological and educational preparation for surgical patients has been recommended for quite some time (Bird 1955; Bernstein and Small 1951; Elman 1951), and since research on the effects of psychoeducational care has been going on since the early 1960s, it seems only reasonable to presume that the usual care provided by nurses and doctors contained some psychoeducational care. It is impossible to gauge the strength of a treatment if one does not know the degree of overlap between experimental and control treatments. Small or absent treatment effects could be due either to ineffective treatments or to a lack of difference between the treatments that experimental and control subjects actually received.

Other reporting problems include unavailability of means and standard deviations for all outcomes measured, not providing base rates on outcomes (like the incidence of medical complications), failure to provide correlation matrices for various outcomes measured, and failure to provide log transformations on outcomes very susceptible to skewness (e.g., length of stay or analgesics usage). Inclusion of this type of information would facilitate meta-analysis in this research domain.

Despite these reporting problems and the limitations of the data set for probing theoretical explanations, this sample of studies was excellent for examining overall treatment effects. Beneficial effects were found on clinically and financially relevant indicators of recovery, pain, and psychological distress. Common threats to internal and construct validity were not plausible alternative explanations of the observed results. Effects continue to be found in the most recent studies and many different types of treatment were found to have beneficial effects.

Studies Included in This Analysis

Notes: The research reported in this chapter was funded by the Russell Sage Foundation and the National Institutes of Health, National Center for Nursing Research (R01 NR01539).
Studies known to be available in multiple forms are listed more than once.

Abegglen, J.
1973   The effect on post-operative pulmonary complications by preoperative teaching of coughing and deep breathing. Unpublished master's thesis, University of Utah.

Aiken, L. H., and T. F. Henrichs
1971   Systematic relaxation as a nursing technique with open heart surgery patients. *Nursing Research* 20:212–217.

Aivazian, E.
1976   The effect of preoperative nursing intervention on anxiety of female patients. Unpublished master's thesis, California State University.

Archuletta, V.; O. B. Plummer; and K. D. Hopkins
1977   *A Demonstration Model for Patient Education: A Model for the Project "Training Nurses to Improve Patient Education."* Boulder: Western State Commission for Higher Education.

Auterman, M. E.
1971   The utilization of preoperative interviews by the operating room nurse as a means to reduce anxiety in surgical patients. Unpublished master's thesis, University of Iowa.

Bafford, D. C.
1977   Progressive relaxation as a nursing intervention: A method of controlling pain for open heart surgery patients. In M. V. Batey, ed., *Communication Nursing Research, Vol 8: Nursing Research: Revising Research Priorities, Choice or Chance.* Boulder, CO: Western Interstate Commission for Higher Education.

Banchik, D.
1974   Effects of a cognitive coping device and preparatory information on psychological stress in surgical patients: A replication. Unpublished master's thesis, Yale University.

Barnett, S.
1981   The effects of structured preoperative instruction on postoperative pain. Unpublished master's thesis, Texas Women's University.

Beadle, R. W.
1976   The value of preoperative teaching/counseling on the postoperative welfare of open heart surgery patients. Unpublished master's thesis, University of North Carolina.

Best, J. K.
1981   Reducing length of hospital stay and facilitating the recovery process of orthopedic surgical patients through crisis intervention and pastoral care. Doctoral dissertation, Northwestern University. *Dissertation Abstracts International* 42:3631B.

Bonilla, K. B.; W. F. Quigley; and W. F. Bowers
1961   Experiences with hypnosis on a surgical service. *Military Medicine* 126:364–370.

Boore, J.
1976   An investigation into the effects of some aspects of pre-operative stress and recovery. Doctoral dissertation, Victoria University of Manchester, England.
1980   Pre-operative information and post-operative recovery. *NAT News* 22:16–17, 19, 22.

Borchman, B.
1977   Reduction of stress through psychological preparation of the surgical patient. Unpublished master's thesis, DePaul University.

Budd, S., and W. Brown
1974   Effect of a reorientation technique on postcardiotomy delirium. *Nursing Research* 23:341–348.

Burns, R. K.
1977   Prevention of post-operative pulmonary complications through a nurse-supervised program of coughing and deep breathing. Unpublished master's thesis, University of Rochester.

Burry, M.
1986   The effects of a structured intervention program in patients with abdominal surgery. Unpublished master's thesis, Georgia State University.

Cappiello, L. M.
1979   The effects of preoperative information upon individuals' responses to decreased visual sensory input. Unpublished master's thesis, University of Virginia.

Carrieri, V.
1975   The effects of an experimental teaching program to postoperative ventilatory capacity. In M. V. Batey, ed., *Communication Nursing Research, Vol 7: Critical Issues in Access to Data.* Boulder: Western Interstate Commission for Higher Education.

Ceccio, C.
1983   The effect of the Jacobson Relaxation Technique on the postoperative level of comfort and analgesic intake in elderly clients with hip fractures. Unpublished master's thesis, Medical College of Ohio.

Chapman, J. S.
1970   Effects of different nursing approaches on· psychological and physiological responses. *Nursing Research Report* 5(1):4–7.

Cohen, F.
1975   Psychological preparation, coping, and recovery from surgery. Doctoral dissertation, University of California. *Dissertation Abstracts International* 37:454B.

Collins, E. M.
1981   An investigation of the effectiveness of preoperative nursing intervention in reducing anxiety and promoting physical and emotional
adjustment in women following hysterectomy. Unpublished master's
thesis, State University of New York at Buffalo.

Collins, N. W., and R. C. Moore
1970   The effect of a preanesthetic interview on the operative use of thio-
pental sodium. *Anesthesia and Analgesia* 49:872–876.

Cook, V.
1984   The effect of structured teaching on the anxiety level of adult preoperative patients. Unpublished master's thesis, Southern Illinois
University.

Crabtree, M. S.
1977   A cost benefit analysis of individual and group preoperative teaching. Unpublished master's thesis, University of Illinois.
1978   Application of cost benefit analysis to clinical nursing practice: A
comparison of individual and group preoperative teaching. *Journal of
Nursing Administration* 8(12):11–16.

Cunning, B. R.
1972   Effects of preoperative programmed instruction on postoperative respiratory complications. Unpublished master's thesis, University of
Iowa.

Daake, D. R.
1985   The effect of pleasant imagery instruction on the control of postsurgical pain. Unpublished master's thesis, Medical College of Georgia.

Darling, L. M.
1983   The effects of preoperative mobilization teaching on postoperative
incidence of early ambulation and its associated pain. Unpublished
master's thesis, University of Iowa.

Davis, D.
1985   The effect of preoperative instruction on pulmonary infection in patients with total hip arthroplasty. Unpublished master's thesis, Arizona State University.

Davis, H. S.
1973   The role of a crisis intervention treatment in the patient's recovery
        from elective surgery. Doctoral dissertation, Northwestern Univer-
        sity. *Dissertation Abstracts International* 36:349OB.

Del Monte, P.
1985   The relationship between preoperative information, environmental
        load, and postcardiotomy delirium. Unpublished master's thesis,
        Adelphi University.

DeLong, R. D.
1971   Individual differences in patterns of anxiety arousal, stress-relevant
        information and recovery from surgery. Doctoral dissertation, Uni-
        versity of California. *Dissertation Abstracts International* 32:554B.

Devine, E. C.; F. W. O'Connor; T. D. Cook; V. A. Wenk; and T. R. Curtin
1988   Clinical and financial effects of psychoeducational care provided by
        staff nurses to adult surgical patients in the post-DRG environment.
        American Journal of Public Health 78:1293–1297.

Dozier, A. M.
1980   Preoperative preparation by nurses to improve the elderly's postop-
        erative recuperation. Unpublished master's thesis, University of
        Rochester.

Dumas, R. D., and B. A. Johnson
1972   Research of nursing practice: A review of five clinical experiments.
        International Journal of Nursing Studies 9:137–149.

Duthler, T. B.
1979   An investigation of the reduction of psychological stress in patients
        facing surgical removal of tumors. Doctoral dissertation, University
        of Missouri. *Dissertation Abstracts International* 40:4477B.

Dziurbejko, M. M., and J. C. Larkin
1978   Including the family in preoperative teaching. *American Journal of
        Nursing* 79:1892–1894.

Egbert, L. D.; G. E. Battit; C. E. Welch; and M. K. Bartlett
1964   Reduction of postoperative pain by encouragement and instruction
        of patients. *New England Journal of Medicine* 270:825–827.

Eldridge, R. A.
1983   Effect of structured preoperative teaching on anxiety and satisfaction
        of cardiac surgery patients and their families. Unpublished master's
        thesis, University of Pittsburgh.

Fell, R.
1983   Relaxation technique to increase comfort level of postoperative pa-
        tients: A partial replication. Unpublished master's thesis, Marquette
        University.

Felton, G; K. Huss; E. A. Payne; and K. Srsic
1976   Preoperative nursing intervention with the patient for surgery: Outcomes of three alternative approaches. *International Journal of Nursing Studies* 13:83–96.

Fidrocki, V. M.
1978   Effect of a pre-operative instruction booklet upon post-operative recovery and satisfaction with care. Unpublished master's thesis, Yale University.

Field, P. B.
1974   Effects of tape-recorded hypnotic preparation for surgery. *International Journal of Clinical and Experimental Hypnosis* 22:54–61.

Fitzgerald, O.
1978   Crisis intervention with open heart surgery patients. Doctoral dissertation, American University. *Dissertation Abstracts International* 39:127A.

Flaherty, G. G., and J. J. Fitzpatrick
1978   Relaxation technique to increase comfort level of postoperative patients: A preliminary study. *Nursing Research* 27:352–355.

Florell, J. L.
1971   Crisis intervention in orthopedic surgery. Doctoral dissertation, Northwestern University. *Dissertation Abstracts International* 32:3633B.

Foreman, M.
1982   Effects of relevant sensory information on recovery from carotid artery surgery. Unpublished master's thesis, Medical College of Ohio.

Fortin, F., and S. Kirouac
1976   A randomized controlled trial of preoperative patient education. *International Journal of Nursing Studies* 13:11–24.

Fowler, J. F.
1973   Preoperative teaching as a method of preventing postoperative atelectasis in the patient undergoing surgery of the gall bladder. Unpublished master's thesis, University of Delaware.

Fraulini, K. E.
1980   Effect of hand and pillow methods of incisional splinting for postoperative deep breathing and coughing on vital capacity and inspiratory pressure in cholecystectomized patients. Unpublished master's thesis, Loyola University, Chicago.

Galczak, S. L.
1980   Effective preoperative information: Does content make a difference. Unpublished master's thesis, University of Northern Illinois.

Gautreaux, M.
1986   The effects of a structured pre-operative teaching program on pulmonary function and hospital length of stay. Unpublished master's thesis, University of Texas, El Paso.

Germon, K.
  1985   The effects of structured preoperative teaching on the outcomes of patients post-operatively cared for in intensive care units. Unpublished master's thesis, Ohio State University.

Gilkey, F.
  1982   Effect of preoperative teaching of trensurethral resection patients on their self concept. Unpublished master's thesis, University of Delaware.

Glick, O.
  1967   The effect of preoperative instruction about coughing and wound splinting on the patient's subjective report of pain with coughing postoperatively. Unpublished master's thesis, University of Iowa.

Goulart, A. E.
  1987   Preoperative teaching for surgical patients. *Perioperative Nursing Quarterly* 3:8–13.

Greenleaf, M.
  1986   Preoperative training in self-hypnosis for surgical patients. Doctoral dissertation, Yeshiva University. *Dissertation Abstracts International* 47:101B, 374.

Griffin, R. B.
  1981   Pain perception and breath control use to relieve postoperative pain. Unpublished master's thesis, Texas Women's University.

Harmon, M. D.
  1975   The effect of preoperative preparation on preoperative anxiety. Unpublished master's thesis, University of Nevada.

Hart, R. R.
  1975   Recovery of open heart surgery patients as a function of a taped hypnotic induction procedure. Doctoral dissertation, Texas Tech University. *Dissertation Abstracts International* 36:5259B.
  1980   The influence of a taped hypnotic induction treatment procedure on the recovery of surgical patients. *International Journal of Clinical and Experimental Hypnosis* 28:324–332.

Hassell, J. M. G.
  1965   The relationship of preoperative preparation to postoperative recovery and satisfaction in patients with intervertebral disc disease. Unpublished master's thesis, University of Utah.

Hawkins, J. E.
  1977   *The efficacy of structured and unstructured preoperative teaching.* Unpublished master's thesis, Emory University.

Hayward, J.
  1975   *Information—A Prescription Against Pain.* London: Whitefriars Press.

Healy, K. M.
   1968   "Does preoperative instruction make a difference?" *American Journal of Nursing* 1:62–67.

Hegyvary, S. T.
   1974   Organizational setting and patient care outcomes: An exploratory study. Doctoral dissertation, Vanderbilt University. *Dissertation Abstracts International* 35:8025A.

————, and P. A. Chamings
   1975a   The hospital setting and patient care outcomes. *Journal of Nursing Administration* 5(3):29–32.
   1975b   The hospital setting and patient care outcomes. *Journal of Nursing Administration*, 5(4):36–42.

Herl, E.
   1975   The effects of a structured and unstructured preoperative teaching plan. Unpublished master's thesis, University of Kansas.

Herrera, R. M.
   1969   The effect of preoperative teaching of deep breathing and coughing techniques to the male patient experiencing lower abdominal surgery. Unpublished master's thesis, Catholic University of America.

Hill, B. J.
   1979   Sensory information, behavioral instructions and coping with sensory alteration surgery. Doctoral dissertation, Wayne State University. *Dissertation Abstracts International*, 40, 2381B.
   1982   Sensory information, behavioral instructions and coping with sensory alteration surgery. *Nursing Research* 31:17–21.

Holden-Lund, C.
   1988   Effects of relaxation with guided imagery on surgical stress and wound healing. *Research in Nursing and Health* 11:235–244.

Hueschen, H. J.
   1980   An investigation of the effect of perioperative nursing practice on stress reduction in the surgical patient. Unpublished master's thesis, University of Kansas.

Hunt, L.
   1978   The effects of preoperative counseling on postoperative recovery-open heart surgery patients. Unpublished doctoral dissertation, University of Toronto.

Hunter, M., and A. Britton
   1973   The value of a preoperative printed teaching tool for abdominal hysterectomy patients. Unpublished master's thesis, University of Michigan, Ann Arbor.

Inman, C.
1973   Relationship between a supportive preoperative interview and post-operative discomfort. Unpublished master's thesis, Texas Woman's University.

Jackson, D. A., and K. S. Vaught
1982   Effects of preoperative visits by recovery room nurse. Unpublished master's thesis, University of Evansville.

Jackson, M. A.
1984   Relationship between structured preoperative instruction and anxiety levels of preoperative cardiovascular surgical patients. Unpublished master's thesis, Texas Woman's University.

Jacobs, N.
1983   Comparison of orthopedic surgical patients who receive a structured versus an unstructured program of pre-operative education as to the number of post-operative analgesics required and the number of post-operative hospitalized days. Unpublished master's thesis, Medical College of Georgia.

Janssen, C.
1980   The effect of relaxation technique on comfort level and analgesic usage of postoperative patients. Unpublished master's thesis, California State University.

Johnson, J. E.
1965   Effect of nurse-patient interaction on the patient's postoperative discomfort. Unpublished master's thesis, Yale University.
1966   The influence of purposeful nurse-patient interaction on the patient's postoperative course. In *Exploring Progress in Medical-Surgical Nursing Practice A.N.A. 1965 Regional Clinical Conferences.* New York: American Nurses' Association.

———; S. S. Fuller; M. P. Endress; and V. H. Rice
1978a   Altering patient's response to surgery: An extension and replication. *Research Nursing and Health* 1:111–121.
1978b   Sensory information, instruction in a coping strategy, and recovery from surgery. *Research in Nursing and Health* 1:4–17.

Johnson, L. I.
1966   Nursing intervention to relieve anxieties which accompany stress of surgical procedures. Unpublished master's thesis, University of Iowa.

Kelly, M.
1985   Relationship of audiovisual instruction to postoperative anxiety. Unpublished master's thesis, Texas Woman's University.

Kelly, M. E.
1983   A study of the effects of preoperative group instruction upon reduction of preoperative anxiety. Unpublished master's thesis, Idaho State University.

Kempe, A. R., and R. Gelazis
1985   Patient anxiety levels: An ambulatory surgery study. *AORN Journal* 41(2):390–391, 394, 396.

King, I., and B. Tarsitano
1982   The effect of structured and unstructured pre-operative teaching: A replication. *Nursing Research* 31:324–329.

Klos, D; K. M. Cummings; J. Joyce; J. Graichen; and A. Quigley
1980   A comparison of two methods of delivering presurgical instructions. *Patient Counseling and Health Education* 2:6–13.

Klueh, T.
1981   Effects of preoperative teaching on anxiety of adult surgical patients. Unpublished master's thesis, University of Evansville.

Koch, F. T.
1971   The effect of preoperative instruction by a professional operating room nurse on postoperative anxiety. Unpublished master's thesis, University of Arizona.

Kosik, S. L., and P. J. Reynolds
1986   A nursing contribution to cost containment: A group preoperative teaching program that shortens hospital stay. *Journal of Nursing Staff Development* 2:18–22.

Langer, E. J.; I. L. Janis; and J. A. Wolfer
1975   Reduction of psychological stress in surgical patients. *Journal of Experimental and Social Psychology* 11:155–165.

Langhans, P. P.
1979   The effectiveness of structured preoperative teaching of cardiovascular surgical patients in reducing the incidence of postoperative pulmonary alterations. Unpublished master's thesis, Texas Women's University.

Lawlis, G. F.; D. Selby; D. Hinnant; and C. E. McCoy
1985   Reduction of postoperative pain parameters by presurgical relaxation instructions for spinal pain patients. *Spine* 10:649–651.

Layne, O. L., and S. C. Yudofsky
1971   Postoperative psychosis in cardiotomy patients: The role of organic and psychiatric factors. *New England Journal of Medicine* 284:512–520.

Lech, P.
1979   An experimental study of the effect of preoperative education utilizing selected Lamaze techniques on male surgical patients' post-operative pain. Doctoral dissertation, University of Pittsburgh. *Dissertation Abstracts International* 40:2125B.

Leigh, J. M.; J. Walker; and P. Janaganathan
1977   Effect of preoperative anaesthetic visit on anxiety. *British Journal of Medicine* 2:987–989.

Levesque, L.; R. Grenier; S. Kerouac; and M. Reidy
  1984   Evaluation of a presurgical group program given at two different times. *Research in Nursing and Health* 7:227–236.

Lewis, W.
  1967   The effect of pre-operative information on the anxiety level of patients with open and closed belief-disbelief systems. Unpublished master's thesis, University of California, Los Angeles.

Lierman, L.
  1981   The effects of psychological preparation and supportive care for mastectomy patients during hospitalization. Doctoral dissertation, University of Utah. *Dissertation Abstracts International* 42:2309B.

Lindeman, C. A., and S. L. Stetzer
  1973   Effect of preoperative visits by operating room nurses. *Nursing Research* 22:4–16.

Lindeman, C. A., and B. Van Aernam
  1971   Nursing intervention with the presurgical patient—The effects of structured and unstructured preoperative teaching. *Nursing Research* 20:319–332.

Lindenberg, V.
  1970   The relationship of an intensive preoperative nursing program to the anxiety level and postoperative outcomes in patients with cholecystectomy surgery. Unpublished master's thesis, University of Utah.

Lubno, M. A.
  1981   The effects of preoperative teaching on the fear/anxiety of selected patients. Unpublished master's thesis, Medical College of Georgia.

Lucas, R. H.
  1975   The affective and medical effects of different preoperative interventions with heart surgery patients. Doctoral dissertation, University of Houston. *Dissertation Abstracts International* 36:5763B.

MacMullen, J. A. S.
  1971   A comparative study of group and individual pre-operative instruction. Unpublished master's thesis, Adelphi University.

Marrow, M.
  1967   Effectiveness of practice of deep breathing exercises preoperatively in affecting pulmonary function postoperatively. Unpublished master's thesis, Case Western Reserve University.

Marsh, G W.
  1979   The use of Lamaze techniques to manage post-operative pain. Unpublished master's thesis, University of Colorado.

Martin, H. R.
  1971   The effects of preoperative preparation on postoperative pain. Unpublished master's thesis, Adelphi University.

McCutchan, J.
1986   Preoperative instruction and anxiety levels in day surgery patients. Unpublished master's thesis, University of Evansville.

McGowan, O. L.
1968   The effects of deliberate preoperative instruction and supervised postoperative activities on the recovery rate of a select group of patients having an abdominal hysterectomy. Unpublished master's thesis, University of Alabama.

McLaren, R.
1982   The effect of preoperative teaching on postcardiotomy delirium. Unpublished master's thesis, California State University.

Megel, M. A.
1972   The effect of including a significant other in the preoperative instruction of an adult surgical patient on the practice of postoperative exercises. Unpublished master's thesis, University of Iowa.

Meyer, J. A.
1975   Effects of preoperative programmed instruction of postoperative respiratory complications. Unpublished master's thesis, University of Iowa.

Mikulaninec, C. E.
1986   A comparison of preoperative teaching methods and effects on learning and level of anxiety. Unpublished master's thesis, University of North Carolina.
1987   Effects of mailed preoperative instructions on learning and anxiety. *Patient Education and Counseling* 10:253–265.

Miremadi, L.
1978   Preoperative instruction in coughing and deep breathing: A comparison of methods. Unpublished master's thesis, California State University.

Moeller-Wiegmann, A.
1980   An examination of the effects of pre-operative relaxation training on the post-operative pain experience of cholecystectomy patients. Unpublished master's thesis, University of Kansas.

Mogan, J.; N. Wells; and E. Robertson
1985   Effects of preoperative teaching on postoperative pain: A replication and expansion. *International Journal of Nursing Studies* 22:267–280.

Namei, S. K.
1975   Inclusion of spouses in group preoperative preparation of patients. Unpublished master's thesis, University of Cincinnati.

Niemeyer, M.
1972   The effect of pre-operative information on general surgery patients' rate of recovery. Unpublished master's thesis, University of Michigan.

Noone, J.
    1985   The effect of sensory information on postoperative pain and stress. Unpublished master's thesis, Adelphi University.

Nosbusch, J. M.
    1981   Benson's relaxation technique as a strategy to reduce patients' perception of incisional pain. Unpublished master's thesis, Marquette University.

Olijnyk, F. J.
    1979   A study of the effects of preoperative psychological preparation of cardiosurgical patients. Unpublished master's thesis, University of Rochester.

Opheim, W. A.
    1972   A study to determine if a method of abdominal muscle relaxation taught preoperatively to cholecystectomy patients will help reduce postoperative pain. Unpublished master's thesis, University of Washington.

Ortmeyer, J. A.
    1978   Anxiety and repression coping styles, and treatment approaches in the interaction of elective orthopedic surgical stress. Doctoral dissertation, Northwestern University. *Dissertation Abstracts International* 38:5536A.

Osthoff, E. G.
    1969   The effect of preoperative teaching of deep breathing and coughing technique. Unpublished master's thesis, Catholic University of America.

Owens, J. F., and C. M. Hutelmyer
    1982   The effect of preoperative intervention on delirium in cardiac surgical patients. *Nursing Research* 31:60–61.

Parker, M. E.
    1968   The effect of preoperative teaching of deep breathing and coughing techniques to a group of patients experiencing upper abdominal surgery. Unpublished master's thesis, Catholic University of America.

Peppers, P. A. H.
    1982   Relationship of preoperative teaching and select recovery variables on abdominal surgical patients. Unpublished master's thesis, Emory University.

Pettigrew, A. C.
    1976   Preoperative instruction of sustained maximal inspiration or yawn: The relationship of changes in pre- and postoperative vital capacity and tidal volume. Unpublished master's thesis, University of Cincinnati.

Pickett, C., and G. Clum
1982   Comparative treatment strategies and their interaction with locus of control in the reduction of postsurgical pain and anxiety. *Journal of Consulting and Clinical Psychology* 50:439–441.

Pollok, C. S.
1972   The effects of preoperative instruction in deep breathing on the rate of postoperative return of vital capacity to the baseline level in selected female surgical patients. Unpublished master's thesis, Virginia Commonwealth University.

Ponder, P. M.
1976   Preoperative instruction in self-care measures for prevention or alleviation of postoperative tympanites. Unpublished master's thesis, Emory University.

Powell, I.
1975   Comparison of the incidence of unusual sensory and thought experience in two groups of cardiac surgical patients: Effects of special preoperative information. Unpublished master's thesis, Case Western Reserve University.

Quinn, J. C.
1986   Another dimension in postanesthesia nursing. *Journal of Post Anesthesia Nursing* 1:26–30.

Quinn, J. R.
1979   Structured preoperative preparation of cardiac surgical patients with inclusion of a "significant other" for a smoother postoperative recovery. Unpublished master's thesis, University of Rochester.

Reading, A. E.
1982   The effects of psychological preparation on pain and recovery after minor gynecological surgery: A preliminary report. *Journal of Clinical Psychology* 38:504–512.

Reichbaum, L. S.
1984   The use of relaxation training to reduce stress and facilitate recovery in open-heart surgery patients. Unpublished doctoral dissertation, University of Pittsburgh.

Ridgeway, V., and A. Mathews
1982   Psychological preparation for surgery: A comparison of methods. *British Journal of Clinical Psychology* 21:271–280.

Risinger, M. A.
1976   The effects of instruction concerning pain on postoperative pain experienced by surgical patients. Unpublished master's thesis, University of Iowa.

Risser, N.; A. Strong; and S. Bither
   1980   The effect of an experimental teaching program on postoperative ventilatory function: A self-critique. *Western Journal of Nursing Research* 2:484–500.

Roth, P. A.
   1970   Preoperative nursing intervention in the reduction of postoperative stress. Unpublished master's thesis, University of Arizona.

Rushton, P.
   1977   Preoperative teaching in spinal surgery patients: Is it really beneficial? Unpublished master's thesis, Brigham Young University.

Ryan, C. R.
   1967   The effect of preoperative teaching of deep breathing and coughing techniques. Unpublished master's thesis, Catholic University of America.

Salkeld, E.
   1975   The effects of structured preoperative teaching on the patient's ability to cough and deep breathe postoperatively. Unpublished master's thesis, University of Arkansas.

Schare, B. L.
   1968   Preoperative nursing intervention for the relief of postoperative pain. Unpublished master's thesis, University of Cincinnati.

Schattner, J.
   1978   The value of systematic relaxation as a nursing intervention in reducing preoperative anxiety from the gynecological cancer patient. Unpublished master's thesis, University of Rochester.

Schimitt. F. E.
   1970   Reduction of anxiety in presurgical patients by a nursing intervention utilizing group and individual interactions. Unpublished master's thesis, University of Iowa.

———, and P. J. Wooldridge
   1973   Psychological preparation of surgical patients. *Nursing Research* 22:108–116.

Scovill, N., and J. Brummer
   1988   Preadmission preparation for total hip replacement patients. Unpublished manuscript.

Seaton, R. M.
   1970   A study to determine whether preoperative instruction in deep breathing and coughing produced more effective deep breathing and coughing postoperatively. Unpublished master's thesis, University of Iowa.

Shekleton, M.
  1973 The effect of preoperative instruction in coughing and deep breathing exercises on postoperative vital capacity and peak expiratory flow rate. Unpublished master's thesis, Case Western Reserve University.
  1983 The effect of preoperative instruction in coughing and deep breathing exercises on postoperative ventilatory function. Doctoral dissertation, Rush University. *Dissertation Abstracts International* 45:04B.

Shimko, C.
  1981 The effect of preoperative instruction on state anxiety. *Journal of Neurosurgical Nursing* 13:318–322.

Sicola, V. R.
  1974 Effect of a preoperative teaching program on postoperative recovery of lumbar laminectomy patients. Unpublished master's thesis, Texas Women's University.

Silver, L. A.
  1983 Effects of structured pre-operative teaching on level of pain. Unpublished master's thesis, University of Illinois at Chicago.

Smith, E. G.
  1979 The relationship of formal versus informal preoperative teaching upon the adult major abdominal surgical patient's level of anxiety due to stress. Unpublished master's thesis, Texas Women's University.

Smith, P. P.
  1984 Individualized, detailed preoperative instruction and its effect on anxiety and recovery for the mastectomy patient. Unpublished master's thesis, University of Texas, El Paso.

Solomon, A. J.
  1973 The effect of a psychotherapeutic interview on the physical results of thoracic surgery. Doctoral dissertation, California School of Professional Psychology. *Dissertation Abstracts International* 34:2319B.

Stewart, K. B.
  1985 The effects of structured preoperative teaching on postoperative recovery. Unpublished master's thesis, West Virginia University.

Streltzer, J., and H. Leigh
  1978 Psychological preparation for surgery: The usefulness of a preoperative psychotherapeutic interview. *Hawaii Medical Journal* 37:139–142.

Surmon, O. S.; T. P. Hackett; E. L. Silverberg; and D. M. Behrendt
  1974 Usefulness of psychiatric intervention in patients undergoing cardiac surgery. *Archives of General Psychiatry* 30:830–835.

Swinford, P. A.
  1986 The effect of relaxation and positive imagery in the control of pain in post-surgical adult patients. Unpublished master's thesis, University of Wisconsin, Oshkosh.

Tessier, J.
   1983   The effects of structured preoperative teaching on the anxiety level and the compliance behavior of postoperative patients. Unpublished master's thesis, Medical College of Georgia.

Thomas, K.
   1980   The effects of structured preoperative teaching on postoperative respiratory complications and hospitalization. Unpublished master's thesis, East Carolina University.

Thompson, D. L.
   1986   The use of guided imagery to reduce acute postoperative pain. Unpublished master's thesis, University of Arizona.

Toney, E.
   1986   Effects of preadmission printed information on anxiety and comfort levels of patients in an ambulatory day surgery. Unpublished master's thesis, University of Alaska.

Twitchell, J.
   1977   At-home preoperative teaching. Unpublished master's thesis, University of Arizona.

Van Pittman, C. G.
   1979   The effectiveness of structured preoperative teaching, as compared to unstructured preoperative teaching, in reducing postoperative pulmonary congestion in cardiovascular bypass surgical patients. Unpublished master's thesis, University of Mississippi.

Van Steenhouse, A. L.
   1978   A comparison of three types of presurgical psychological intervention with male open heart surgery patients. Doctoral dissertation, Michigan State University. *Dissertation Abstracts International* 39:1449A.

Voshall, B. A.
   1976   The effects of preoperative teaching on postoperative pain. Unpublished master's thesis, University of Kansas.
   1980   The effect of preoperative teaching on postoperative pain. *Topics in Clinical Nursing* 2:39–43.

Vraciu, J. K., and R. A. Vraciu
   1977   Effectiveness of breathing exercises in preventing pulmonary complications following open heart surgery. *Physical Therapy* 57:1367–371.

Walker, E.
   1977   The influence of structured preoperative instruction on preoperative anxiety in the adult client undergoing major surgery of the hip. Unpublished master's thesis, State University of New York at Buffalo.

Wallace, L. M.
   1984   Psychological preparation as a method of reducing the stress of surgery. *Journal of Human Stress* 10:62–76.

1986   Communication variables in the design of pre-surgical preparatory information. *British Journal of Clinical Psychology* 25:111–118.

Weiss, O. F.; M. Weintraub; K. Sriwatanakul; and L. Lasagna
1983   Reduction anxiety and postoperative analgesic requirements by audiovisual instruction. *Lancet* 1:43–44.

Wells, J. K.; G. S. Howard; W. F. Nowlin; and M. J. Vargas
1986   Presurgical anxiety and postsurgical pain and adjustment: Effects of a stress inoculation procedure. *Journal of Consulting and Clinical Psychology* 54:831–835.

Wells, N.
1982   The effect of relaxation on postoperative muscle tension and pain. *Nursing Research* 31:236–238.

Wendland, D.
1971   The evaluation of effectiveness of a crisis approach to concerns of hysterectomy patients as shown by patient interviews and behavioral observations of anxiety. Unpublished master's thesis, University of California, Los Angeles.

White, C.
1983   Effect of structured preoperative teaching on postoperative anxiety levels of intensive care unit patients. Unpublished master's thesis, Georgia State University.

White, J. S.
1966   The effect of psychiatric nursing care on the reduction of postoperative complications. Unpublished master's thesis, University of Utah.

Whitted, D. R., and M. Doyle
1961   Investigation of the effects of preoperative interviews on selected surgical patients. Unpublished master's thesis, Adelphi University.

Williamson, K. C.
1980   The effects of specific nursing intervention measures on the postoperative pain response of adult abdominal surgery patients. Unpublished master's thesis, University of Kansas.

Wilson, E.
1983   The effect of progressive relaxation in the anticipation phase of the pain experience on the actual postoperative pain sensation with increased activity. Unpublished master's thesis, Marquette University.

Wilson, J. F.
1977   Determinants of recovery from surgery: Preoperative instruction, relaxation training and defensive structure. Doctoral dissertation, University of Michigan. *Dissertation Abstracts International* 38:1476B.
1981   Behavioral preparation for surgery: Benefit or harm? *Journal of Behavioral Medicine* 1:79–102.

Witney, L.
  1976   Preoperative teaching: Group approach. Unpublished master's the-
          sis, University of Illinois at Chicago.

Wong, J., and S. Wong
  1985   A randomized controlled trial of a new approach to preoperative
          teaching and patient compliance. *International Journal of Nursing Stud-
          ies* 22:105–115.

Yakubik, R. A.
  1987   An investigation of the effect of a localized relaxation technique on
          the sensory and affective components of pain in elderly postopera-
          tive laparotomy and inguinal herniorrhaphy patients: A pilot study.
          Unpublished master's thesis, University of Texas, San Antonio.

Young, L., and M. Humphrey
  1985   Cognitive methods of preparing women for hysterectomy: Does a
          booklet help? *British Journal of Clinical Psychology* 24:303–304.

Yvans, E.
  1966   A study to explore the effect of a planned, pre-operative nursing
          visit, with post-operative reinforcement, on the amount of analgesic
          used post-operatively by cholecystectomy patients. Unpublished
          master's thesis, University of Washington.

4

Juvenile Delinquency Treatment: A Meta-Analytic Inquiry into the Variability of Effects

Mark W. Lipsey

One need not look beyond the daily newspaper to establish that crime is a matter of considerable concern in our society. Far less obvious is what should be done about it. As with almost any important matter, this is one on which opinions can differ sharply, not only in the political arena but among social science researchers and criminological experts as well.

Among the many approaches to crime prevention that have been advocated are punishment (deterrence), amelioration of social conditions that produce crime, target hardening, community prevention (e.g., neighborhood watches), and a host of other such notions. Of particular interest here are the options articulated for dealing with an actual or potential perpetrator once he or she (usually he) is identified. The two major schools of thought have been labeled "just desserts" (i.e., punishment proportionate to the offense) and "rehabilitation" (i.e., treatment aimed at reforming the miscreant and preventing future criminal behavior) (Cullen and Gilbert 1982). While these approaches are not necessarily mutually exclusive in practice, they differ so greatly in philosophy that they do not easily coexist within a given criminal justice program.

One domain within which rehabilitation has particular appeal is that of juvenile crime. While juvenile crime is often serious, and unquestionably represents a large proportion of the total criminal activity in a community, the nature of adolescence is generally seen as justifying special handling, a concept institutionalized in the separate juvenile courts in which minors are tried. The most important feature of adolescence, of course, is that it is a formative period marked by behavior that will not necessarily be continued into adulthood. A rehabilitative strategy that shapes the delinquent offender toward more prosocial behavior during this formative stage, therefore, is particularly attractive. Also, since youth have a potentially long adulthood before them, the payoff in reduced criminality over a lifetime resulting from effective preventive intervention at an early age can be substantial.

The potential benefits of rehabilitation for juvenile delinquents can be attained, of course, only if effective intervention is applied. Unfortunately, the question of whether delinquency intervention in general or various specific varieties of intervention are in fact effective has not been convincingly resolved despite decades of research by behavioral scientists. It was the goal of the study described in this chapter to make the most comprehensive and probing attempt to date to synthesize and interpret the large body of research on the effects of preventive or rehabilitative treatment for delinquency. Moreover, given the history of controversy and uncertainty about the results of this body of research, it was deemed important to also attempt to determine why it has proven so difficult to interpret.

This chapter provides an overview of a large meta-analytic survey of the delinquency treatment research conducted over the last four decades. Before turning to a description of the methods and results of this investigation, a brief look at the history of previous attempts to review the delinquency treatment literature will provide some useful context.

Previous Research Reviews

Best known among the reviews of the criminal rehabilitation literature is Lipton, Martinson, and Wilks's broad survey of research on correctional treatments (1975). They made a detailed examination of 231 separate studies involving interventions for both juveniles and adults. Martinson's widely quoted conclusion was that "with few isolated exceptions, the rehabilitative efforts that have been reported so far have had no appreciable effect on recidivism" (1974, p. 25). Greenberg, who updated the Lipton et al. review, echoed that pessimism: "The blanket assertion that 'nothing works' is an exaggeration, but not by very much" (1977, p. 141).

Reviews that have focused exclusively on delinquency treatment have reached similar conclusions. Romig (1978), for example, attempted to identify the characteristics of successful treatment of delinquents, cataloging the available studies with a level of detail rivaling that of Lipton et al. In each category of treatment, however, he found relatively few convincing positive results to report. More critical stances on delinquency treatment research were taken by Wright and Dixon (1977) and Lundman, McFarlane, and Scarpitti (1976). They too reported finding little evidence of significant effects on juvenile crime.

The predominantly negative reviews of rehabilitation that dominated the 1970s were not without challenge. Palmer (1975, 1983), for example, argued that they overlooked many positive instances of success in their haste to generalize and gave little attention to the issues of fit between the type of juvenile and the type of treatment. In a similar vein, Gendreau and Ross (1979) offered "bibliotherapy" to discouraged professionals in the form of a summary of correctional treatments that had produced positive evaluation results. Even more pointed remarks came from Gottfredson (1979), who satirically itemized the "treatment destruction techniques" that critics could use to discredit any promising treatment concept.

The tone for the 1980s was set by the reports of the National Academy of Science's (NAS) Panel on Research on Rehabilitative Techniques (Sechrest, White, and Brown 1979; Martin, Sechrest, and Redner 1981). The first report combed through the available treatment evaluation research, and reviews of that research, and reluctantly concluded that there was indeed little evidence of successful treatment for either juveniles or adults: "Although a generous reviewer of the literature might discover some glimmers of hope, those glimmers are so few, so scattered, and so inconsistent that they do not serve as a basis for any recommendation other than continued research" (Sechrest, White, and Brown 1979, p. 3). *But* the NAS panel emphasized the possibility that the problem was the nature of the evidence rather than the failure of the concept. In particular, they identified a variety of factors essential to credible evaluation research that they found lacking in this literature—well-controlled designs, sensitive measures, strong and well-implemented treatments, and the like.

For purposes of making a broad assessment of delinquency treatment effects, the most important development since the National Academy of Science's report has been the rise of meta-analysis as a technique for aggregating the continuously growing research literature. Unfortunately, the meta-analyses to date have been limited efforts that, in many ways, have raised more questions than they have answered.

The most extensive of these meta-analyses was conducted by Garrett (1984, 1985), who focused on adjudicated delinquents placed in residential facilities, either community or institutional. She examined 111 studies of the effects of treatment programs in such facilities on a wide range of outcome variables. On the other end of the spectrum, Kaufman (1985) restricted his meta-analysis to "prevention" treatment of preadjudicated at-risk juveniles. He looked only at delinquency outcome measures used in randomized research designs using a sample of 20 studies. The meta-analysis by Gottschalk et al. (1987) fell somewhere between these other two efforts. Their sample of 90 studies included only treatment of adjudicated delinquents, but was not restricted to treatment in residential facilities; indeed, only about half the sample was residential.

All three of these meta-analyses found a positive grand mean effect size for the better-designed studies, averaged over studies and outcome measures. The magnitude of this overall effect ranged from around one-fourth to one-third of a standard deviation superiority for the mean treatment group outcome compared with the mean control group outcome. These three studies differed, however, in their assessment of the statistical significance of this effect (Garrett did not test; Kaufman reported significance; Gottschalk et al. reported nonsignificance) and on most other topics examined. In particular, results were inconsistent with regard to the relative efficacy of different treatment modalities, the role of amount or frequency of treatment, and the relationship of research design to study outcome.

More recently, Whitehead and Lab (1989) conducted a meta-analysis of 50 delinquency treatment studies published in journals since 1975 that used control groups and dichotomous recidivism measures. They adopted the Phi coefficient for an effect size index and chose, apparently arbitrarily, a value of .20 as the minimum necessary to consider an effect worthwhile (equivalent to an effect size of about .41 in standard deviation units). A simple count, using no statistical analysis, showed a minority of studies yielding effects as large as .20 and Whitehead and Lab concluded that, therefore, treatment was not effective. From their summary table, the mean Phi coefficient can be computed as .12, equivalent to a difference of about .25 standard deviation units between treatment and control (Cohen 1988). Despite the authors' disparaging conclusions, therefore, this meta-analysis also yielded a positive mean effect of about the same order of magnitude as the previous efforts.

Andrews et al. (1990) responded to the negative conclusions of the Whitehead and Lab meta-analysis with their own reanalysis, augmented by additional studies. They distinguished between appropriate correctional services, defined as those delivered to high-risk cases using modes of treatments matched with client learning styles, and various categories of inappropriate services. They found a mean Phi coefficient of .30 for appropriate services (equivalent to .63 standard deviation units), which was significantly larger than the mean values for inappropriate services. The grand mean over all the studies in their analysis was an effect size of .21 standard deviation units, quite comparable to those found in virtually all the previous meta-analyses.

While meta-analytic reviews of delinquency treatment are reaching somewhat more favorable conclusions than earlier conventional reviews, the efforts of this sort to date have been quite circumscribed. Moreover, the different approaches and restrictions adopted by the various meta-analysts make it difficult to compare their results or find interpretable patterns across them.

The meta-analysis reported here was designed to improve on previous reviews of delinquency treatment research, both conventional and meta-analytic, in the following ways: (1) broadening the coverage of the literature by making an exhaustive search for relevant studies, both published and unpublished; (2) coding sufficient detail from each eligible study to support a probing analysis of the correlates of measured treatment effects, including those stemming from the research methods used as well as from the nature and circumstances of treatment; (3) applying state of the art statistical analysis for meta-analytic data to properly assess the magnitude of the effects found in these studies and the sources of variability in those effects.

This chapter is a preliminary report of the major results from that meta-analytic investigation. It focuses primarily upon the variability in the delinquency treatment effects found in the research literature and identification of the major sources of that variability.

Methods

Eligibility Criteria

Research reports were defined as eligible for inclusion in this meta-analysis according to a set of detailed criteria specified prior to the search for relevant studies and periodically revised to incorporate new distinctions required by ambiguous instances that had to be resolved during the search. In abbreviated form, these criteria were as follows:

1. There had to be some intervention or treatment, broadly defined, that had as its aim (explicitly or implicitly) the reduction, preven-

tion, treatment, remediation, and so forth, of delinquency or antisocial behavior problems similar to delinquency. Delinquency was defined as behavior chargeable under applicable laws whether or not apprehension occurs or charges are brought; antisocial behavior was defined as actions that are threatening, disruptive, or damaging to property, to other persons, or to self. The large category of treatments targeted solely on substance abuse and no other component of antisocial behavior, however, was excluded.

2. The majority of the subjects to whom the treatment was applied had to be juveniles, defined as persons age 21 or younger. To exclude childhood behavior problems without legal implications, however, studies involving juveniles below the age of 12 were not included unless the antisocial behaviors treated were clearly of a type chargeable as delinquent offenses.

3. There had to be measured outcome variables with quantitative results reported that included at least one delinquency measure. Additionally, there had to be some comparison that contrasted one or more designated treatments with one or more designated control conditions on those outcome variables.

4. The treatment versus control comparison groups used in the research had to be based on random assignment of subjects to conditions or, if assignment was nonrandom, there had to be both premeasures and postmeasures on the outcome variable; some evidence of matching prior to treatment; or a range of measures of such characteristics as prior delinquency history, sex, and age which allowed some assessment of the similarity of the treatment and control groups prior to treatment. Pre-test-post-test studies with no control group and post-test-only comparisons between nonrandom groups with no information about group equivalence were not eligible.

5. To maintain some homogeneity in cultural context and social meaning of delinquency, studies had to be set in the United States or a substantially similar English-speaking country (e.g., Canada, Britain, Australia) and reported in English. The juvenile subjects in the study, however, were not required to be English-speaking or "Anglo"; for example, studies of Latino delinquents set in the United States qualified.

6. To restrict the studies to the relatively modern era with regard to criminal justice practices and conceptions of delinquency, studies were eligible only if the date of reporting or publication was 1950 or later, that is, post–World War II.

Identification and Retrieval of Eligible Research Reports

Bibliographic citations for potentially eligible research reports were obtained primarily from three sources. One initial source was the bibliographies of previous literature reviews and meta-analyses—for example, those cited earlier in this chapter. The major source was a comprehensive search in the bibliographic databases of the Dialog system. For this purpose an extensive set of keywords was developed around alternative expressions of the concepts "research," "delinquency," and "treatment." Keyword searches for studies with title, abstract, or index terms representing conjunctions of these three concepts were then conducted in all the Dialog databases that were judged potentially relevant. Appendix 4.A lists the databases examined. In a few instances where keyword search was judged less than optimal, it was supplemented by manual searches through the bibliographic volumes themselves.

A third source of bibliographic information was citations within the reports that were identified by the above procedures and subsequently retrieved and screened for eligibility. These and other incidental sources produced more than 8,000 citations for which available information indicated potential relevance to the meta-analysis. The bibliography is quite complete through 1986 and is currently being updated to include more recent material. It should be noted that the procedures for generating this bibliography included no restrictions according to type of report or nature of publication. Thus books, technical reports, conference papers, theses, and dissertations, as well as published journal articles, were included.

As much as possible of the material identified in this bibliography was located at university libraries in the southern California area, through interlibrary loan or through bulk purchases of microfiche from relevant services. Reports that could not be located through these channels (mostly technical reports and conference papers) were pursued by writing directly to authors whenever addresses could be found either in the original citation or in membership directories for such organizations as the American Psychological Association, American Society of Criminology, and American Sociological Association. At the time of this writing, search and retrieval activities are still continuing but the preponderance of identified material has been either located and screened for eligibility or declared unretrievable after persistent effort.

Coding of the Studies

Each eligible report was coded by a doctoral student in psychology who had been trained in the task through study of a detailed coding manual and supervised practice coding. The coding scheme consisted of 154 items that a coder completed on the basis of the text of the selected report. Additionally, certain of these items were repeated if a study had multiple outcome measures or breakdowns of the results for subsets of subjects or different times of measurement. Table 4.3 (presented later) lists the major variables coded. A brief description of the major categories of information extracted by the coding follows.

EFFECT SIZE. The major treatment group versus control group comparison was selected for each study and all quantitative outcome variables contrasting those two groups were identified. These variables were then divided into those that indexed delinquent behavior and those that represented other behavior or characteristics: for example, school grades, self-esteem. For each such outcome variable, a coding was made of the direction of the effect—that is, whether it favored the treatment group, control group, or neither. Studies without direction of effect information for at least one outcome variable were dropped from further consideration.

Where sufficient quantitative information was reported, an effect size estimate was then computed for each outcome variable. The effect size index used for this purpose was Cohen's d (Cohen 1988), defined generally as the difference between the treatment group mean score and the control group mean score divided by the pooled standard deviations of those scores. The resulting effect size was given a positive value if treatment group performance was superior or "better" than control group performance and a negative value if control group performance was superior. Effect size was thus represented as the number of standard deviation units by which the treatment group outperformed the control group on the identified outcome variable. This is the basic formulation developed and elaborated for meta-analysis by Glass and Hedges (Glass, McGaw, and Smith 1981; Hedges 1981; Hedges and Olkin 1985).

When means and standard deviations were not reported for an outcome variable, which occurred with unfortunate frequency, the effect size was estimated, if possible, from whatever statistical information was reported—p, t, or F values, contingency tables, and the like. A common, but not universal, form for reporting delinquency outcome in the studies was recidivism rate: the proportion of subjects in each experimental group who were rearrested, reconvicted, or whatever subsequent to treatment. Such proportion and percentage data were converted to effect size estimates using the arcsine transformation described in Cohen (1988).

Effect size estimates were computed using the statistics available for the comparison of treatment and control groups on each outcome variable without attempting to adjust for any lack of comparability between the groups at the time of measurement. These effect sizes, of course, may be biased upward or downward by such factors as nonrandom designs that yield initial nonequivalence between the experimental groups and attrition from either or both groups after assignment to experimental conditions. The approach that was taken to this problem was to code separately the information that was available in each study regarding these matters, that is, nonequivalence and attrition. That information was subsequently used in the analysis to determine the nature of its relationship with effect size and, where relationships were found, to partial them out statistically before considering what effects might be attributable to treatment. Similarly, the details of the various measures—for example, source, type, and period covered for delinquency measures—were coded so that differences in effect size that were related to the metric used could also be statistically controlled in later analyses.

In addition to the overall "aggregate" comparison between treatment and control groups on each outcome variable, many of the studies also reported results for subgroups: for example, males versus females. When possible, effect size estimates were also computed for these "breakdown" groups separately: for example, effect size for males and effect size for females. Moreover, both "aggregate" and "breakdown" comparisons sometimes had follow-up measures, that is, outcome variables measured at more than one time subsequent to treatment. When possible, effect size estimates were separately computed for each follow-up comparison.

The full coding on effect size, therefore, represented delinquency and nondelinquency outcome variables for the aggregate treatment versus control comparison, any breakdowns of that comparison, and any follow-up comparisons of either the aggregate or the breakdown comparison. This chapter reports only on delinquency outcome for the aggregate comparison at the first point of measurement subsequent to treatment. It is these data that give the broadest overview of the effects of delinquency treatment.

METHOD VARIABLES.   To enable inquiry into the relation between the methodology used in a study and the effects found in that study, a wide range of information was coded about study design, measures, samples, attrition, and the like. Particular attention was given to the extent of initial equivalence between treatment and control groups prior to application of the treatment.

STUDY CONTEXT.   When available, information was coded about the year and form of publication of each study, country in which it was conducted, source of funding, and characteristics of the researcher—for instance, institutional affiliation and discipline.

NATURE OF TREATMENT.   An important set of issues, of course, has to do with the characteristics of the treatments used in the various studies and their relationships to the study outcome. Accordingly, information was coded regarding the treatment type, setting, sponsorship, duration, intensity, and a wide range of other such features.

NATURE OF SUBJECTS.   Study outcome may also vary with the nature of the juvenile subjects treated. The coding scheme recorded, where available, information about the demographic characteristics of the juveniles (e.g., race, sex, age), prior delinquency history, and other such matters.

Results

The analyses presented here focus on the distribution of measured delinquency effects in the studies coded for this meta-analysis; that is, on the effect size indices for the treatment versus control group comparisons on delinquency outcome variables. Results on other (nondelinquency) outcomes will be reported in later papers. At the time of this writing 443 studies were coded and available for analysis and the results that follow are based on that set.

Effect Size Distribution for Delinquency Outcomes

Many of the studies had multiple delinquency outcome variables. Creating a distribution of effects for all of them would have overrepresented studies that reported more variables and underrepresented those that reported fewer variables. This distortion was judged undesirable because of both the statistical dependencies created by using multiple effects from a single study and the potential misrepresentation of the pattern of outcomes across studies. Instead, an analysis of all delinquency measures was first done to identify the types of variables that were most commonly used in this literature. Then, when multiple delinquency outcomes were available in a study, a single one was selected for analysis according to criteria designed to identify that representing the category most widely used in other studies. This selection was done blindly with regard to the effect size on the various candidate measures. Since the most common delinquency outcome measure was some variation on the concept of recidivism—rearrest, reconviction, and so on—this process acted to favor measures of that sort. The measures selected in this manner will be referred to as "primary" delinquency measures. (Table 4.3, which will be discussed in more detail later, provides descriptive information about these measures, especially in items 32 through 37.)

After selection of the primary delinquency measure for each study, the effect size for that measure was weighted by the coefficient developed by Hedges (1981) to correct for bias in estimation. Effect sizes based on small samples tend to run larger than the population values that they estimate and must be reduced proportionately. The specific weighting coefficient used for all effect sizes in this study was $1 - (3/(4n_t + 4n_c - 9))$ where n_t is the sample size for the treatment group and n_c is the sample size for the control group (Hedges 1981; Hedges and Olkin 1985). Where it is necessary to be specific, this formulation of the effect size will be referred to as the "n-adjusted effect size."

DIRECTION AND SIZE OF EFFECTS.   We are now in a position to examine the effect size distribution for evidence regarding the efficacy of delinquency treatment. The treatment effect information that covers the largest number of studies is the coding of the simple direction of the difference between the treatment and control groups. A difference favoring treatment indicates that the treatment group had a better outcome (less delinquency) than the control group; a difference favoring the control group indicates that the treatment group had a worse outcome; a difference favoring neither group indicates that they were exactly equal or that the original study reported no significant difference without providing actual values. Table 4.1 shows the breakdown of direction of effect for the primary delinquency measures, one from each of the 443 studies in the present analysis.

If, in the general case, treatment is not effective in reducing delinquency we would still expect some positive and negative differences due to sampling error, but would expect them in equal proportions. If

Table 4.1 **Direction of Treatment versus Control Group Differences on Primary Delinquency Outcome Measure for All Studies**

	N	%
Favors Treatment	285	64.3
Favors Control	131	29.6
Favors Neither	27	6.1
Total	443	

Binomial test (by z approximation) that population proportions are .50/.50: $z = 7.32$ $p<.001$ (hypothesis rejected).

we take the relatively few cases in which neither treatment nor control group is favored and divide them evenly between the other two categories, the proportions can be tested by the normal distribution approximation to the binomial to determine if they depart from the expectation of a 50-50 split (Siegel 1956). As indicated in Table 4.1, the large skew toward differences that favor the treatment group is statistically significant.

To get information about the magnitude (not just the direction) of the differences between treatment and control groups in these studies, we must turn to the computed effect size values. Since not all studies in the database provided sufficient information for computation of effect size, a smaller set is available for this analysis ($n = 397$). Direction of effect breakdowns for this subset of studies (not shown) were virtually identical with those shown in Table 4.1 for all studies.

Figure 4.1 presents the distribution of n-adjusted effect size values on the primary delinquency outcome measures and summary statistics for the distribution. The mean and median effect size values are positive, though numerically modest, showing that on average these studies found lower delinquency for treatment groups than control groups. The unweighted mean of .172 is the value most directly comparable to the results of the other delinquency meta-analyses reviewed in the introduction to this chapter. Recall that those values were in the range of .20 to .33 but represented more highly selected sets of studies.

It should be noted that the effect size values represented in Figure 4.1 give equal representation to each study irrespective of its sample size (other than the slight correction already applied for biased estimation in small samples). Thus one effect size is contributed to this distribution by a study of $n = 10$ and, similarly, only one is contributed by a study of $n = 1,000$. Since effect size estimates based on large

Figure 4.1   Distribution of Unweighted *n*-Adjusted Effect Sizes for Primary Delinquency Measures

Count	ES	
2	−1.20	xx
2	−1.10	xx
0	−1.00	
1	−.90	x
1	−.80	x
4	−.70	xxxx
6	−.60	xxxxxx
7	−.50	xxxxxxx
9	−.40	xxxxxxxxx
11	−.30	xxxxxxxxxxx
26	−.20	xxxxxxxxxxxxxxxxxxxxxxxxxx
33	−.10	xxxxxxxxxxxxxxxxxxxxxxxxxxxxxxxxx
49	.00	xxxxxxxxxxxxxxxxxxxxxxxxxxxxxxxxxxxxxxxxxxxxxxxxxxx
20	.00	xxxxxxxxxxxxxxxxxxxx
48	.10	xxxxxxxxxxxxxxxxxxxxxxxxxxxxxxxxxxxxxxxxxxxxxxxxxx
33	.20	xxxxxxxxxxxxxxxxxxxxxxxxxxxxxxxxx
38	.30	xxxxxxxxxxxxxxxxxxxxxxxxxxxxxxxxxxxxxx
18	.40	xxxxxxxxxxxxxxxxxx
25	.50	xxxxxxxxxxxxxxxxxxxxxxxxx
13	.60	xxxxxxxxxxxxx
14	.70	xxxxxxxxxxxxxx
12	.80	xxxxxxxxxxxx
4	.90	xxxx
8	1.00	xxxxxxxx
1	1.10	x
2	1.20	xx
1	1.30	x
3	1.40	xxx
0	1.50	
3	1.60	xxx
1	1.70	x
2	1.80	xx

Summary Statistics:

Number of cases	397
Unweighted mean	.172
Median	.100
Standard deviation	.438
Variance	.192

samples are statistically more reliable than those based on small samples, some accommodation of the sample size differences must be made in order to test the statistical significance of the means of the effect size distribution.

Hedges and Olkin (1985) have shown that the optimal procedure is to weight each effect size inversely by its variance (which reflects sam-

Table 4.2   Statistical Tests for Effect Size Means and Homogeneity

A. *n*-Adjusted Effect Sizes for All Studies
   Inverse-variance weighted ES mean .103   (*n* = 397)
   .99 confidence interval for mean .083 to .123
   Inverse-variance weighted ES variance .089
   Homogeneity test statistic H = 1319.00 df = 237
   Chi-square .01 critical value 273.78
B. *n*-Adjusted Effect Sizes for Studies with Random Assignment
   Inverse-variance weighted ES mean .110   (*n* = 294)
   .99 confidence interval for mean .086 to .134
   Inverse-variance weighted ES variance .080
   Homogeneity test statistic H = 904.14   df = 293
   Chi-square .01 critical value 351.46
C. *n*-Adjusted Effect Sizes for Studies with Random Assignment
   and No Appreciable Attrition from Experimental Groups
   Inverse-variance weighted ES mean .140   (*n* = 78)
   .99 confidence interval for mean .094 to .186
   Inverse-variance weighted ES variance .090
   Homogeneity test statistic H = 281.08   df = 77
   Chi-square .01 critical value 107.98

ple size). This was done in the present data with the restriction that treatment and control *n* were separately Windsorized at 300 to prevent a few very large studies from dominating the results.

With inverse-variance weighted effect sizes, a confidence interval can be determined and the statistical significance of the mean effect size for the distribution in Figure 4.1 can be assessed. Table 4.2 (part A) summarizes the statistical information for this procedure: the inverse-variance weighted mean, the confidence interval, and some homogeneity statistics that will be discussed later.

The inverse-variance weighted effect size mean shown in Table 4.2 (part A) is positive and very similar to the median value for the unweighted effect sizes shown in Figure 4.1. The .99 confidence interval does not include zero; thus the positive mean effect displayed here is statistically significant.

One might wonder, however, if the positive value of the mean effect size is simply a reflection of biased results in the studies represented. In particular, since many of these studies did not use randomly assigned control groups, the positive mean effect size may indicate only initial nonequivalence between the treatment and comparison group reappearing as a pseudo-treatment effect in the outcome measures. If, for example, the juveniles selected for treatment were less delinquency prone, on average, than those selected for comparison groups, post-treatment outcome measurement would be expected to show differences favoring treatment.

Table 4.2 (parts B and C), reports the results of two tests of this possibility. First, the mean effect size and confidence interval were determined for that subset of studies which reported random assignment to experimental conditions ($n = 294$). Second, and more probing, the mean effect size and confidence interval were computed for the subset of studies which reported both random assignment and no (or trivial) attrition from the experimental groups between assignment and outcome measurement ($n = 78$). This latter case excludes studies that used random assignment but, prior to outcome measurement, may have lost that initial equivalence because of differential attrition from the treatment and control groups.

As Table 4.2 shows, the mean effect size for each of these selected subsets of studies is positive and, indeed, very similar to the mean for all the studies together. The modest differences are in the positive direction; that is, the better controlled studies yielded slightly larger mean effect sizes than the general mix of studies. Moreover, the confidence intervals indicate that the mean effect sizes for the selected subsets of studies are statistically significant. It does not appear that the overall positive effect size mean can be attributed to bias resulting from inclusion of results from poorly controlled studies.

The answer to the general question "Does treatment reduce delinquency?" therefore appears to be "Yes, on average there is a positive effect." But, while positive and statistically significant, the mean effect sizes found here appear relatively modest. If we take the inverse-variance weighted mean for the distribution of effects on the primary delinquency measures as the standard, treated juveniles showed about .10 standard deviation units less delinquency subsequent to treatment than did the control juveniles. At first impression, this sounds quite trivial.

This figure is more meaningful if we translate it into something more directly relevant than standard deviation units. Since the modal measure represented in these data is a rearrest recidivism rate, one alternative is to express the mean effect in those terms. If we assume that control groups without treatment recidivate at the rate of 50 percent, which is about the mean value for those studies that used simple dichotomous recidivism measures ($n = 208$), we can convert the treatment-control difference from standard deviation units to percentages using the arcsine transformation from Cohen (1988). This procedure shows that .10 standard deviation units is equivalent to a decrease of 5 percentage points from a 50 percent baseline. In other words, the mean treatment effect of .10 standard deviation is equivalent to a reduction in average recidivism from 50 to 45 percent. This formulation of the effect is much more interpretable and, while it still shows that the result is modest, does reveal that it is not trivial. A reduction in recidivism of 5 percentage points from a baseline of 50 percent amounts to a 10 percent decrease in recidivism (5/50). While a 10 percent average drop may not be spectacular, it cannot be said to be obviously negligible.

Moreover, the true effect represented in these studies is almost certainly larger than these figures indicate. Effect sizes are attenuated by the unreliability of the study outcome measures upon which they are calculated. With few exceptions, the measures in the present collection of studies represent some aspect of officially recorded delinquency—arrests, probation violations, reconvictions, and the like. It is well known that officially recorded contacts represent a small proportion of the total number of delinquent behaviors in which a juvenile engages (Williams and Gold 1972). As a result, it is largely a matter of chance whether a particular delinquent act eventuates in an officially recorded contact with an agent of law enforcement or the juvenile justice system. This large chance component makes such delinquency measures very unreliable. Lipsey (1982, 1983) estimated their reliability to be around .20–.30.

We can correct the inverse-variance weighted mean effect size of .10 for the attenuation that would result from the low reliability of the delinquency outcome measures at issue simply by dividing by the square root of the reliability (Hedges 1981). If we assume reliability of .25, the resulting deattenuated mean effect size is .20; that is, it doubles. Translating this into simple dichotomous recidivism terms, we find that it is equivalent to a decrease in a treatment group of 10 percentage points from a control baseline of 50 percent recidivism. Or, in overall percentage terms, it is equivalent to a 20 percent decrease in recidivism (10/50). Without the masking effect of highly unreliable delinquency measures, therefore, the overall treatment effect found in this meta-analysis could be quite large enough to have practical significance.

Despite this relatively positive finding with regard to delinquency treatment, some care must be taken in the conclusions drawn at this point. What we have shown is that the average treatment versus control difference in these studies favors the treatment group. The extent to which that difference reflects the efficacy of the treatments employed, rather than some other feature of these studies—for example, some methodological characteristic—is still in some doubt (though Table 4.2 seems to rule out one of the more obvious possibilities). The grand mean effect size averaged over so many diverse studies is rather like a main effect in a complex analysis of variance design: It generalizes over all the other factors and interactions that may be influencing the outcome to make a crude overall comparison. Before interpreting that main effect, we need to determine whether it is equally representative of the results of all the types of studies in the database. This, in turn, requires testing of the homogeneity of the effect size distribution.

HOMOGENEITY TESTS. If the values in the effect size distribution are tightly clustered around the mean—for example, varying no more than would be expected by sampling error—that mean is a reasonable representation of the outcome of each and all of the studies. If the variation is great, however, the mean may not represent any distinct group of studies and may be quite misleading. Of particular concern in the present context is the possibility that methodologically low-quality studies would spuriously yield larger effect sizes than higher-quality studies, thus biasing the distribution upward and overstating the magnitude of the actual effects of treatment.

Hedges (1982) has developed a test of the homogeneity of effect sizes that is useful in this regard. It requires computation of a term, H, that can be tested with the chi-square distribution. Table 4.2 reports the summary statistics for the homogeneity tests on the distributions of effect sizes for all studies and for the selected subsets of studies. All three distributions show significant heterogeneity. Indeed, for the full set of studies the H statistic, which is a sum of squares term, shows more than three times as much heterogeneity as would be expected on the basis of sampling error alone.

The task to which we now turn is attempting to identify the sources of the variability in the effect size distributions. In particular, we want to try to determine the extent to which variation in study methodology contributes to effect size variation in contrast to differences among studies on the substantive factors of treatment and subject type.

Analyzing Effect Size Variability

If some of the variability in effect sizes is systematically related to differences in the studies from which they originate, we should be able to find a pattern of correlations between the relevant study characteristics and effect size. Our ability to investigate such correlations is limited by the availability of variables representing study characteristics in the meta-analysis which, in turn, is limited by what authors report when they write up their studies. Table 4.3 lists the major study characteristics that were coded in the present meta-analysis, reports the frequency breakdown on each for the 443 studies in the present database, and indicates the proportion of studies for which information on the item was unavailable.

For purposes of analyzing effect size variability, study characteristics were grouped into 11 clusters (all but "Outcome" on Table 4.3). These clusters, in turn, represent three larger categories—study context, method, and treatment. The clusters are listed descriptively below with a shorthand label for each. They are sequenced from the more fundamental and general methodological issues to the more study-specific issues of treatment and study context. The items included in each are marked with an asterisk or double asterisk in Table 4.3.

Method
    Experimental groups, sample size, sampling (Samples)
    Initial equivalence of experimental groups (Equivalence)
    Attrition from experimental groups (Attrition)
    Characteristics of the control condition (Control)
    Characteristics of the delinquency outcome measures (Measures)
    Information about the effect size computation (ES Info)

Treatment
    Characteristics of subjects/clients treated (Subjects)
    Amount or intensity of treatment (Dosage)
    Characteristics of the condition (Treatment)
    Treatment philosophy and context (Tx Philos)

Study Context
    Country, publication year, author's discipline, etc. (Context)

A straightforward approach to analyzing the variability of a single dependent variable (effect size in this case) as a function of various independent variables (such as those in the above clusters) is multiple regression. To employ this technique, however, a number of procedural and conceptual issues must be faced.

One problem, noted earlier, is the uneven sample sizes upon which the effect sizes are based. A study with a large sample should be given more weight in the analysis than one with a small sample since it represents information about the response of more people and yields more reliable results. Hedges and Olkin (1985) have shown that the same inverse-variance weights that were used earlier to compute effect size means, confidence intervals, and homogeneity statistics can be applied

Table 4.3 Descriptive Data for Major Variables Coded

	N	%		N	%
STUDY CONTEXT					
1. Country of Study			Conference paper	7	1.6
United States	407	91.9	Missing	0	0.0
Canada	12	2.7	6. Year of Publication		
Britain	15	3.4	1950–1959	5	1.1
Other	7	1.6	1960–1969	58	13.1
Missing	2	0.5	1970–1979	207	46.7
2. Author's Discipline			1980–1987	166	37.5
Psychology	135	30.5	Missing	7	1.6
Criminal justice	68	15.3			
Sociology	43	9.7	METHOD		
Education	43	9.7	*Experimental Groups,*		
Social work	24	5.4	*Sample Size, Sampling*		
Psychiatry/medicine	12	2.7	*(Samples)*		
Political science	8	1.8	7. Number of Treat-		
Other	7	1.6	ment Groups in		
Missing	103	23.3	Design**		
3. Author's Affiliation			One	364	82.2
Academic	246	55.5	Two	52	11.7
Government agency	53	12.0	Three	18	4.1
Program agency	88	19.9	More	6	1.4
Research firm	30	6.8	Missing	3	0.7
Other	2	0.5	8. Number of Control		
Missing	24	5.4	Groups in Design		
4. Source of Research			One	341	77.0
Funding			Two	79	17.8
Agency/organiza-			Three	15	3.4
tion	127	28.7	More	2	0.4
Federal	126	28.4	Missing	6	1.4
State/local govern-			9. Post-Test Total		
ment	56	12.6	Sample Size**		
Funded, unknown			1–25	38	8.6
source	15	3.4	26–50	59	13.3
No funding indi-			51–75	47	10.6
cated	117	26.4	76–100	39	8.8
Missing	2	0.5	101–150	65	14.7
5. Type of Publication			151–200	46	10.4
Journal/book chap-			201–300	47	10.6
ter	168	37.9	301–500	30	6.8
Technical report	192	43.3	501–800	21	4.7
Dissertation/thesis	44	9.9	801+	27	6.1
Book	32	7.2	Missing	24	5.4

Table 4.3 *(Continued)*

	N	%			N	%
10. Method Quality: Representativeness of Sampling**			Matched groupwise		22	5.0
Low	140	31.6	Random with serious degradation		27	6.1
Moderate	164	37.0	Individual selection			
High	138	31.2	(e.g., by need)		32	7.2
Missing	1	0.2	Convenience comparison group		28	6.3
11. Method Quality: Statistical Power**			Missing		1	0.2
Low	227	51.2	14. Confidence/Explicitness of Assignment Procedure*			
Moderate	113	25.5				
High	103	23.3	Very low		1	0.2
Missing	0	0.0	Low		13	2.9
Initial Equivalence of Experimental Groups (Equivalence)			Moderate		36	8.1
			High		111	25.1
			Very high		279	63.0
12. Unit on Which Assignment to Experimental Groups Based			Missing		3	0.7
			15. Method Quality: Treatment/Control Group Comparability*			
Individual	409	72.3				
Intact group	20	4.5	Low		88	19.9
Program area	10	2.3	Moderate		202	45.6
Missing	4	0.9	High		153	34.5
13. Procedure for Assignment to Groups**			Missing		0	0.0
			16. Rating: Overall Similarity of Treatment and Control*			
Random after matching	61	13.8	Very similar	1	12	2.7
Random, no matching	134	30.2		2	114	25.7
Regression discontinuity	4	0.9		3	141	31.8
				4	75	16.9
Wait list control	12	2.7		5	58	13.1
Nonrandom, matched on pretest	14	3.2		6	27	6.1
			Very different	7	3	0.7
Nonrandom, matched on individual features	37	8.4	Missing		13	2.9
			17. Confidence/Explicitness of Group Similarity*			
Nonrandom, matched on demographics	71	16.0	Very low		2	0.5
			Low		15	3.4
			Moderate		190	42.9

	N	%		N	%
High	188	42.4	Favors control	71	16.0
Very high	37	8.4	Favors neither	66	14.9
Missing	11	2.5	Missing	212	47.9
18. Researcher's Comparison of Treatment/Control Equivalence*			22. Direction of Treatment/Control Ethnicity Difference*		
No comparisons made	95	21.4	Favors treatment	64	14.4
No statistically significant differences	96	21.7	Favors control	67	15.1
Significant differences unimportant	27	6.1	Favors neither	50	11.3
Significant differences uncertain	51	11.4	Missing	262	59.1
Significant differences important	26	5.9	23. Direction of Treatment/Control Delinquency History Difference*		
Descriptive differences unimportant	71	16.0	Favors treatment	66	14.9
Descriptive differences uncertain	46	10.4	Favors control	59	13.3
Descriptive differences important	20	4.5	Favors neither	32	7.2
Missing	11	2.5	Missing	286	64.6
19. Direction of Treatment/Control Pre-Test Difference*			24. Direction of Treatment/Control Delinquency Typology Difference**		
Favors treatment	64	14.4	Favors treatment	24	5.4
Favors control	53	12.0	Favors control	22	5.0
Favors neither	7	1.6	Favors neither	23	5.2
Missing	319	72.0	Missing	374	84.4
20. Direction of Treatment/Control Sex Difference**			*Attrition from Experimental Groups*		
Favors treatment	53	12.0	*(Attrition)*		
Favors control	55	12.4	25. Treatment Group N Change from Pre- to Post-Test**		
Favors neither	99	22.3	Gain	9	2.0
Missing	236	53.3	Loss	108	24.4
21. Direction of Treatment/Control Age Difference**			No difference	241	54.4
			Missing	85	19.2
			26. Control Group N Change from Pre- to Post-Test**		
			Gain	11	2.5
			Loss	96	21.7
Favors treatment	94	21.2	No difference	246	55.5

Table 4.3 *(Continued)*

	N	%		N	%
Missing	90	20.3	31. Number of Delin-		
27. Method Quality:			quency Outcome		
Attrition			Measures Not		
Problems**			Codable*		
Low	136	30.7	None	330	74.5
Moderate	183	41.3	One	43	9.7
High	115	26.0	Two	24	5.4
Missing	9	2.1	Three	12	2.7
Characteristics of the Con-			Four	8	1.8
trol Condition (Control)			Five	6	1.4
28. Type of Control			More	11	2.4
Condition**			Missing	9	2.0
No treatment	57	12.9	32. Weeks After Treat-		
Wait list	17	3.8	ment Begins When		
Minimal contact	32	7.2	Primary Measure		
Treatment as usual	307	69.3	Taken**		
Placebo	18	4.1	1–13	79	17.8
Other	6	1.4	14–26	114	25.7
Missing	6	1.4	27–52	111	25.1
29. Confidence/Explicit-			53–112	61	13.8
ness of Control			113+	32	7.2
Condition*			Missing	46	10.4
Very low	0	0.0	33. Period Covered in		
Low	4	0.9	Primary Delin-		
Moderate	45	10.2	quency Measure-		
High	129	29.1	ment, Weeks**		
Very high	255	57.6	1–13	60	13.5
Missing	10	2.3	14–26	131	29.6
Characteristics of the De-			27–52	130	29.3
linquency Outcome			53–112	52	11.7
Measures (Measures)			113+	30	6.8
30. Number of Delin-			Missing	40	9.0
quency Outcome			34. Type of Delinquency		
Measures Coda-			Represented in		
ble**			Primary Measure*		
One	164	37.0	Antisocial behavior	24	5.4
Two	86	19.4	Unofficial delin-		
Three	65	14.7	quency	19	4.3
Four	38	8.6	School disciplinary	12	2.7
Five	27	6.1	Arrests/police con-		
More	47	10.6	tact	195	44.0
Missing	16	3.6	Probation contact	35	7.9

	N	%			N	%
Court contact	80	18.1	sure Demonstrated?			
Parole contact	25	5.6	Yes		16	3.6
Institutional disciplinary	15	3.4	No		427	96.4
Institutionalization	28	6.3	39. Reliability of Primary Delinquency Measure Demonstrated?			
Catchment area indicator	4	0.9				
Missing	6	1.4	Yes		22	5.0
35. Range of Offenses Covered in Primary Measure*			No		421	95.0
			40. Sensitivity of Primary Delinquency Measure Demonstrated?			
All offenses	385	86.9				
Status offenses only	10	2.3				
Other restricted	37	8.4	Yes		1	0.2
Missing	11	2.5	No		442	99.8
36. Type of Scaling of Primary Delinquency Measure*			41. Rating: Overlap of Measure with Content of Treatment**			
Dichotomous recidivism	247	55.8	Very low	1	137	30.9
Summed dichotomy	9	2.0		2	82	18.5
Frequency or rate	141	31.8		3	58	13.1
Severity index	11	2.5	Moderate	4	61	13.8
Event timing	6	1.4		5	39	8.8
Rating of amount	8	1.8		6	25	5.6
Other	8	1.8	Very high	7	37	8.4
Missing	13	2.9	Missing		4	0.9
37. Source of Data for Primary Delinquency Measure**			42. Rating: Potential for Social Desirability Bias*			
Self-report, juvenile	33	7.4	Very low	1	311	70.2
Therapist, teacher, etc.	14	3.2		2	62	14.0
School records	14	3.2		3	20	4.5
Police records	127	28.7	Moderate	4	9	2.0
Probation records	41	9.3		5	14	3.2
Court records	114	25.7		6	9	2.0
Institutional records	52	11.7	Very high	7	14	3.2
Other records	5	1.1	Missing		4	0.9
Missing	43	9.7	43. Confidence/Explicitness re Overlap and Social Desirability*			
38. Validity of Primary Delinquency Mea-						

Table 4.3 *(Continued)*

	N	%			N	%
Very low	2	0.5		Missing	7	1.6
Low	6	1.4	47. Confidence/Explicit-			
Moderate	46	10.4	ness re Delin-			
High	218	49.2	quency Risk*			
Very high	164	37.0	Very low	1	0.2
Missing	7	1.6	Low	4	0.9
44. Method Quality:			Moderate	59	13.3
Psychometric			High	160	36.1
Properties of Pri-			Very high	213	48.1
mary Measure**			Missing	6	1.4
Low	286	64.6	48. Proportion of Juve-			
Moderate	106	23.9	niles with Prior			
High	51	11.5	Offense History*			
Missing	0	0.0	None	16	3.6
45. Method Quality:			Some	62	14.0
Blinding in Collec-			Most	68	15.3
tion of Outcome			All	206	46.5
Data*			Some, can't esti-			
Low	287	64.8	mate	50	11.3
Moderate	90	20.3	Missing	41	9.3
High	55	12.4	49. Predominant Type of			
Missing	11	2.5	Prior Offenses			
			No priors	18	4.1
TREATMENT			Mixed	149	33.6
Characteristics of Subjects/			Person crimes	6	1.4
Clients Treated (Sub-			Property crimes	91	20.5
jects)			Status offenses	39	8.8
46. Level of Delinquency			Other	11	2.3
Risk/Involvement**			Missing	129	29.1
Nondelinquent,			50. Aggressive History			
normal	3	0.7	of Juveniles*			
Nondelinquent,			No	116	26.2
symptomatic	26	5.9	Yes, some juveniles	91	20.5
Predelinquents	64	14.4	Yes, most juveniles	7	1.6
Delinquents	155	35.0	Yes, all juveniles	7	1.6
Institutionalized,			Some, can't esti-			
nonjuvenile justice	7	1.6	mate	69	15.6
Institutionalized,			Missing	153	34.5
juvenile justice	87	19.6	51. Sex of Juveniles*			
Mixed, low end	37	8.4	No males	10	2.3
Mixed, high end	33	7.4	Some males	26	5.9
Mixed, full range	24	5.4	Mostly males	188	42.4

	N	%
All males	154	34.8
Some, can't estimate	18	4.1
Missing	47	10.6
52. Average Age of Juveniles at Time of Treatment**		
6–11	8	1.8
12	7	1.6
13	38	8.6
14	92	20.8
15	100	22.6
16	83	18.7
17	22	5.0
18	15	3.4
19	19	4.3
20–21	6	1.4
Missing	53	12.0
53. Predominant Ethnicity of Juveniles**		
Anglo	143	32.3
Black	52	11.7
Hispanic	8	1.8
Other minority	2	0.5
Mixed, none >60%	70	15.8
Mixed, can't estimate	32	7.2
Missing	136	30.7
54. Rating: Overall Heterogeneity of Treated Juveniles**		
Very homogeneous 1	2	0.5
2	98	22.1
3	142	32.1
Moderately heterogeneous 4	82	18.5
5	67	15.1
6	23	5.2
Very heterogeneous 7	4	0.9
Missing	25	5.6

	N	%
55. Confidence/Explicitness re Information on Heterogeneity		
Very low	2	0.5
Low	25	5.6
Moderate	209	47.2
High	179	40.4
Very high	2	0.5
Missing	26	5.9
56. Source of Clients for Treatment**		
Voluntary, family	14	3.2
Non–criminal justice agency	33	7.4
Criminal justice agency, voluntary	142	32.1
Criminal justice agency, mandatory	201	45.4
Multiple sources	14	3.2
Researcher solicits	30	6.8
Missing	9	2.1
Amount or Intensity of Treatment (Dosage)		
57. Duration, Weeks from First to Last Treatment Event**		
1–6	69	15.6
7–13	60	13.5
14–26	108	24.4
27–39	51	11.5
40–52	52	11.7
53–78	9	2.0
79–112	18	4.1
113+	10	2.3
Missing	66	14.9
58. Frequency of Treatment Contact**		
Continuous	71	16.0
Daily	55	12.4
2–4 per week	48	10.8
1–2 per week	151	34.1

Table 4.3 *(Continued)*

		N	%			N	%
Less than weekly		45	10.2		5	90	20.3
Missing		73	16.5		6	70	15.8
59. Mean Hours Contact				Substantial	7	21	4.7
per Week*				Missing		35	7.9
Less than 1		45	10.2	63. Rating: Intensity of			
1–2		108	24.4	Treatment Event**			
3–5		44	9.9	Weak	1	11	2.5
6–10		30	6.8		2	54	12.2
11–20		12	2.7		3	108	24.4
21–30		9	2.0	Moderate	4	119	26.9
31–50		10	2.3		5	75	16.9
51–100		4	0.9		6	27	6.1
Continuous		70	15.8	Strong	7	9	2.0
Missing		111	25.1	Missing		40	9.0
60. Mean Total Number				64. Confidence/Explicit-			
of Hours of Con-				ness re Ratings of			
tact**				Amount/Intensity*			
1–10		65	14.7	Very low		9	2.0
11–20		32	7.2	Low		34	7.7
21–40		42	9.5	Moderate		191	43.1
41–100		40	9.0	High		162	36.6
101–200		37	8.4	Very high		15	3.4
201–1,000		35	7.9	Missing		32	7.3
1,000+		8	1.8	65. Evidence of Degra-			
Continuous		71	16.0	dation in Treat-			
Missing		113	25.5	ment Delivery**			
61. Confidence/Explicit-				Yes		132	29.8
ness of Informa-				Possible		68	15.3
tion on Treatment				No		195	44.0
Amount*				Missing		48	10.8
Very low		7	1.6	66. Method Quality: In-			
Low		46	10.4	tegrity of Treat-			
Moderate		111	25.1	ment Implementa-			
High		127	28.7	tion*			
Very high		95	21.4	Low		194	43.8
Missing		57	12.8	Moderate		158	35.7
62. Rating: Amount of				High		87	19.6
Meaningful Con-				Missing		2	0.5
tact**				*Characteristics of the*			
Trivial	1	15	3.4	*Treatment Condition*			
	2	59	13.3	*(Treatment)*			
	3	82	18.5	67. Role of Researcher in			
Moderate	4	71	16.0	Treatment**			

	N	%		N	%
Delivered treatment	28	6.3	Group/family counseling	33	7.4
Planned, controlled	162	36.6	Other counseling	14	3.2
Influential, no direct role	57	12.9	Behavioral therapy	24	5.4
Independent of treatment	157	35.4	Skill/employment training	36	8.1
Missing	39	8.8	Service broker, multimodal	29	6.5
68. Treatment Modality; Therapy Type**			All other	5	1.1
Juvenile Justice Interventions			Missing	0	0.0
Probation, regular	2	0.5	69. Confidence/Explicitness re Treatment Modality*		
Probation, added counseling	36	8.1	Very low	0	0.0
Probation, restitution	12	2.7	Low	0	0.0
Probation, other enhancement	37	8.4	Moderate	47	10.6
Parole, regular	2	0.5	High	140	31.6
Parole, enhanced	15	3.4	Very high	251	56.7
Institutionalization, regular	4	0.9	Missing	5	1.1
Institutionalization, added counseling	43	9.7	70. What the Treatment Attempts to Change		
Institutionalization, community residential	13	2.9	Broadband delinquency	238	53.7
			Status offenses	21	4.7
Institutionalization, other enhancement	33	7.4	Other specific offenses	16	3.6
Deterrence, shock contact	11	2.5	School performance	22	5.0
All other juvenile justice interventions	7	1.6	Psychological attribute	52	11.7
			Social attribute	52	11.7
Non-Juvenile Justice Interventions			Skill level	27	6.1
Residential, camp	21	4.7	Other	10	2.3
School, added counseling	17	3.8	Missing	5	1.1
School, other enhancement	26	5.9	71. Who Administers the Treatment**		
Individual counseling	23	5.2	Criminal justice personnel	112	25.3
			School personnel	19	4.3
			Public mental health personnel	44	9.9
			Private mental health personnel	77	17.4

Table 4.3 *(Continued)*

	N	%		N	%
Non mental health			Theoretical Devel-		
counselors	44	9.9	opment**		
Laypersons	85	19.2	Black box label	60	13.5
Researcher	14	3.2	Action strategy	134	30.2
Other	16	3.6	Conceptual ratio-		
Missing	32	7.2	nale	140	31.6
72. Format of Treatment			Hypothesis testing	40	9.0
Sessions**			Integrated theory	68	15.3
Juvenile alone	22	5.0	Missing	1	0.2
Juvenile and pro-			77. Treatment Etiological		
vider	123	27.8	Orientation**		
Juvenile group	180	40.6	Individual	163	36.8
Juvenile with family	47	10.6	Individual, mixed	106	23.9
Mixed	44	9.9	Sociological, micro	88	19.9
Other	10	2.3	Sociological, macro	32	7.2
Missing	17	3.8	Labeling	23	5.2
73. Treatment Site a			Sociological, mixed	24	5.4
Public Facility**			Missing	7	1.6
Yes, criminal justice	138	31.2	78. Program Age*		
Yes, non–criminal			New (<2 years)	277	62.5
justice	86	19.4	Established	155	35.0
No, private	132	29.8	Defunct	5	1.1
Mixed	31	7.0	Missing	6	1.4
Other	17	3.8	79. Program Sponsor-		
Missing	39	8.8	ship**		
74. Treatment Site a			Researcher, one		
Residential/Insti-			cohort	112	25.3
tutional Set-			Researcher, multi-		
ting**			ple cohorts	34	7.7
Yes	123	27.8	Independent private	41	9.3
No	302	68.2	Public, non–crimi-		
Mixed	7	1.6	nal justice	85	19.2
Missing	11	2.5	Public, criminal		
75. Formal Setting*			justice	165	37.2
Yes	311	70.2	Missing	6	1.4
No	65	14.7	80. How Fully Treatment		
Mixed	36	8.1	Is Described*		
Missing	31	6.8	Detailed	62	14.0
Treatment Philosophy and			General	166	37.5
Context			Descriptive label	188	42.4
(Tx Philos)			No description	22	5.0
76. Treatment Level of			Missing	5	1.1

	N	%
OUTCOME		
Descriptive Outcome		
81. Tone of Report		
Positive	315	71.1
Neutral	102	23.0
Negative	22	5.0
Missing	4	0.9
82. Author's Interpretation of Study Result		
Success	226	51.0
Mixed	123	27.8
Failure	60	13.5
No conclusion	20	4.5
Missing	14	3.2
Statistical Outcome, Primary Delinquency Measure		
83. Direction of Treatment/Control Difference at Post-Test		
Favors treatment	285	64.3
Favors control	131	29.6
Favors neither	18	4.1
Missing	9	2.0
84. Statistical Significance of Post-Test Difference		
Significant	97	21.9
Not significant	177	40.0
Missing	169	38.1
85. Unadjusted Post-Test Effect Size		
−2.00 to −1.00	4	0.9
−0.99 to −0.50	14	3.2
−0.49 to −0.25	25	5.6
−0.26 to −0.01	79	17.8
0.00	21	4.7
+0.01 to +0.25	111	25.1
+0.26 to +0.50	72	16.3
+0.51 to +1.00	54	12.2

		N	%
+1.00 to +2.00		17	3.8
Missing		46	10.4
Statistical Information re Effect Sizes/Outcomes (ES Info)			
86. Confidence/Explicitness of Information for Post-Test Effect Size**			
Highly estimated	1	4	0.9
Moderate estimation	2	5	1.1
Some estimation	3	12	2.7
Slight estimation	4	33	7.4
No estimation	5	334	75.4
Missing		55	12.4
87. Type of Post-Test Means Reported**			
Arithmetic		167	37.7
Median		2	0.5
Proportion		242	54.6
Other		9	2.0
Missing		23	5.2
88. Type of Post-Test Variances Reported			
Standard deviation		125	28.2
Variance		1	0.2
Standard error		4	0.9
Proportion		215	48.5
Other		5	1.1
Missing		93	21.0
89. Type of Statistical Test Researcher Used for Post Difference			
No report		103	23.3
t, F, z		107	24.2
Chi-square		93	21.0
Nonparametric		16	3.6
ANCOVA		15	3.4

Table 4.3 *(Continued)*

	N	%		N	%
Blocked	2	0.5	Low	105	23.7
Other	3	0.7	Moderate	206	46.5
Missing	104	23.5	High	130	29.3
90. Method Quality:			Missing	2	0.5
Controls for Sub-			92. Confidence/Explicit-		
ject Heterogeneity			ness for Overall		
Low	209	47.2	Method Quality		
Moderate	151	34.1	Ratings**		
High	83	18.7	Very low	2	0.5
Missing	0	0.0	Low	8	1.8
91. Method quality:			Moderate	58	13.1
Appropriateness			High	303	68.4
of Statistical			Very high	71	16.0
Analysis*			Missing	1	0.2

*Variables included in initial cluster definitions for multiple regression analyses.
**Variables included in pared-down clusters for hierarchical multiple regression analysis.

to this situation. The approach used here to analyze the variation in effect sizes, therefore, is a weighted multiple regression in which the contribution of each case (study) to the analysis is weighted by the inverse variance of the effect size.

Additionally, not all the potential predictor variables for these analyses were in the form of graduated or continuous measures appropriate for correlational analysis; many were categorical. Categorical variables with more than two categories were recoded into a rank order sequence that reflected the natural progression of the categories if there was one. If there was not, conceptually similar and small *n* categories were aggregated and each case was dummy coded, 1 or 0, to index membership in each of the resulting categories.

Another issue that arises in this analysis is how to handle the missing values, since a fair number are sprinkled throughout the items in the predictor clusters. A two-step procedure was used to resolve this matter. First, a missing value indicator for each potential predictor variable was created in dichotomous form: 1 if a value was present and 0 if it was missing. These dichotomies were then correlated with the effect size dependent variable to determine if there was any relationship between missing data in a set of studies and the effect sizes found in those studies. Then, in the regression analyses, the means of nonmissing values on a predictor were substituted for each missing value in order to keep the number of cases up. If the proportion of missing values was under 10 percent, no further adjustments were made. If the proportion of missing values was greater than 10 percent, however, the correlation for the missing value dichotomy was examined. If nonsignificant, no further adjustments were made; if significant, the missing value dichotomy was itself entered into the regression equation along with the original variable from which it was derived. With this procedure, all cases could be used in the analysis, but any information about effect size carried by the fact that information on an item was missing for some studies was retained. Although a number of these "missing value" codes were involved in the preliminary regression analysis, none proved sufficiently strong as predictor variables to be retained in the final regression model.

Finally, some attention must be paid to the multicollinearity of the predictor variables and variable clusters, that is, the correlations and confoundings among the predictors themselves. To the extent that there are appreciable correlations, especially among the clusters of variables that are the primary focus in the present analysis, decisions must be made about where to allocate the confounded variance and in what sequence the clusters should be entered as predictors into the analysis.

One cluster—study context—proved to have no predictive power beyond that available in the other clusters and was dropped from further consideration. Once the specifics of the method and treatment used in a study were accounted for, items such as discipline of the author and year of publication that constituted the study context cluster added nothing else. This is not surprising since we would not expect the author's training and other such matters to have influence on study results except by way of the character of the specific treatments and methods employed in the study.

CLUSTERS OF PREDICTOR VARIABLES.   We first examine the relationship of each individual cluster of variables to effect size. This is done by constructing a single weighted multiple regression for each cluster in which only the variables from that cluster are entered as predictors. The question to be answered here is whether any of these clusters show notable correlations with effect size and thus potentially explain some of its variance. More particularly, we would like to know if the variability in effect sizes primarily reflects differences in methodology used in the various studies or if it primarily reflects differences in the treatments and treatment circumstances under investigation. If the former,

Table 4.4   Multiple Correlation of Predictor Clusters with Effect Size (Diagonals) and with Each Other (Off-Diagonals)

	Samp	Equi	Attr	Cont	Meas	ESIn	Subj	Dosa	Trea	TxPh
Method										
Samples	.20*									
Equivalence	.08	.28*								
Attrition	.11	.10	.22*							
Control	.01	.16*	−.14*	.08						
Measures	.04	.27*	.09	.16*	.28*					
ES Info	.06	.02	−.07	.05	.15*	.10				
Treatment										
Subjects	.11	.04	.02	.12*	.08	.12	.19			
Dosage	.03	.07	.05	−.01	.09	.04	.07	.24*		
Treatment	.12	.11	.16*	.02	.09	.19*	.11	.09	.40*	
Tx Philos	−.01	.09	.06	.07	.11	.17*	.04	.01	.18*	.20*

| Samp | Equi | Attr | Cont | Meas | ESIn | Subj | Dosa | Trea | TxPh |
| | | Method | | | | | Treatment | |

$*p < .05$

we have methodological bias that must be accounted for; if the latter, we have interesting differences in the effectiveness of treatment that bear further investigation.

The diagonals of the matrix in Table 4.4 report the multiple correlations between each cluster of independent variables and effect size. All the clusters having to do with treatment produced relatively large multiple correlations, as did some of the method clusters. In particular, clusters having to do with the sampling, equivalence between experimental groups, attrition, and characteristics of the delinquency outcome measures used were moderately correlated with effect size.

We must, however, consider the possibility of confoundings among the variables represented in the clusters. Dosage, for example, might correlate with effect size because studies that used high dosage also happen to frequently use a design that biases effect sizes upward. The off-diagonal correlations in Table 4.4 show the relationships among the clusters. They are obtained by using the regression equation for each cluster to compute a predicted effect size for each case and then correlating those predicted values. Some of those correlations are quite low, showing little relationship between the variables in one cluster and those in another, but others are large enough to raise a question about the independence of the relationship of the respective clusters with effect size. In particular, there are four statistically significant correlations showing confoundings between a method cluster and a treatment cluster. What appear to be relationships between the nature of the treatment and the resulting effect may therefore only reflect confounded method artifacts.

The next step in the analysis was to use hierarchical multiple regression with these clusters to examine their conjoint relationship with effect size. To conserve degrees of freedom, variables were dropped from each cluster if neither the zero-order correlation with effect size nor the beta coefficient in the multiple regression equation for the cluster reached .10, so long as this did not make the cluster size smaller than three variables or omit a variable of unusual conceptual interest (e.g., whether subjects were randomly assigned). The variables remaining in these pared-down clusters are marked with a double asterisk in Table 4.3.

HIERARCHICAL MULTIPLE REGRESSION. The pared-down clusters were stepped into the hierarchical weighted multiple regression in the order indicated on Table 4.4 and the listing above. Entering all the method clusters before any of the treatment clusters made it possible to examine the independent contribution of treatment characteristics beyond those explainable by methodological characteristics with which they were confounded. Within the method category the sequence allowed investigation of the successive influence of the nature of the samples, initial group equivalence, subsequent attrition, the nature of the control group, particulars of outcome measurement, and particulars of the effect size information. This sequence was chosen to reflect the approximate temporal sequence of the major methodological steps in mounting experimental research. That is, samples are drawn before assignment to groups, assignment precedes attrition, and so forth. Thus where there are confoundings between clusters, the contested variance in the effect size distribution is assigned to the methodological step that comes earliest in the sequence.

Within the treatment category, any number of reasonable sequences might be adopted. The sequence that was chosen (subjects, dosage, treatment, treatment philosophy) was designed to be conservative about attributing effects to specific treatment modalities if they could be accounted for by more general factors. Stepping the subject cluster into the analysis as the first of the treatment clusters, for example, ensured that no effects would be attributed to dosage and treatment modality if they could alternatively be accounted for by differences among types of subjects in their responsiveness to treatment. Similarly, entering dosage before treatment modality ensured that no effect would be attributed to specific treatment types that might only be a general function of the amount or intensity of treatment delivered, irrespective of type. Treatment philosophy, on the other hand, is a general factor (philosophy, nature of setting, etc.), but it was stepped in last on the presumption that these matters should have only indirect influence on treatment outcomes. The only interesting aspect of treatment philosophy, in other words, is what influence it might have that cannot be explained by the specifics of the subjects, dosage, and treatment type.

In addition to examining the relative influence of the different variable clusters themselves, it is interesting to consider the possibility of interactions among the clusters. The regression weights from preliminary analysis were used to construct factors combining the individual variables within each cluster into a single composite variable. The cross-products of these factors could then be entered as additional predictor variables in the regression analysis to examine the influence of cluster level interactions. Since the total number of cross-product terms for 11 clusters is quite large, testing of interactions was limited to two-way interactions (e.g., dosage by treatment modality) and, further, to those cross-products that seemed most promising in preliminary analysis.

A cluster of cross-product terms representing interactions among the method clusters was entered in the analysis after the last method cluster but before the first treatment cluster. Similarly, a cluster of cross-products representing interactions between method clusters and treatment clusters was entered after the last treatment cluster. Finally, a cluster representing interactions among treatment clusters was entered after everything else.

Table 4.5 reports the summary results for this stepwise regression procedure and indicates the variance accounted for by each cluster as it is added to the regression equation. The method clusters and method interactions altogether have a multiple correlation of .50 with effect size, accounting for 25 percent of the variability in effect size. Of the method clusters, all but the one representing the nature of control groups (Control) and the one encoding effect size information (ES Info) made statistically significant contributions to predicting effect size.

Even more interesting, perhaps, is the strong relationship of the treatment clusters to effect size above and beyond what could be attributed to the method variables. Addition of these clusters and their interactions increased the multiple correlation from .50 to .68 and accounted for an additional 22 percent of the variance in effect size. All of the treatment clusters made statistically significant contributions to predicting effect size except those dealing with subject characteristics (Subjects). Most of the contribution of treatment variables came from the cluster having to do with the treatment modality (Treatment). The

Table 4.5   Summary Table for Stepwise Hierarchical Inverse-Variance
Weighted Multiple Regression Using All Clusters to Predict Effect
Size on the Primary Delinquency Measure

Step	Variable Cluster	Cumulative Multiple R	Cumulative R-Square	R-Square Change	Change as Proportion of Total R-Square
	Method			.25	.53
1	Samples	.20	.04	.04*	.09
2	Equivalence	.31	.10	.06*	.12
3	Attrition	.36	.13	.03*	.07
4	Control	.40	.16	.03	.06
5	Measures	.44	.20	.04*	.08
6	ES Info	.46	.21	.01	.03
7	Meth x Meth	.50	.25	.04*	.09
	Treatment			.22	.47
8	Subjects	.51	.26	.01	.02
9	Dosage	.53	.29	.03*	.07
10	Treatment	.63	.40	.11*	.24
11	Tx Philos	.65	.42	.02*	.04
12	Tx x Meth	.68	.46	.04*	.09
13	Tx x Tx	.68	.47	.01	.02

$*p < .05$

effect size found in a delinquency treatment study thus depends substantially upon the methodological characteristics of the study, but it is
also importantly influenced by the nature and circumstances of the
treatment under study, as indeed we would expect.

Overall, therefore, the clusters of predictor variables included in this
analysis accounted for nearly 50 percent of the variability in the effect
size distribution. Of that, the largest share (53 percent) was associated
with methodological variables, but the independent contribution of
treatment variables was also considerable.

At this point, we can ask how well the multiple regression model
performed in accounting for the total variability in effect size among
the studies. As shown in Table 4.2, the variance of the distribution of
n-adjusted effect sizes was calculated to be .089, a value more than
three times as great as expected from sampling error alone. The variance of the residuals from the multiple regression was .047, or 53 percent of the total (consistent with an $R^2 = .47$). Testing those residuals
for homogeneity yielded $H = 798.61$ (df = 311), to be compared with

an alpha = .01 critical chi-square value of 371.17. While substantially reduced, significant heterogeneity still remained in the effect size distribution after fitting the multiple regression model.

Despite its statistical significance, however, it seems unlikely that the variation in the effect sizes not accounted for by the model was meaningful or important. The variance of the residuals, .047, includes a portion of approximately .024 (27 percent of total variance) attributable to sampling error (computed using techniques from Hedges 1984). Additionally, it almost certainly includes measurement error in the effect size values themselves, many of which were estimated from limited statistical information available in the study reports and subject, further, to whatever errors that coders may have made in computations with the information. A recoding of 25 studies (approximately every fifteenth) yielded a correlation of about .90 between effect size estimates for different coders, but this does not reflect the error inherent in estimating effect size from incomplete statistical information as was sometimes done. If the overall measurement error in effect size is as high as 20 percent of the nonsampling error variance (i.e., reliability coefficient = .80), then another 15 percent of the total variance must be measurement error (.20 (.089−.024)/.089). With 47 percent of the variance accounted for by the multiple regression model, 27 percent by sampling error, and 15 percent by measurement error in the effect size estimates, only about 11 percent is left unaccounted for. Little of the variability remaining after fitting the multiple regression model, therefore, is likely to be meaningful despite its statistical significance.

CLUSTER-LEVEL RELATIONSHIPS WITH EFFECT SIZE. Detailed discussion and interpretation of the weightings of the individual predictor variables in each cluster that resulted from the multiple regression exceeds the scope of this chapter. Moreover, some refinement of the coding and categorization beyond the present preliminary form is doubtless necessary before such detailed scrutiny will be fully rewarding. It is possible, however, to give a general characterization of the relationship between each major cluster of variables (excluding interactions) and the distribution of effect sizes. A summary of those relationships is presented in Table 4.6.

METHOD. The method cluster that accounted for the largest proportion of variance in effect size was that dealing with the pre-treatment equivalence of the treatment and control groups used in the study (Equivalence). Not surprisingly, the greater the magnitude and number of differences between the treatment and control groups prior to treat-

Table 4.6 General Nature of the Multiple Regression Results for Each Major Variable Cluster

Cluster	R^2 Change	
Method		
Samples	.04	Larger studies with larger sample sizes were associated with smaller effect sizes.
Equivalence	.06	Specific dimensions of initial nonequivalence between treatment and control groups (e.g., sex, delinquency type) were associated with larger or smaller effect sizes. Overall method of subject assignment (e.g., random vs. nonrandom), however, was not associated with effect size.
Attrition	.03	Greater attrition from either treatment or control group was associated with smaller effect sizes.
Control	.03	Control groups receiving some contact, e.g., "treatment as usual" in the juvenile justice system, were associated with smaller effect sizes than "no treatment" controls except for probation treatment as usual.
Measures	.04	Large number of delinquency outcome measures, long spans of time covered in those measures, and weak reliability and validity were associated with smaller effect sizes.
ES Info	.01	Less explicit reporting of statistical results was associated with larger effect sizes as was more explicit reporting of general methodological procedures.
Treatment		
Subjects	.01	Juveniles with more indication of delinquency (higher "risk") were associated with larger effect sizes.
Dosage	.03	Longer duration treatment and that judged to provide larger amounts of meaningful contact were associated with larger effect sizes.
Treatment	.11	(1) Treatment provided by the researcher or situations where the researcher was influential in the treatment setting were associated with larger effect sizes.

Table 4.6   *(Continued)*

Cluster	R^2 Change	
		(2) Treatment in public facilities, custodial institutions, and the juvenile justice system were associated with smaller effect sizes.
		(3) Behavioral, skill-oriented, and multimodal treatment was associated with larger effect sizes than other treatment approaches.
Tx Philos	.02	Treatment judged to have a more sociological and less psychological orientation was associated with larger effect sizes.

ment, the greater were the delinquency differences subsequent to treatment. More surprising was the finding that the nature of subject assignment to groups (random versus nonrandom), often viewed as synonymous with design quality, had little relationship to effect size. What mattered far more was the presence or absence of specific areas of nonequivalence—for example, sex differences—whether they occurred in a randomized design or not.

Loss of equivalence between treatment and control groups can also occur after a study begins via attrition. While the Attrition cluster played a smaller role in effect size than initial nonequivalence, it was appreciable nonetheless. Curiously, attrition from both the treatment and control groups appears to suppress effect sizes. This is the result that would occur if more amenable juveniles tended to drop out of treatment groups and/or more delinquent juveniles tended to drop out of control groups.

Other important design issues were sample size (Samples) and the type of control group selected (Control). Studies with larger samples tended to have smaller effect sizes. On first blush, this may appear to be a reflection of the upward bias known to occur in estimation of effect sizes from small samples (Hedges 1981; Hedges and Olkin 1985). Statistical adjustments were applied to the effect size values in order to control that bias, however. More likely, there is a general size of study effect here—small studies may be done more carefully, have more consistently delivered treatments, and the like. It is notable in this regard that studies having more outcome variables and more experimental groups also showed smaller effect sizes.

Whatever the study size, control groups that received some attention—for example, "treatment as usual" in a juvenile justice setting—showed less contrast with treatment groups (smaller effect size) than those control groups that received no treatment at all. Since the treatments studied in juvenile justice contexts are often augmentations to services that can already be extensive (e.g., custodial care), this is not surprising. The one exception, "treatment as usual" for probation services, is consistent with this pattern since probation contact is usually quite minimal.

The remaining method variable cluster of consequence was that dealing with the nature of the delinquency outcome measurement (Measures). Although collectively these variables were correlated with effect size, no readily interpretable pattern was evident. Other than number of delinquency measures, which was probably part of the study-size effect discussed above, the strongest relationship was a tendency for delinquency measures covering a longer period of time post-treatment to be associated with smaller effect sizes.

TREATMENT.   Of primary interest in Table 4.6 are those clusters that show an important influence of the type and circumstances of treatment upon delinquency. Since all the method clusters were stepped into the regression analysis prior to any of these clusters, we can have some confidence that any relationships that emerge represent characteristics of effective treatment rather than confoundings with influential method variables.

The cluster of variables representing subject characteristics (Subjects) was stepped into the analysis first among the treatment clusters to test the possibility that certain juveniles were especially responsive to treatment, whatever its nature. While there was a slight tendency for studies of juveniles with higher risk levels—that is, greater involvement with delinquency—to show larger effect sizes, the overall influence of this cluster was small and statistically nonsignificant. The prospect of such a relationship, however, deserves further scrutiny in later analysis. Targeting high-risk juveniles was one of the criteria for "clinically relevant" treatment in the Andrews et al. (1990) meta-analysis cited in the introduction to this chapter.

In similar spirit, the cluster of variables dealing with the amount or intensity of treatment (Dosage) was entered into the analysis next. This permitted consideration of the possibility that the size of the treatment dose was more important than the specific nature of the treatment administered. As the National Academy of Science's review of correctional treatment observed, weak and incompletely delivered treatments cannot be expected to have meaningful effects (Sechrest, White, and Brown 1979).

The regression analysis did show a modest positive relationship between effect size and the duration, frequency, and amount of treatment. The relationship seems to be weakened, however, by an unexpected confounding. Some of the treatment dosage variables are such that they are naturally higher for juveniles in institutional care—for example, frequency of contact. As a category, treatment in institutional context seems to be associated with smaller effect sizes. This results in a somewhat curvilinear relationship in which effect size increases with amount of treatment up to amounts associated with institutional care (i.e., "continuous" frequency of contact) and then declines. Subsequent analysis of this relationship will require more refined breakdowns among treatment categories than those used in the present analysis.

By far the strongest relationship with effect size was found for the cluster of variables representing treatment modality and the nature of the treatment provider (Treatment). These relationships showed three different facets. First, treatments that were delivered by the researcher, or in which the researcher had a considerable influence, showed larger effect sizes. A cynical interpretation of this pattern might suggest that these larger effects stemmed from some interest on the part of the researcher in making the treatment look good. It is at least equally plausible, however, that treatment delivered or administered by the researcher for research purposes was better implemented and monitored than the typical practices of service agencies. If such is the case, the "researcher involvement" variable becomes a more general indicator for treatments mounted with enthusiasm and careful administrative control—a circumstance that may well lead to larger effects.

The second facet of the variables in the treatment cluster was an association between smaller effect sizes and treatments provided in public facilities, or within the juvenile justice system or custodial institutions. Since these findings overlap considerably with the pattern of findings for specific treatment modality, we turn to them now.

The most influential variables in the Treatment cluster were those that dummy-coded various specific treatment types separately for juvenile justice and non–juvenile justice sponsors. A rather consistent pattern emerged which is most easily seen by looking at the mean effect size for each category of treatment. Since we want to examine treatment effects unconfounded by method effects, the mean effect sizes for each treatment category were computed from the multiple regression residuals after all method clusters were removed (adding back the grand mean, of course). To make these mean effect sizes more interpretable, each was also translated into the equivalent reduction it represented in a dichotomous recidivism rate when compared with a hypothetical control group with 50 percent recidivism. Table 4.7 reports the results.

Treatment modality is often described rather crudely in the source studies upon which this analysis relies, often by no more than a label or phrase. It is correspondingly difficult to code into a meta-analysis in any definitive way. It would be a mistake, therefore, to focus on any particular category in Table 4.7 and draw a general conclusion about the efficacy of treatments offered under such various conventional labels as "restitution" or "counseling." This would also contribute to the unfortunate tendency in delinquency treatment to advocate a "magic bullet," a specific treatment concept alleged to be a superior approach to delinquency. Moreover, the categories in Table 4.7 include instances of varying efficacy ranging above and below the category mean and they overlap considerably for those many treatments with multiple elements.

A more appropriate approach to interpreting Table 4.7 is to examine the broader patterns in the ranking of more and less effective treatment modalities. Viewed this way, there is striking consistency in both the juvenile justice and non–juvenile justice treatments. In both cases, the more structured and focused treatments (e.g., behavioral, skill-oriented) and multimodal treatments seem to be more effective than the less structured and focused approaches (e.g., counseling). It will be the task of subsequent analysis of these data to better tease apart the various treatment parameters that account for this ranking.

It is noteworthy that the best of the treatment types, both inside and outside the juvenile justice system, show delinquency effects of meaningful practical magnitude, in the range of 10–20 percentage points reduction in recidivism. Since these are reductions from a presumed 50 percent control group baseline, they represent decreases of 20–40 percent (i.e., 10/50 to 20/50). It is also interesting that the treatment types that show this large order of effects are, with few exceptions, those defined as most "clinically relevant" in the Andrews et al. review (1990).

Finally, it should be noted that a number of treatment approaches were associated with mean effect sizes of virtually zero. This family of treatments simply may not work, as many critics have charged. Further, a couple of treatment categories appear to produce negative ef-

Table 4.7   Residualized Effect Size Estimates After Removal of Method
Variance for Different Treatment Modalities

Treatment Modality	Effect Size	Equivalent Recidivism Change from 50% Control
Juvenile Justice		
Employment (4)	.37	−.18
Multimodal (12)	.25	−.12
Behavioral (8)	.25	−.12
Institutional, other (9)	.20	−.10
Skill-oriented (15)	.20	−.10
Community residential (12)	.16	−.08
Any other juvenile justice (5)	.14	−.07
Probation/parole, release (16)	.11	−.05
Probation/parole, reduce caseload (11)	.08	−.04
Probation/parole, restitution (13)	.08	−.04
Individual counseling (20)	.08	−.04
Group counseling (39)	.07	−.03
Probation/parole, other enhancement (7)	.07	−.03
Family counseling (6)	.02	−.01
Vocational (9)	−.18	+.09
Deterrence (9)	−.24	+.12
Non–Juvenile Justice		
Skill-oriented (17)	.32	−.16
Multimodal/broker (29)	.21	−.10
Behavioral (31)	.20	−.10
Group counseling (17)	.18	−.09
Casework (7)	.16	−.08
Family counseling (29)	.10	−.05
Advocacy (4)	.10	−.05
Other counseling (5)	.06	−.03
School class/tutor (14)	.00	−.00
Individual counseling (24)	−.01	+.00
Any other non–juvenile justice (3)	−.01	+.00
Employment/vocational (22)	−.02	+.01

Note: The number of studies in each category is reported in parentheses.

fects—most notably, deterrence treatments. This category includes shock
incarceration and the "scared straight" program model that received
considerable publicity a few years ago.

Whatever patterns one discerns in these results, they do indicate

that the specifics of what is done in delinquency treatment are important. No generalized placebo or Hawthorne effect is likely to be able to
account for the differential outcomes of different approaches.

The final cluster of treatment-related variables to be entered in the
regression represented those that indicated something about the treatment philosophy: its etiological orientation, level of theory development, and related matters. This cluster was only weakly related to effect size. It appears that there is little in the reported treatment
philosophy, above and beyond the characteristics of its subjects, dosage, and treatment type, that influences the size of effects.

Conclusions

What is presented here is only the most general analysis of the measured effects from delinquency treatments studies. While it was demonstrated that the grand mean of those effects is positive, indicating at
least modest overall treatment effects, the primary focus of this phase
of the investigation has been upon the variability of effects. This variability was shown to be far in excess of what would be expected simply
on the basis of sampling error. It follows that there must be some circumstances in which studies yield large effects and others in which
they yield small effects. The challenge is to discover the nature of those
circumstances.

If research results are shaped primarily by the methods chosen, we
should know which aspects of the methods are most important and
investigate the bias they introduce. If, on the other hand, some substantial portion of the variability in measured delinquency effects stems
from the nature of the treatments applied and the characteristics of the
juvenile recipients of those treatments, then it behooves us to discover
which treatment circumstances produce the largest effects and put that
information to practical use.

The analyses presented in this chapter are less concerned with the
details of these issues than with charting the overall domain. The results indicate that both method and treatment influence the effects of
delinquency treatment studies. Although method variables collectively
seem to play a somewhat greater role, the largest single category of
influences is the nature of the treatment itself. Subsequent work using
this database will focus on closer specification of the details of method
and treatment, and their interaction, that are most important in shaping study outcome.

The pattern of the general results presented here throws some light on the checkered history of research reviews in delinquency treatment. The grand mean effect size is perilously close to zero. While not so close as to justify the "nothing works" rhetoric of the 1970s, convincing positive effects would be difficult to discern in any sample from this literature. This would be especially true if the sample was of modest size and if the review primarily used "box score" techniques that keyed on the statistical significance of individual study findings. The sample sizes used in this literature (median around 60 in each experimental group) do not yield sufficient statistical power for an individual study to find statistical significance for effect sizes in the range of .10–.20 standard deviation units.

Moreover, the wide variability in effects found in this literature means that different reviews that sampled different portions of it could come, quite honestly, to rather different conclusions. On the high end of the distribution are studies that show impressively large effects, as Gendreau and Ross (1979), Palmer (1975), Andrews et al. (1990), and others have asserted. On the low end of the distribution, and even in the middle, a considerable number of studies can be produced that show insignificant and even apparently negative effects, as Martinson (1974), Whitehead and Lab (1989), and others have insisted. If the heterogeneity of the distribution of effects in delinquency treatment research is as large as an elephant, perhaps it is no wonder that each reviewer, grasping here a tail of the distribution and there a hump, describes the beast so differently.

Appendix 4.A Bibliographic Databases Used in Search

Arts and Humanities Citation Index
Books in Print
British Books in Print
British Education Index
Child Abuse and Neglect
Criminal Justice Periodical Index
CRISP: National Institute of Mental Health
Dissertation Abstracts Online
ERIC (Educational Resources Information Center)
Family Resources
Federal Research in Progress
Library of Congress Books
Medline
Mental Health Abstracts
National Criminal Justice Reference Service
National Technical Information Service
PAIS International (Public Affairs Information Service)
Psychological Abstracts
Social Science Citation Index
Sociological Abstracts
SSIE Current Research (Smithsonian Science Information Exchange)
U.S. Government Printing Office Publications
U.S. Political Science Documents

Notes: The research reported in this paper was funded by the National Institute of Mental Health, Antisocial and Violent Behavior Branch (MH39958 and MH42694), and the Russell Sage Foundation.

There were 443 studies involved in the analysis presented in this paper. The full bibliography of studies can be obtained from the author at the Psychology Department, Claremont Graduate School, Claremont, CA 91711.

5

Do Family and Marital Psychotherapies Change What People Do? A Meta-Analysis of Behavioral Outcomes

William R. Shadish, Jr.

What best distinguishes family and marital therapies from other therapies is the belief that the crucial forces in a person's life are his or her interactions with family members (Nichols 1984). Thus family members are usually included in therapy to help change family interaction patterns.

Perhaps the most common way to categorize different kinds of marital and family therapies is through their theoretical orientation; that is, the system that therapists use for guidance in dealing with clients. Much controversy exists in the psychotherapy literature about whether orientation makes any difference to therapy outcome. Examining the magnitude of orientation effects, and exploring variables that might moderate or explain such effects, is the major purpose of this chapter.

This study considers the relative effectiveness of six theoretical orientations adapted from Wamboldt, Wamboldt, and Gurman (1985): (1) behavioral, (2) systemic, (3) humanistic, (4) psychodynamic, (5) eclectic, and (6) other. Extended descriptions of these orientations are available in Jacobson and Gurman (1986) and Nichols (1984). Behavioral orientations use concepts and interventions from learning theory—for example, role playing, reciprocal contracts, or behavioral skills training. Behavioral treatments usually focus on present behavioral interactions or cognitions rather than on historical matters such as childhood experiences. They pay little attention to unconscious motivations or other

Note: Studies used in this analysis are indicated by a †.

underlying causes of symptoms, preferring instead to deal directly with symptoms themselves. Treatments involve thoroughly assessing and then modifying these presenting symptoms through modifying the stimuli and contingencies that maintain them. The behavioral category includes such interventions as parent management training, behavioral marital therapy, the psychoeducational models, the McMaster model, and Alexander's functional model.

Systemic orientations share many of the pragmatic, present-oriented tendencies of the behavioral orientations, but offer very different theoretical explanations for problems and provide somewhat different interventions. Theoretically, as one might guess, systemic orientations conceive of marriage or the family as a relatively stable system governed by the behaviors and thoughts of system members, by interaction patterns among them, and by salient environmental matters such as education of children or occupational choices. Unlike behavioral treatments, treatment focus is usually not on directly modifying the symptoms themselves. Symptoms are viewed as stemming from problems in this larger system, as when a husband's criticisms of his wife are prompted by his inability to cope with job stresses. The therapist's goal is usually to change various aspects of the system until a more satisfactory equilibrium can be found. Unlike behavioral interventions, systemic interventions may have little obvious direct connection to the symptom, so that any way to "shake up the system" might receive serious consideration. Systemic orientations include MRI brief therapy, the strategic models, Zuk's triadic model, the structural models, and the Milan systemic model.

The theoretical underpinnings of humanistic therapies include a commitment to personal freedom, self-determination, and the fulfillment of personal needs. The emphasis is on helping family members to experience the "here-and-now" feelings associated with their family and on facilitating the spontaneous, open, and creative expression of those feelings. Attention is focused more on helping individual family members to meet their personal needs than on the needs of the family system itself. The presumption is that the family will do better as each of its individual members does better. Examples of humanistic orientations include Rogerian therapy, Gestalt therapy, and symbolic-experiential (e.g., Satir) therapy.

Psychodynamic theories emphasize the role of individual motivations, emotions, beliefs, and other mental forces in shaping family interactions. The best-known example, of course, is psychoanalysis, where individual behavior is seen to be caused by the internal dynamics of the id, ego, and superego, some of which are partly unconscious and so not easily available to analyze. These dynamics are determined early in childhood, the most famous example of which is the Freudian Oedipus complex. Symptoms are a function of these internal psychodynamics. Such theories are adapted to family and marital therapies by stressing the role of family members as objects through which these internal dynamics are made manifest. The goal of psychodynamic therapies is to understand the internal personality dynamics of each family member, and then to change these dynamics so that family members deal with each other as they really are rather than on the basis of dynamics developed in childhood. Examples of psychodynamic therapies include psychodynamic marital, psychoanalytic family, multigenerational (e.g., Framo, Baker, Williamson), contextual, and Bowen family systems approaches.

Eclectic therapies adapt concepts and interventions from other orientations. Such therapists tend to believe that no single orientation has the "best" answer, but that each orientation has something valuable to offer depending on the client's presenting problem. The guiding premise is to select an intervention that seems best tailored to the problem. In general, these therapies were of two kinds in the present study: therapies that explicitly claimed to combine two or more of the preceding four categories, or therapies that were specifically labeled as eclectic and were described as using interventions borrowed from multiple orientations. Finally, we included a sixth category for therapies that did not fit any of the first five classifications.

In psychotherapy research generally, including marital and family therapy, a persistent finding is that these orientation differences make little difference to outcome once one controls for differences in the ways studies are conducted. But this assertion has been very controversial, particularly among behaviorists (Wilson and Rachman 1983). Behavioral orientations frequently yield larger raw effect sizes than other orientations, but these differences fade once one adjusts raw effect sizes for the influence of such variables as reactivity of measurement (Smith, Glass, and Miller 1980) or experimenter allegiance (Berman, Miller, and Massman 1985). For example, several meta-analyses have found that experimenter allegiance to a theoretical orientation increases the effect size produced by the therapy (Berman et al. 1985; Robinson, Berman, and Neimeyer 1990; Smith et al. 1980). Allegiance carries the connotation of particularly high devotion, loyalty, fidelity, and even ardor for the particular orientation. The presumed causal chain seems to be that allegiance increases the experimenter's belief that the therapy will be effective; such a belief is communicated to the client, which engenders a belief in the client that therapy will work; and the client belief acts like a placebo effect to increase effectiveness. The finding has been demonstrated many times and may account for differences previously thought due to orientation. Similarly, Smith et al. (1980) found that behavioral treatments had larger raw effect sizes, but that this may have been due to their use of more reactive measures:

> reactive instruments are those that reveal or closely parallel the obvious goals or valued outcomes of the therapist or experimenter; which are under the control of the therapist, who has an acknowledged interest in achieving predetermined goals; or which are subject to the client's need and ability to alter his scores to show more or less change than what actually took place. (pp. 66–67)

After adjusting effect size for use of reactive measures, the apparent superiority for behavioral treatments again disappeared.

Most of these matters have not yet been explored in the marital and family therapy literatures. Hahlweg and Markman (1988), for example, meta-analyzed the effects of behavioral marital therapy, but did not compare it with other orientations. This chapter makes such orientation comparisons and then explores artifactual and theoretical explanations for any observed effects. First, we focus on describing overall therapy effectiveness and exploring whether any single variable can explain the diversity of therapy outcomes. These variables fall into four general categories: study methodology (e.g., whether subjects were matched prior to randomization), dependent variable characteristics (e.g., reactivity of measurement), client and treatment inputs and context (e.g., therapist experience, client presenting problem), and therapy process (e.g., use of communication training). To no one's surprise, we will find that no single variable—including orientation—is capable of explaining outcome fully. Second, we explore whether differences in therapy outcome can be explained by a multivariate set of these same predictors in a regression model and, if so, which variables best account for therapy outcome. The key issue here will be whether orientation effects remain significant after accounting for other possible explanations. Third, after briefly exploring some methodological problems that might complicate substantive interpretation of results, we try to determine if conclusions about orientation effects would change under several statistical models that have not been widely applied to meta-analytic data, including univariate and multivariate random effects models, multiple equation path models, and latent variable models.

The chapter will conclude by discussing the nature of meta-analytic explanation and its use with regard to orientation effects.

Methods

Data in this study are drawn from a larger project that included 163 randomized trials (Shadish 1989b; Shadish et al. 1989). From this larger project, only effect sizes computed on behavioral dependent variables are included. This selection was made using a variable that categorized the dependent variable as (1) affect (e.g., anger, depression), (2) behavior (e.g., number of marital fights), (3) cognition (e.g., mental ability, obsessive thoughts), (4) physiology (e.g., EEG readings), or (5) compound measures combining more than one of the above (e.g., happiness, general dissatisfaction). Selection on category (2) of this variable still leaves considerable heterogeneity among studies since behavioral dependent variables can vary on such dimensions as self-report versus observer ratings of behavior, the specificity versus generality of the behaviors, and the degree of blindedness of the experimenter to data gathering on the dependent variable. Nonetheless, by selecting on behavioral outcomes, this chapter can explore the degree to which therapy changes the actual behaviors of therapy clients, and what factors might influence those changes.

Developing a Coding Manual

Our 24-page coding manual was adapted from Smith et al. (1980) by including variables reflecting our particular hypotheses, elaborating instructions to minimize coding errors, and including appropriate variables suggested by family and marital therapy experts. The manual codes over 100 variables associated with (1) general study characteristics, (2) presenting problem, (3) circumstances of and surrounding treatment, (4) outcome, and (5) effect size. Reliability of codings from a penultimate version of the coding manual is reported later in this chapter. The manual was then revised, and five graduate students in clinical psychology were trained for several weeks in its use, including practice doing codings. Codings were subsequently monitored in regular team and individual meetings throughout the study.

Coding Effect Sizes

The effect size, d, is computed by differencing group means, dividing the difference by the pooled standard deviation, and applying a correction for small sample bias (Hedges and Olkin 1985, p. 81, equation 10). A positive effect size means that the treatment group did better than the control group. We developed methods for computing effect sizes when means, standard deviations, and sample sizes were not reported (manual available on request). Some of these methods yield exactly the same effect size estimate as d; others are inexact in that they do not yield the same estimate as d. When researchers reported a finding only as nonsignificant, we coded the effect size as zero.

Meta-analyses of the same studies can yield different conclusions, if the rules used for coding effect sizes are different (Matt 1989). We coded all plausible outcome variables at all post-tests and follow-ups except when a variable was mentioned in the methods section but never again. When both total scores and subscale scores were reported, and when the same effect size method was available to code both, we coded subscale scores and excluded the total score. Coding both would introduce linear dependencies among the measures. While total scores may be more reliable than subscales, they may also be less reliable if poorly constructed, and they may lose more specific information if therapy causes differential responses in subscales. We did not compute an effect size for any computation that did not involve a family or marital therapy, such as individual therapy versus a control group.

All analyses in this chapter are performed on effect sizes aggregated to the study level since computation yielded significant, positive intraclass correlations within treatment comparisons and within study (ranging from .22 to .73). Since the intraclass correlation is less than 1, this strategy is conservative, so total aggregation underestimates the effective sample size just as treating the individual effect sizes as the units overstates the sample size.

Literature Search

Computer searches of all years of *Psychological Abstracts* (yielding 1,262 abstracts/references), *Dissertation Abstracts International* (198), *Social Science Citation Index* (178), *National Center for Family Research* (150), *National Clearinghouse for Mental Health* (601), and *Mental Health Abstracts* (300) used a 22-step combination of relevant substantive terms (e.g., family) with relevant research terms (e.g., outcome). We also reviewed the bibliographies of 46 reviews of the literature (Gurman, Kniskern, and Pinsof 1986), and reviewed the tables of contents and abstracts of relevant journals in the personal and university libraries in our vicinity. Finally, we sought the advice of several nationally known consultants who were specialists in the area. We retained 163 studies for our meta-analysis, including 59 dissertations. Studies met the following criteria:

1. Subjects were randomly assigned to conditions, including assignment to different treatments or to treatment and control groups. We included seven studies in which assignment was haphazard rather than formally random, as when subjects were assigned in alternating order to two treatments and no obvious selection bias seemed present.
2. Only studies of clinically distressed clients were included; analogue or enrichment studies were excluded.
3. We included studies aimed at changing family interaction even if not explicitly labeled "family therapy," such as studies in which parents are trained to change parent-child interactions. We also included studies that used a family or marital therapy to address an "individual" problem such as agoraphobia.
4. We excluded studies of physiologically oriented sex therapies, but included studies with partly sexual presenting problems using traditional marital therapy.
5. When we located a study in both dissertation and published form, we coded the former on the assumption that dissertations report more complete results and thus yield more accurate estimates of population effect sizes.

These procedures yielded 163 studies, from which the present chapter uses 106 studies reporting behavioral outcomes. Descriptive characteristics of the 106 studies are presented in Table 5.1. Most of that table is self-explanatory, but several points are worth noting. First, the number of comparisons includes both post-tests and follow-ups. However, most studies did not include follow-ups, and we will exclude follow-ups from the analyses in this chapter. Second, the number of measures per study refers to distinct measures, but not whether that measure was administered at both post-test and follow-up since the latter information was already coded at the comparison level. Third, the effect size is the lowest unit of coding in the meta-analysis, reflecting a unique combination of measure and comparison. (Note that these studies contained about 11 effect sizes per study on behavioral outcomes.) Fourth, 75 percent of the studies were conducted by psychologists. This might imply that we were differentially effective in locating studies across disciplines. But this may also reflect differential frequency of this research across areas, since the computer search procedures we used had no obvious disciplinary biases in them.

Table 5.1   Some Descriptive Characteristics of 106 Studies with Behavioral Dependent Variables

Study Characteristics	Mean (Range)	Total
Year Published	1979 (1967–1988)	
Form of Publication		
Journal article		57
Book/chapter		2
Dissertation		45
Unpublished manuscript		2
Number of Studies		106
Family studies		70
Marital studies		36
Number of Comparisons		208
Treatment-control comparisons		89
Treatment-treatment comparisons		119
Number of Comparisons/Study	1.96 (1–9)	
Number of Effect Sizes		1,203
Number of Measures/Study	11.26 (1–86)	
Professional Affiliation of First Author		
Psychology		80
Social work		4
Psychiatry		4
Education		16
Nursing		1
Unknown		1

Results

Descriptive Results

Before reviewing orientation effects, it is instructive to examine overall outcome. Figure 5.1 is a histogram of study-level effect sizes for all treatment-control comparisons taken within three weeks of termination of therapy. Effect sizes range from $-.15$ to 4.76. They are positively skewed, with only four studies reporting negative average effect sizes, providing little evidence for significant negative effects of these therapies. Over these 58 studies, the ordinary least squares (OLS) mean effect size (i.e., the straight average) is .83 with a standard error of .11 and a 95 percent confidence interval of $.61 < d < 1.05$. However, studies with smaller sample sizes should produce less accurate estimates of population effect sizes than studies with larger sample sizes, so one should give more weight to estimates of d from larger sample studies

Figure 5.1   Study-Level Effect Sizes for Treatment-Control Comparisons
($n = 58$)

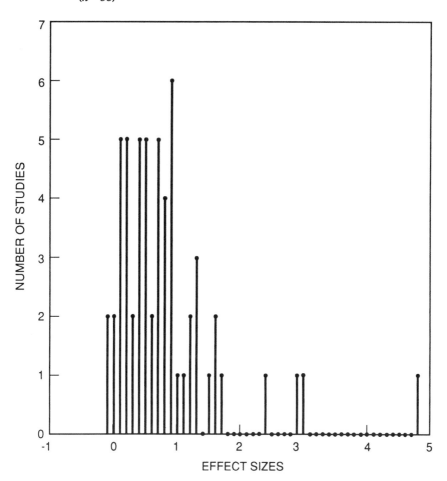

when combining estimates over studies. Hedges and Olkin (1985) suggest a weighted least squares (WLS) estimator, d_+, that can be computed from knowledge of d and the sample size per group (p. 111, formula 6).[1] Using those weights, the mean effect size $d_+ = .70$ with

---

[1]Other meta-analysts have justified use of weighted least squares estimates using a rationale that may be only partially correct. For example, Robinson et al. (1990) used sample size to predict effect size, and then saved, squared, and correlated residuals with sample size. They found that smaller sample sizes were significantly associated with larger resid-

a standard error of .054 and a 95 percent confidence interval of $.59 < d_+ < .81$.

INTERPRETING EFFECT SIZES.   One interpretation of effect sizes appeals to the notion that they are standard scores that can be used to assess overlap between treatment and control distributions. Specifically, an effect size of .70 implies that the mean of the treatment distribution is .70 standard deviations above the mean of the control distribution. Reference to a unit normal table then shows that an effect size of .70 implies that a family or marital therapy client at the mean was better off than 76 percent of the control clients. A second use of this table is to compute the probability that a randomly chosen treatment response is greater than a randomly chosen control response. The Gaussian (unit normal) lookup is at $z = ES/\sqrt{2}$, where ES is the effect size indicator. With $d_+ = .70$, this yields $z = .4949$, and a probability of .69 that a randomly chosen family/marital therapy response will have a better outcome than a randomly chosen control response.

Another interpretation is to translate the average effect size back into an original metric. For example, Jacobson (1977)[†] used the Marital Adjustment Scale (MAS), a commonly used, brief, self-report scale of marital satisfaction (Locke and Wallace 1959). The range on this scale is from 2 to 158, with a score around 100 reflecting positive adjustment if one spouse's report is used (Jacobson et al. 1984), and a range and mean about double that if two spousal reports are combined. Jacobson (1977)[†] used the combined score and found standard deviations of 25.8 to 40.2; similarly, Baucom (1984)[†] found standard deviations of 20.74 to 44.58. Multiplying these figures by the effect size of .70 suggests that treated couples did about 14 to 31 combined spousal MAS points better than control couples—or about 5–10 percent of the range of the scale.

An additional way to interpret effect sizes is to convert them into correlation coefficients. Assuming an average sample size of ten per group, which we will see later is accurate for the present sample of studies, an effect size of .70 converts to a simple correlation of .35 (Hedges and Olkin 1985, p. 77), so that family and marital treatments account for about 12 percent of the variance in outcome. This latter figure may sound modest, but one can make a case for its importance in two ways. First, Rosnow and Rosenthal (1989) point out that this

---

uals, and so used weighted least squares methods of a kind similar to those used here. But the WLS analysis could profitably be used even if—as was true in the present data— residuals were *uncorrelated* with sample sizes, because WLS still yields more accurate estimates of effect size and yields smaller standard errors.

correlation would be quite large compared with some commonly accepted "important" effects in medical research. For example, in 1988 a medical group reported that a randomized experiment found the risk of heart attack was cut significantly by every-other-day doses of aspirin. The findings were viewed as so compelling that the experiment was terminated since it was deemed unethical to continue giving the placebo to the control group (Rosenthal 1989). Yet the relevant effect size for this finding measured as a correlation was a mere $r = .034$. Rosenthal (1989) lists many similar examples. Second, Rosenthal and Rubin (1982) point out that a correlation can be related to a treatment success rate by computing $.50 + r/2$. Thus an effect size of .70 and accompanying correlation of .35 implies a success rate of about 68 percent in marital and family therapies compared with only 32 percent in control groups—roughly doubling the number of positive outcomes that would occur spontaneously without therapy.

HOMOGENEITY/SPECIFICATION TESTS.   As an adjunct to d_+, Hedges and Olkin (1985, p. 123) describe Q, a test of the homogeneity of the effect sizes being averaged. Rejection of Q implies that the effect sizes from these 58 studies may not measure the same population parameter. Q is a test of model specification; that is, whether or not the category being used is sufficient to account for all systematic variance in effect size (also true for the test of model specification for multiple regression, Q_e, reported later in this chapter).[2] In the present case, $Q = 222.77$

[2]Q is a novel statistic and could easily be misunderstood, particularly because calling Q a homogeneity test may remind readers of homogeneity of variance tests under OLS approaches. Tests of homogeneity of variance in OLS ANOVA are similar to and different from Q. They are different in that even when homogeneity is rejected under Q, all descriptive statistics and inferential tests remain accurate. By contrast, OLS ANOVA is not robust to extreme violations of homogeneity of within-column sampling variance, or violations in the presence of greatly unequal sample sizes, in which case the accuracy of ANOVA inferential tests may be problematic. On the other hand, rejection of Q and homogeneity of variance complicates the interpretation of the average effect size in *both* cases. With rejection of Q, the investigator has to qualify the meaning of the average effect size to note that the underlying data points may not estimate the same population parameter, or alternatively, that more predictors are needed to account for nonrandom effect size variance. The same problem occurs with rejection of homogeneity of variance under OLS approaches, since the rejection indicates the presence of other variables that are causing an interaction between the levels of the ANOVA factor and the effect sizes (Bryk and Raudenbush 1988). However, an advantage of the WLS approach is that transformations of data to meet homogeneity assumptions are not necessary under the WLS approach used in this chapter; indeed, there is some question if such transformations are the optimal strategy for OLS either, despite their common use (Bryk and Raudenbush 1988).

(df = 57), rejected at the .01 level. The simplest interpretation is that marital and family therapy is not just one, uniform therapy producing one, uniform effect (plus random error); rather, these are diverse therapies producing diverse outcomes that must be more finely subdivided to be properly understood. One might, for instance, subdivide them into different orientations. In general, one could either use available substantive or methodological variables to construct subgroups of studies that may yield homogeneous effect sizes, or construct multiple regression models that account more adequately for nonhomogeneity.

CATEGORICAL TESTS. Table 5.2 presents the first option—categorical tests that examine what variables might account for variability in marital and family therapy (Table 5.2 also presents random-effects model estimates

Table 5.2   Effect Sizes as a Function of Various Predictor Variables Under Different Statistics Models

Variable	OLS Statistics		WLS Fixed-Effects Statistics			WLS Random-Effects Statistics		
	d	se	d_+	se	Q_h	$\sigma^2(\Delta)$	Δ	se
Study Methodology								
Effect Size Method								
Exact ($n = 40$)	.87*	.15	.79*	.07	161.36*	.57*	.76*	.14
Approximation (25)	1.06*	.15	.79*	.08	88.15*	.31*	.95*	.15
	F (1,63) = .73		Q_b (1) = .21					
			Q_w (63) = 249.51*					
Form of Publication								
Publication (31)	1.09*	.18	.75*	.08	108.82*	.65*	.95*	.17
Dissertation (27)	.53*	.10	.66*	.07	113.27*	.04*	.61*	.09
	F (1,56) = 6.79*		Q_b (1) = .68					
			Q_w (56) = 222.09*					
Blindedness to Treatment								
No influence likely (4)	.24*	.05	.18	.21	4.22	.00	.23	.21
Indirect influence (18)	.99*	.19	.96*	.09	96.46*	.40*	.93*	.19
Direct influence (25)	.89*	.21	.59*	.09	56.81*	.72	.72*	.20
	F (2,44) = 1.10		Q_b (2) = 16.27*					
			Q_w (44) = 157.48*					
Source of Clients								
Experimenter-solicited (31)	1.01*	.19	.69*	.08	109.46*	.75*	.86*	.18
Other/self-referred (27)	.62*	.10	.71*	.07	113.26*	.00	.75*	.08
	F (1,56) = 3.14		Q_b (1) = .05					
			Q_w (56) = 222.72*					

	OLS Statistics		WLS Fixed-Effects Statistics			WLS Random-Effects Statistics		
Variable	d	se	d_+	se	Q_h	$\sigma^2(\Delta)$	Δ	se
University-Based Clients								
Mostly university (5)	.82*	.23	.64*	.19	9.21	.00	.71*	.19
Some university (4)	1.01*	.16	.78*	.22	14.80*	.00	.92*	.23
No university (48)	.82*	.13	.70*	.06	197.27*	.55*	.72*	.13
	$F\,(2,54)=.08$		$Q_b\,(2)=.23$					
			$Q_w\,(54)=221.28*$					
Assignment to Conditions[a]								
Random (52)	.84*	.12	.70*	.06	215.86*	.50*	.73*	.12
Haphazard (7)	.70*	.15	.67*	.18	9.33	.00	.68*	.19
	$F\,(1,57)=.18$		$Q_b\,(1)=.02$					
			$Q_w\,(57)=225.19*$					
Matching[a]								
Matching Occurred (13)	.43*	.11	.43*	.12	18.29	.00	.45*	.12
No Matching (46)	.93*	.14	.76*	.06	200.66*	.52*	.82*	.13
	$F\,(1,57)=3.60$		$Q_b\,(1)=6.26*$					
			$Q_w\,(57)=218.95*$					
Client and Treatment Context and Inputs								
Problem Category								
Family (39)	.62*	.09	.63*	.06	161.25*	.10*	.60*	.09
Couple (19)	1.27*	.26	.91*	.11	56.76*	.90*	1.10*	.25
	$F\,(1,56)=8.33*$		$Q_b\,(1)=4.75*$					
			$Q_w\,(56)=218.02*$					
Locus of Presenting Problem								
Individual child (23)	.63*	.14	.48*	.10	57.53*	.15	.55*	.13
Individual adult (3)	.59	.42	.51*	.21	2.52	.15	.53	.30
Couple (19)	1.27*	.26	.91*	.11	56.76*	.90*	1.10*	.25
Family (7)	.64*	.14	.57*	.15	28.45*	.00	.69*	.15
Extrafamilial (6)	.57*	.23	.96*	.12	57.66*	.16*	.68*	.22
	$F\,(4,53)=1.98$		$Q_b\,(4)=15.37*$					
			$Q_w\,(53)=202.92*$					
Problem Type								
Behavioral (31)	.61*	.11	.66*	.07	124.82*	.11*	.58*	.10
Nonbehavioral (27)	1.08*	.20	.75*	.08	97.14*	.72*	.94*	.19
	$F\,(1,56)=4.56*$		$Q_b\,(1)=.77$					
			$Q_w\,(56)=221.96*$					
Experimenter Allegiance								
Yes (54)	.84*	.12	.61*	.06	161.06*	.47*	.72*	.12
No (10)	.80*	.16	1.03*	.11	55.46*	.03*	1.02*	.14
	$F\,(1,62)=.02$		$Q_b\,(1)=12.23*$					
			$Q_w\,(62)=216.52*$					
Treatment Location								
University (35)	.99*	.17	.69*	.08	110.53*	.67*	.85*	.17

Table 5.2 *(Continued)*

Variable	OLS Statistics		WLS Fixed-Effects Statistics			WLS Random-Effects Statistics		
	d	se	d_+	se	Q_h	$\sigma^2(\Delta)$	Δ	se
Nonuniversity (18)	.54*	.11	.47*	.10	31.65*	.00	.48*	.10
	$F\,(1,51) = 3.25$		$Q_b\,(1) = 2.97$					
			$Q_w\,(51) = 142.18^*$					
Therapist Gender								
Male (9)	.77*	.32	.57*	.15	19.48*	.61	.64*	.31
Female (6)	.72*	.29	.55*	.21	8.19	.15	.56*	.27
Both Male/Female (29)	.67*	.09	.58*	.08	81.54*	.01	.65*	.08
	$F\,(2,41) = .09$		$Q_b\,(2) = .02$					
			$Q_w\,(41) = 109.21^*$					
Therapist Degree								
Professional Degree (15)	.97*	.22	.70*	.12	45.83*	.42*	.84*	.21
In Degree Training (34)	.75*	.15	.58*	.08	77.80*	.42	.64*	.14
No Training/Degree (3)	.30*	.07	.30	.44	1.46	.00	.27	.32
	$F\,(2,49) = .92$		$Q_b\,(2) = 1.20$					
			$Q_w\,(49) = 125.08^*$					
Therapist Experience								
Experienced (21)	.97*	.26	.58*	.10	80.55*	1.02*	.80*	.25
Inexperienced (28)	.65*	.09	.60*	.08	41.89*	.00	.61*	.08
	$F\,(1,47) = 1.64$		$Q_b\,(1) = .03$					
			$Q_w\,(47) = 122.44^*$					
Treatment Process								
Treatment Orientation								
Behavioral (36)	.98*	.16	.69*	.07	99.21*	.63*	.84*	.16
Systemic (11)	.52*	.18	.38*	.14	21.93*	.09	.42*	.17
Humanistic (6)	.75*	.21	.60*	.18	21.93*	.06	.72*	.21
Eclectic (11)	.76*	.25	.93*	.10	82.66*	.42*	.78*	.24
Other (6)	.81*	.23	.73*	.14	28.45*	.09*	.85*	.21
	$F\,(4,65) = .68$		$Q_b\,(4) = 11.29^*$					
			$Q_w\,(65) = 254.18^*$					
Behavioral Orientation								
Behavioral (36)	.98*	.16	.69*	.07	99.21*	.63*	.84*	.16
Nonbehavioral (32)	.72*	.12	.72*	.07	162.02*	.19*	.70*	.11
	$F\,(1,66) = 1.63$		$Q_b\,(1) = .16$					
			$Q_w\,(66) = 261.23^*$					
Treatment Standardization								
High (43)	.84*	.13	.63*	.07	116.43*	.46*	.72+	.13
Partial (11)	.97*	.27	.93*	.10	88.86*	.54*	.89*	.26
Unstandardized (4)	.40	.28	.41	.23	9.40*	.09	.44	.27

Variable	OLS Statistics		WLS Fixed-Effects Statistics			WLS Random-Effects Statistics		
	d	se	d_+	se	Q_h	$\sigma^2(\Delta)$	Δ	se
	$F(2,55)=.64$		$Q_b(2)=7.88^*$					
			$Q_w(55)=214.69^*$					
Treatment Implementation								
Documented (18)	.85*	.16	.90*	.09	100.56*	.18*	.85*	.15
Partially Documented (36)	.86*	.16	.59*	.07	106.32*	.64*	.72*	.16
Undocumented (7)	.45*	.14	.44*	.16	11.28	.00	.45*	.16
	$F(2,58)=.69$		$Q_b(2)=10.18^*$					
			$Q_w(58)=218.16^*$					
Time Focus of Therapy								
Present (54)	.87*	.12	.74*	.06	208.12*	.45*	.77*	.11
Present/Historical (4)	.30	.22	.27	.18	7.50	.03	.31	.21
	$F(1,56)=1.69$		$Q_b(1)=6.00^*$					
			$Q_w(56)=215.62^*$					
Communication Training								
Sole Emphasis (19)	.82*	.15	.63*	.10	65.36*	.15	.76*	.14
Partial Emphasis (28)	1.05*	.22	.87*	.08	135.48*	.97*	.90*	.21
No Emphasis (19)	.77*	.15	.60*	.10	46.76*	.14*	.67*	.14
	$F(2,63)=.67$		$Q_b(2)=6.42^*$					
			$Q_w(63)=247.60^*$					
Dependent Variable Characteristics								
Unit Described by Measure								
Child (25)	.68*	.14	.53*	.09	57.80*	.23	.59*	.14
Adult (11)	.90*	.22	.66*	.13	37.59*	.24	.82*	.21
Couple (20)	1.29*	.25	.92*	.11	58.06*	.83*	1.12*	.24
Family (18)	.33*	.10	.31*	.10	44.14*	.00	.35*	.10
Extrafamilial (5)	.40	.33	.94*	.12	58.02*	.34*	.53	.31
	$F(4,74)=4.00$		$Q_b(4)=26.75^*$					
			$Q_w(74)=255.61^*$					
Outcome Mode								
Self-Report (24)	.83*	.22	.53*	.09	71.27*	.83*	.70*	.21
Ratings by Others (47)	.75*	.10	.71*	.06	190.52*	.17*	.70*	.09
	$F(1,69)=.16$		$Q_b(1)=2.85$					
			$Q_w(69)=261.79^*$					
Smith et al. Reactivity[b]								
Low (Categories 1–3) (31)	.81*	.15	.74*	.07	166.87*	.39*	.73*	.14
Medium (Category 4) (37)	.75*	.14	.56*	.07	93.64*	.47*	.65*	.14
High (Category 5) (9)	.60*	.16	.54*	.15	9.98	.00	.57*	.15
	$F(2,74)=.25$		$Q_b(2)=4.01$					
			$Q_w(74)=270.49^*$					

Table 5.2 *(Continued)*

Variable	OLS Statistics		WLS Fixed-Effects Statistics			WLS Random-Effects Statistics		
	d	se	d_+	se	Q_h	$\sigma^2(\Delta)$	Δ	se
Blindedness on Measure								
Blind (45)	.76*	.11	.71*	.06	190.06*	.27*	.70*	.11
Not Blind (17)	.89*	.27	.60*	.11	40.75*	.88	.73*	.26
	$F\,(1,60)=.26$		$Q_b\,(1)=.72$					
			$Q_w\,(60)=230.81^*$					
Measure Specificity								
Tailored to Treatment (53)	.99*	.20	.66*	.06	176.59*	1.71*	.84*	.19
General Marital/Family (14)	.62*	.22	.43*	.13	38.37*	.37*	.54*	.21
	$F\,(1,65)=.50$		$Q_b\,(1)=2.62$					
			$Q_w\,(65)=214.96^*$					
Measure Manipulability								
Not Very Manipulable (5)	1.09*	.31	.87*	.20	38.24*	.15	1.08*	.28
Moderately Manipulable (36)	.75*	.13	.70*	.07	148.43*	.31*	.66*	.12
Very Manipulable (38)	.73*	.14	.55*	.07	90.83*	.46*	.63*	.14
	$F\,(2,76)=.43$		$Q_b\,(2)=3.73$					
			$Q_w\,(76)=178.50^*$					
Who Completed Measure								
Wife (14)	.62*	.21	.49*	.12	42.89*	.36*	.55*	.01
Husband (8)	.48*	.16	.50*	.15	14.04	.02	.52*	.15
Child (17)	.64*	.20	.49*	.11	36.82*	.39	.55*	.19
Couple Jointly (23)	1.14*	.23	.79*	.10	69.85*	.86*	.98*	.23
Family Jointly (6)	.51*	.18	.41*	.16	21.63*	.00	.52*	.16
Other (23)	.66*	.12	.77*	.08	108.72*	.09*	.70	.11
	$F\,(5,85)=1.55$		$Q_b\,(5)=11.33^*$					
			$Q_w\,(85)=293.95^*$					

Note: Asterisks by d, d_+, $\sigma^2(\Delta)$, or Δ indicate the statistic is significantly different from zero at $p<.05$. Significance is computed by multiplying standard errors by $+/-$ 1.96 to obtain 95 percent confidence intervals around effect size estimates; intervals that do not include zero are significant. Asterisks by numbers under column Q_h indicate rejection of the test of homogeneity of effect size within category; those by Q_b indicate significant differences among effect sizes between categories; those by Q_w indicate rejection of model specification, suggesting the categories are insufficient to explain effect size variation. Numbers in parentheses beside categories are number of study-level effect sizes in category; number in parentheses beside Q statistics are degrees of freedom. Finally, 13 negative variance components in this table were truncated to .00.
[a]Sums to 59 studies because one study used multiple conditions, one of which was a control group haphazardly assigned and matched.
[b]Smith et al.'s reactivity categories 1–3 (1980) are combined in this analysis since their category 1 had only one study and their category 3 had no studies.

that will be discussed later in this chapter).[3] Those variables are divided into four categories: study methodology, client and treatment context and inputs, treatment process, and dependent variable characteristics. Orientation effects are included under therapy process; Table 5.2 suggests that they are statistically significant. Not only are effect sizes for all orientations significantly different from zero, but the Q_b test suggests between-categories differences are significant, as well. Hedges and Olkin do not outline a posteriori follow-up tests for a significant Q_b. However, with five intervals one could use the Bonferroni inequality to obtain a corrected alpha of .01, and a two-tailed confidence interval defined by $z = +/-2.57$. This very conservative procedure suggests that no orientation differs reliably from any other, although systemic and eclectic very nearly do so.

Table 5.2 suggests many other interpretations as well—for example, that presenting problems regarding couples yield higher effect sizes than those regarding families and that matching subjects reduces effect sizes. Each of these other variables may be confounded with orientation effects. Hence interpretation of these simple categorical tests is equivocal at best and is further complicated by the fact that Q_w was rejected for every variable, suggesting that no single variable—including orientation—accounts adequately for variation in effect sizes. To remedy this, one can further subdivide categories until Q_w is not rejected or pursue multivariate regression models. The two approaches are conceptually similar, but the regression approach is probably more efficient and so is reported subsequently. First, however, consider a methodological matter in Table 5.2.

---

[3] In Table 5.2, the total sample of studies equals 58, but the number of study-level effect sizes is more than 58 for some variables. Although each study contributed only one effect size estimate to each category, it sometimes contributed an effect size to more than one category (for example, if a study included both specific and general outcome measures). This violates the assumption of independence of observations underlying OLS and WLS categorical tests. In the case of OLS F tests, Robinson et al. (in press), adapting the work of Kenny and Judd (1986), argue that the F test error term in unbiased if only one observation from a study enters each cell and that the numerator of the F test is too small (conservatively biased) if the intraclass correlation is positive and a study contributes an observation to more than one cell. The same argument generalizes to the WLS statistics in Table 5.2. Since only one observation per study enters into each category, Q_w is unbiased. When a study contributes an observation to more than one category, Q_b is conservatively biased since the intraclass correlations mentioned earlier were all positive. An alternative to this procedure is to delete observations from studies so that each study contributes only one observation to the entire analysis for each variable. Such deletions yield unbiased tests; but such deletions lose information and degrees of freedom within categories. The trade-offs between these two alternatives require further research.

ORDINARY VERSUS WEIGHTED LEAST SQUARES ANALYSES.   Table 5.2 contrasts two different ways of analyzing this data: ordinary least squares analyses (OLS), which were traditionally used in the past by Smith et al. (1980) and others, and weighted least squares analyses (WLS), which apply the weighting procedure developed by Hedges and Olkin (1985). To the best of our knowledge, results of an extensive empirical contrast between these two approaches has not been published before on real data. Inspection of Table 5.2 shows that the choice of OLS versus WLS technique can make a very large difference to interpretation. For example, 90 percent of the standard errors reported in Table 5.2 decreased under WLS analysis. On average, WLS standard errors were 26 percent smaller than the OLS standard errors. This decrease had little effect on the significance of treatment-control effect sizes, and the direction and magnitude of effect sizes were comparable with d_+ being about 13 percent smaller than d on average. The overall correlation between d and d_+ was high ($r = .69$, $p < .001$), but the pattern of effect sizes over categories within variables changed in 12 of 28 cases. OLS F tests yielded only 4 significant effects over 28 variables; WLS Q tests yielded 12 significant effects. Hence the overall interpretation of results from the two analyses is very different. Homogeneity of variance assumptions were violated in 22 of 28 OLS analyses, often severely enough or with disparate enough sample sizes to require transformation of raw effect sizes that would make their interpretation more difficult. These many differences suggest that since WLS is theoretically preferable, continued use of OLS analyses may be of questionable wisdom.

WLS and OLS estimates differ by a function of study sample size. Hence, an important source of differences between OLS and WLS estimates in Table 5.2 should be differential sample sizes in the 58 studies coded in this meta-analysis. Figure 5.2 presents a histogram of the average sample size per group in the studies. Sample sizes are generally small and disparate over studies, ranging from 4 to 119, with a mean of 14.79 and a mode of 10. Ninety percent of the studies had per-group sample sizes of 22 or less.

Regression Models

The present data contain many variables that may be confounded with orientation effects—not only those in Table 5.2, but also many continuous variables not reported in that table such as attrition rate and therapy dose. One way to examine the possibility of such confounding is to use multiple regression to partial out such effects. Such partialing will be incomplete and inaccurate for reasons outlined in the conclu-

Figure 5.2   Sample Size per Group ($n = 58$)

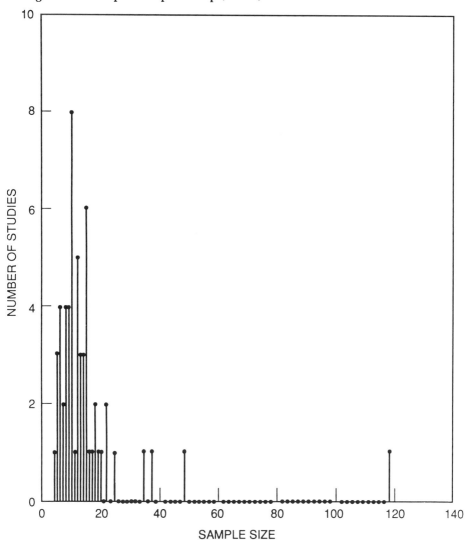

sions to this chapter. But regression is still informative because it allows us to examine whether effects persist under a model that is prone to eliminate redundant effects. A technical problem with regression, however, is that the number of studies ($n = 58$) is small in comparison to the number of predictors. As a rough guide, one might include no more predictors than the square root of the number of studies, about seven or eight predictors in this case. Reducing the number of predictors could be accomplished many ways, two of which are explored here: theoretical selection, and empirical selection.

THEORETICAL SELECTION.   When this research first began, several hypotheses were explicitly drawn from past psychotherapy research:

1. Published research would report higher effect sizes than unpublished research.
2. Behavioral treatments would yield higher effect sizes that nonbehavioral treatments.
3. Experimenter allegiance to a therapy would result in higher effect sizes for that therapy.
4. Studies conducted in universities (or with university clients) would yield higher effect sizes than nonuniversity studies.
5. Reactive measures would yield higher effect sizes than nonreactive measures.
6. Studies with high differential attrition would yield different effect sizes from studies with lower differential attrition (a nondirectional hypothesis).
7. Studies with more results reported simply as nonsignificant, and therefore assigned an effect size of zero, would yield lower effect sizes than other studies.

These hypotheses suggest eight variables (two variables for the fourth hypothesis) to be entered into regression.

Following the weighted least squares procedures outlined in Hedges and Olkin (1985, chap. 8), we computed such an equation. The resulting multiple R was .57. The test for significance of the predictor set was $Q_r(8) = 47.89$, $p < .005$.[4] Two predictors had significant beta weights. Specifically, publications had higher effect sizes than dissertations (beta = .34), and experimenter allegiance to a therapy was associated with lower effect sizes (beta = − .57). Finally, the test for model specification was rejected ($Q_e = 99.14$, df = 50, $p < .005$), suggesting that nonrandom variance in effect size still remained unaccounted for by these predictors. Note that effects for behavioral orientation were not significant in this regression.

One could include interaction terms in the equation, but the eight

[4] Hedges and Olkin's example is computed in SAS, which includes the intercept in this test; the present regression is computed in SPSSx, which excludes the intercept. Therefore one must add a degree of freedom to the degrees of freedom for the Q_e test (Hedges and Olkin 1985, p. 174).

predictors yield 28 possible first-order interactions for a total of 36 predictors, far more than desirable. I know of no theoretical rationale in the psychotherapy literature to help select among those 28 interactions. Hence selection is partly empirical and partly intuitive. The two empirically significant predictors in the first equation are retained along with their interaction. Also reactivity of measurement and behavioral/nonbehavioral treatment orientation are retained because they suggest intuitively interesting interactions:

1. Allegiance may be more powerful in behavioral than nonbehavioral treatments.
2. Allegiance may affect reactive measures more than nonreactive measures.
3. Behavioral treatments may display higher effects on reactive than nonreactive measures.

This regression equation has four raw predictors and four interactions among predictors, and yields a multiple R of .63 ($Q_r = 58.89$, df $= 8$, $p < .005$). Reactive measures yielded higher effect sizes (beta $= .85$), and reactivity interacted with allegiance. However, in the latter case the beta weight exceeded allowable bounds (beta $= -1.27$), probably owing to collinearity with reactivity. To solve this problem the equation was run again with tolerance $= .40$, equivalent to excluding variables with a variance inflation factor (VIF) exceeding 2.5 (since tolerance $=$ 1/VIF; Neter, Wasserman, and Kutner 1983). The resulting multiple R of .54 was significant ($Q_r = 42.09$, df $= 4$, $p < .005$). The high tolerance solved the collinearity problem but allowed only four predictors into the equation, of which only two had beta weights that were significantly different from zero. Specifically, experimenter allegiance to therapy decreased effect size (beta $= -.57$), and publications had higher effect sizes than unpublished works (beta $= .33$). These were the same variables that were significant in the model without interactions; so it seems that these particular interactions will do little to account for the remaining variation in effect size. Model specification was rejected for this model ($Q_e = 104.92$, df $= 54$, $p < .01$). We conclude, then, that neither main effects for behavioral orientation nor selected interactions with orientation add significantly to our ability to predict effect size.

EMPIRICAL SELECTION.   Empirical selection of variables simply enters all variables and retains the significant ones. However, because the number of predictors and the sample size are about the same, we first develop a regression equation for each of four subsets based on the four sets of variables in Table 5.2—study methodology, dependent variable characteristics, client and treatment inputs and context, and treatment process—and then retain significant predictors from each set for inclusion in a final equation. This capitalizes on chance but gives another way of looking at the data.

The 11 variables assessing *study methodology* were (1) proportion of results reported only as nonsignificant, and so coded zero; (2) random versus haphazard assignment of subjects to condition; (3) matching of subjects; (4) year the study was published; (5) differential mortality from groups; (6) blinding of therapist to conditions; (7) whether or not the study was conducted in a university setting; (8) number of outcome measures used in the study; (9) experimenter allegiance to treatment; (10) published versus unpublished status; and (11) whether the method used to compute the effect size was exact or approximate. The resulting multiple R was .64 ($Q_r = 60.59$, df = 11, $p < .005$). The more measures used, the lower the overall effect size (beta = $-.32$); therapist allegiance decreased effect size ($-.47$); and unpublished works had lower effect sizes than published works ($-.28$). Model specification was rejected ($Q_e = 86.43$, df = 47, $p < .005$). Setting tolerance to .40 resulted in no notable changes to this equation.

The 13 variables assessing *dependent variable characteristics* included were (1) experimenter blindedness to the dependent variable; (2) reactivity of the dependent variable; (3) manipulability of the dependent variable; (4) specificity of the dependent variable; (5) whether the dependent variable was a self-report or a rating by someone else; (6) whether or not the dependent variable was completed by the wife/mother, (7) by the husband/father, (8) by a child, (9) by a family jointly; (10) whether or not the dependent variable described an individual child, (11) an individual adult, (12) a couple, (13) a family. The resulting multiple R was .56 ($Q_r = 46.19$, df = 13, $p < .005$). Effect sizes were lower if the dependent variable described a child (beta = $-.41$) or the family ($-.45$). Model specification was rejected ($Q_e = 100.82$, df = 45, $p < .005$). Running this equation with tolerance equal to .40 resulted in only minor changes. The multiple R was .54 ($Q_r = 43.65$, df = 10, $p < .005$; $Q_e = 103.36$, df = 48, $p < .005$). Three predictors were significant: The two above with beta weights of $-.34$ and $-.41$, respectively, and specific measures led to higher effect sizes compared with general measures (.22). Three predictors were dropped for collinearity (dependent variable described couple; dependent variable was completed by husband; reactivity of dependent variable), but they did not have significant beta weights in the previous equation.

The 12 variables assessing *inputs into therapy* (client characteristics, therapist characteristics, therapy context) were (1) male therapists; (2) female therapists; (3) Smith et al.'s (1980) coding.of years of therapist experience; (4) whether or not therapists had a professional mental health degree; (5) whether or not therapists had prior therapeutic experience; (6) whether therapists solicited clients versus whether clients referred themselves or were referred by others; (7) the degree to which clients came from university settings; (8) behavioral or nonbehavioral presenting problem; (9) whether or not the identified patient was a child, (10) an adult, (11) a couple, (12) a family. The multiple R was .52 ($Q_r = 40.29$, df $= 12$, $p < .005$). Lower effect sizes were associated with child presenting problems (beta $= -.57$), adult presenting problems ($-.43$), family presenting problems ($-.49$), male therapists ($-.29$), behavioral presenting problems ($-.24$), and therapists without professional mental health degrees ($-.29$). Model specification was rejected ($Q_e = 106.72$, df $= 46$, $p < .005$). Running this equation with tolerance of .40 yielded minor changes. The multiple R was .48 ($Q_r = 33.40$, df $= 10$, $p < .005$; $Q_e = 113.61$, df $= 48$, $p < .005$). The first four significant predictors listed above were again significant (betas of $-.41$, $-.33$, $-.32$, $-.27$, respectively), but behavioral presenting problems was no longer significant. In addition, owing to collinearity, both couple presenting problems and whether therapists had professional degrees were dropped. This latter variable had been significant in the regression without a stringent tolerance level.

Nine predictors associated with *therapy process* were (1) whether treatment was standardized using a manual or training; (2) whether standardization was checked with implementation checks during therapy; (3) the extent to which communication training was emphasized in therapy; (4) the extent to which therapy attended primarily to current matters or to both current and historical matters; (5) therapy dose (number of minutes per session times number of sessions); (6) whether treatment was systemic in orientation, (7) behavioral in orientation, (8) humanistic in orientation, or (9) eclectic in orientation. The multiple R was .55 ($Q_r = 43.72$, df $= 9$, $p < .005$). Lower effect sizes were associated with studies in which treatment implementation was not checked to ensure that it was delivered as intended (beta $= -.29$), and higher effect sizes were associated with eclectic (.50) and behavioral (.43) treatments. Model specification was rejected ($Q_e = 103.29$, df $= 49$, $p < .005$). Running the equation with tolerance of .40 yielded some changes. Multiple R was .52 ($Q_r = 39.77$, df $= 8$, $p < .005$; $Q_e = 107.24$, df $= 50$, $p < .005$). Treatment implementation was again a significant predictor (beta $= -.23$), but only systemic orientation was significant in this equation, associated with lower effect sizes ($-.21$). The behavioral treatment predictor was dropped for collinearity.

SUMMARY EQUATION.   A summary regression equation included the predictors that were significant in previous runs, except that only the code for behavioral versus nonbehavioral treatment orientation was entered into this equation since 53 percent of the orientations studied were behavioral. This selection of predictor variables capitalizes on chance, but provides a more succinct summary. The resulting multiple R was .73 ($Q_r = 77.34$, df = 13, $p < .005$). Studies with more outcome measures had lower effect sizes (beta = $-.22$), studies with individual adult presenting problems had lower effect sizes ($-.28$), and therapist allegiance decreased effect size ($-.49$). Model specification was rejected ($Q_e = 69.67$, df = 45, $p < .01$). Running the equation with tolerance set to .40 yielded the same three predictors having significant beta weights of $-.24$, $-.30$, $-.48$, respectively (multiple R = .71; $Q_r = 74.89$, df = 12, $p < .005$; $Q_e = 72.12$, df = 46, $p < .05$).

INTERPRETING RESULTS.   Does orientation make a difference? The results so far would lead us to think not. Although behavioral orientations did better and systemic orientations did worse than other orientations in the treatment process regression, this result did not persist in the summary equation when other variables such as number of measures, allegiance, and presenting problem were partialed out. Each of these latter variables should be regarded as a confounding variable rather than a substantively explanatory variable. Proponents of various orientations rarely if ever claim that using many measures or being allegiant is an inherent part of their work. While some therapies often are limited to certain kinds of problems—for example, using behavioral treatments for bed-wetting—orientations as a whole also rarely claim they are effective with only certain problems.

However, we might want to consider why these three variables have the relationship with study effect size that we observed. The effect for number of measures can be explained two ways. Some authors may not report all their nonsignificant results, yielding studies with fewer measures and higher overall effect sizes; and studies using large numbers of outcome measures may include peripherally relevant measures (Dush, Hirt, and Schroeder 1989). The finding about individual adult presenting problems may have occurred because only 3 of the 58 studies had adult presenting problems and those problems were schizophrenia (Kopeikin, Goldstein, and Marshall 1983)[†], alcoholism (Orchen

1983),[†] and substance abuse (Steier 1983),[†] all of which are particularly difficult problems. The therapist allegiance effect is troublesome because the direction of this finding is *opposite* past meta-analyses (e.g., Berman et al. 1985; Robinson et al. 1990; Smith et al. 1980). Other authors' findings that allegiance increases effect size are intuitively plausible and fit better with past research on experimenter expectancy effects. Two explanations are worth exploring. First, some past researchers used ordinary least squares analyses rather than weighted least squares analyses; the two approaches yielded different results for allegiance in the present study (Table 5.2). Second, the present study coded allegiance differently than past studies. Berman et al., Robinson et al., and Smith et al. all used a *relative* rating in which allegiance was rated as a preference for one therapy over other therapies. For example, Berman et al. coded allegiance as positive for cognitive therapy if it was preferred to systematic desensitization therapy. The present study used an *absolute* rating in which allegiance was positive if the experimenter expressed positive sentiments toward the therapy no matter what the experimenter's sentiments were toward other therapies. Absolute ratings yield more positive allegiance ratings for multiple therapies in a study since an experimenter may *prefer* one therapy but still express allegiance toward both. Perhaps relative ratings should be called preference rather than allegiance. The present study can approximate a relative preference rating by comparing treatments rated high on experimenter allegiance to those rated low on experimenter allegiance *within the same study comparison*. Doing so replicates past findings, with the OLS $d = .27$ ($n = 31$, se $= 10$, $p < .05$), and the WLS $d_+ = .23$ (se $= .06, p < .05$; $Q = 75.16$, $p < .05$)—that is, experimenter preference increases effect size for the preferred treatment.

Note that the discrepant findings in the present study came from two very different kinds of analyses. The finding that allegiance decreases effect size was obtained from between-studies treatment-control comparisons, with multiple regression being used to try to adjust statistically for confounding variables across studies. Although one might correctly object at this point that the model was not well specified, this argument loses some but not all of its force in the next section when we see that it is easily made well specified after a minor adjustment. Conversely, the finding that preference increases effect size was obtained from within-study treatment-treatment comparisons in which subjects were randomly assigned to conditions. When those conditions differ partly according to allegiance, as they did in the above 31 treatment-treatment comparisons, then subjects are randomly assigned to preference. Whereas the between-studies treatment-control analysis relies on regression to adjust for extraneous variables statistically, the within-study treatment-treatment analysis relies on experimentally induced group (partial) equivalence to control for extraneous variables (the equivalence is only partial for reasons discussed in the conclusion of this chapter). This discrepancy between regression and experimental models is reminiscent of a similar debate about the relative accuracy of econometric selection bias models versus experimental models (Heckman, Hotz, and Dabos 1987; LaLonde 1986; Stromsdorfer 1987). That debate has not yielded any easy answers and does not apply fully to the present case. Still, experimental models tend to be preferred when a choice has to be made. If so, this discrepancy calls into question naive reliance on the adequacy of model specification tests in between-studies correlational analyses of meta-analytic data. We will return to this matter in the discussion section of this chapter.

WINSORIZING OUTLIERS.   One might, however, argue that these regressions still have not fairly tested orientation effects. The reason is that model specification (Q_e) was rejected in all preceding equations, implying that the existing variables do not sufficiently account for effect size variation. One solution would be to add more variables that are related to outcome to the equation, something that we cannot do since we did not code more. To the extent that such variables do exist, however, beta weights may be incorrect. However, an alternative explanation might be that outliers in Figure 5.1 are so extreme as to be not predictable. If so, an option is to Winsorize the distribution of study-level effect sizes; that is, to reduce the effect size for one or more outliers to some smaller number. This procedure is commonly recommended in statistics and even in natural sciences such as physics (Hedges 1987; Hedges and Olkin 1985). The rationale is that such extremely large effect sizes may be due to chance and, if so, should be reduced to a more plausible large value. A test for outliers (Hoaglin, Mosteller, and Tukey 1983, pp. 39–40) locates only three in Figure 5.1, at the upper end of the distribution at 3.13, 3.66, and 4.76. The highest outlier is from a study by Bogner and Zielenbach-Coenen (1984).[†] Reducing this effect size to 3.7, just a bit higher than the two remaining outliers in Figure 5.1, resulted in a model that fit the data (multiple $R = .76$; $Q_r = 75.79$, df $= 13$, $p < .005$; $Q_e = 54.97$, df $= 45$, $p > .10$). The same three predictors were significant with the same interpretation as before. Hence, orientation makes no difference even in a well-specified model.

Some analysts would argue that one should not try to substantively justify Winsorizing outliers. To them the justification is purely statistical, that such an extreme effect size is so unlikely as to be plausibly a result only of chance. Seeking a substantive justification risks constructing a specious interpretation of a chance effect. Nonetheless, this procedure is not widely used or accepted among most behavioral scientists. Hence it is worth examining this study more closely to demonstrate that this procedure can well be warranted. Close examination of the study suggests a possible explanation for the outlier. A self-report of marital conflict in this study yielded two extreme effect sizes (13.73, 6.32). Simply Winsorizing these two effect sizes to 3.89, the value of the next highest effect size in the Bogner and Zielenbach-Coenen study, results in a well-specified model with no change in interpretation of predictors (multiple $R = .76$; $Q_r = 75.81$, df $= 13$, $p < .005$; $Q_e = 55.16$, df $= 45$, $p > .10$). Bogner and Zielenbach-Coenen report standard deviations that are quite a bit lower (RTE post-test SD $= 2.4$; RTEF $= 0.6$; WLCG $= 0.9$) on this measure compared with the normative sample gathered by Hahlweg et al. (1984).† If one substituted the pooled standard deviation (2.39) from the much larger ($n = 190$) Hahlweg et al. normative sample for the pooled standard deviations obtained in the small ($n = 24$) Bogner and Zielenbach-Coenen sample, the effect sizes drop from 13.73 to 4.70 and from 6.32 to 4.07. This change alone yields a well-specified model with no change in interpretation (multiple $R = .76$; $Q_r = 75.93$, df $= 13$, $p < .005$; $Q_e = 56.06$, df $= 45$, $p > .10$). (This Winsorized summary equation is the one that will be used for replication purposes later in this chapter.) We can conclude, then, that meta-analysts ought to inspect outliers carefully to ensure that no unusual study characteristics might have contributed to generating such extreme effect sizes.

Meta-Analysis and Missing Data

The analyses to this point treat orientation effects as a product of how the study was conducted—for instance, that they result from different clients being selected or different measures being used. But a major alternative explanation is that these effects could be a product of how the meta-analysis is conducted. In this section and the next we explore two such possibilities: the treatment of missing data and coder reliability/confidence.

MISSING DATA ON THE PREDICTORS.   Accurate tests of predictors require that one be able to code the predictor from the report of the original study. For example, it might be difficult to test the effects of theoretical orientation if orientation was not reported in the primary study and so could not be coded. Fortunately, this did not happen often in the present study. Coders were told to complete all codes if possible and to rate the confidence they had in their decision (1 = guess, 2 = more likely than not; 3 = certain or almost certain; Orwin and Cordray 1985). It usually proved possible to make those guesses, leaving very little missing data on predictors. Of the 45 variables entered into the previous regression analyses, 34 had complete data, 5 had data in 57 of 58 cases, 2 had data in 56 of 58 cases, and 4 had complete data in 45–48 of the 58 cases (these were the three therapist experience variables and the variable concerning blinding of the experimenter to treatment). In previous regressions, missing data were replaced by the variable mean, thus keeping all 58 studies.

We might, however, try alternative procedures to see if the previous conclusions we reached still replicate. One alternative is listwise deletion—to delete cases with missing data on any variable. Doing so, the multiple R of .74 was significant ($Q_r = 39.07$, df = 14, $p < .005$), and the model was well specified ($Q_e = 31.93$, df = 32, $p > .50$). With fewer cases in this model, however, none of the beta weights were individually significant, but the same three significant variables in the Winsorized regression closely approached significance in this equation, with the same interpretation. Another option is using mean substitution and a missing data index for each variable with missing data. The missing data index is scored as 1 if the variable is missing in a study and 0 otherwise, following Cohen and Cohen (1983); and it is entered into the regression equation along with the predictors to see if the tendency to have missing data on a variable is systematically related to effect size. Recomputing the Winsorized regression equation with this option made almost no difference to regression statistics, and none of the missing data indicators were significantly related to effect size.

MISSING DATA ON THE DEPENDENT VARIABLE.   The dependent variable is effect size, and missing data on this variable obviously precludes accurate assessment of therapy outcome. The key problem is the 130 (of 620) treatment-control effect sizes coded as zero because they were reported only as nonsignificant. This problem might also preclude accurate assessment of predictor variables if some predictors are more prone to have missing effect sizes than others. For example, if authors using systemic orientations are differentially more likely to report results only as nonsignificant, this would bias the outcomes of such studies toward zero. All the analyses reported to this point include those zero effect size estimates as legitimate dependent variables to be predicted. This procedure may bias average effect size estimates since the distribution of observed effect sizes in Figure 5.1 suggests that nonsignificant effects may be slightly larger than zero. Figure 5.3 presents a histogram excluding such effect sizes. Compared with Figure 5.1, it shows a tendency toward higher effect sizes; but study-level effect sizes were still available on all 58 studies. Mean d_+ for Figure 5.3 was .81 with standard error of .05 and 95 percent confidence interval of $.71 < d_+ < .91$ ($Q = 225.51$, df $= 57$, $p < .005$). If d_+ with all effect sizes is a lower bound estimate, the effect size might be between .70 and .81 if all effect sizes were computable.

Figure 5.3   Study-Level Effect Sizes Excluding Findings Reported Only as Nonsignificant (*n* = 58)

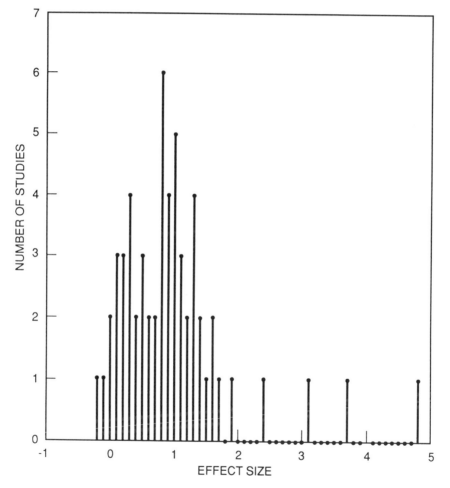

Excluding these effect sizes also makes little difference to the Winsorized regression equation, except that the model is no longer well specified ($Q_e = 69.17$, df $= 45$, $p < .01$; multiple R $= .72$; $Q_r = 72.71$, df $= 13$, $p < .005$). As before, high experimenter allegiance (beta $= -.46$), adult presenting problems ($-.34$), and use of more outcome measures ($-.21$) were significantly associated with lower effect sizes. In addition, unpublished work had lower effect sizes than published work (beta $= -.34$), possibly a function of a nonsignificant trend ($p = .11$) for publications to report more findings only as nonsignificant. Further Winsorization modestly improved the fit of this model. Specifically, the three outliers in Figure 5.3 (Bogner and Zielenbach-Coenen 1984; Jacobson 1977; Roberts et al. 1978)† were set to 2.5, a level just above that of the fourth highest study. This model yielded a significant multiple R $= .76$ ($Q_r = 71.37$, df $= 13$, $p < .005$), the same interpretation of predictors, and a nonsignificant specification test ($Q_e = 52.65$, df $= 45; .25 > p > .10$).

DISCUSSION OF MISSING DATA.   This tentative exploration suggests that our conclusions about marital and family therapy are not much biased by the problem of missing data. Even so, the problem is worth further attention for many reasons. One gets modestly different answers if one codes missing effect sizes as zero versus excluding them altogether. Coding these effect sizes as zero has two flaws: zero may be an inaccurate estimate of true effect size and these 130 nonsignificant effect sizes are likely to be distributed around some value rather than all take on the same value. Eliminating these effect sizes altogether effectively substitutes the study mean effect size for nonsignificant findings. That, in turn, probably overestimates outcome because study effect sizes are usually based on (roughly) the same sample size, so power theory suggests that the nonsignificant effects are probably somewhat smaller than the significant ones. Fortunately, methods are being developed to apply to this problem. Hedges and Olkin (1985) outline maximum likelihood procedures to estimate these effect sizes. But the problem is not just with missing data on dependent variables. Little and Rubin (1987) review several techniques for dealing with missing data that might also apply to independent variables; although not aimed specifically at meta-analysis, their techniques should apply. Rubin's multiple imputation model (1987) for nonresponse in surveys seems particularly worth further exploration in this regard.

Effects of Reliability and Confidence on Results

Conclusions about marital and family therapy might also be biased if different variables are coded at different levels of reliability or confidence. In particular, less reliably coded variables would have attenuated relationships with effect size. Hence we also need to assess if our previous conclusions change if we try to take reliability and confidence into account.

To examine the issue of whether confidence level makes a difference to results, we first recomputed the weighted least squares results reported in Table 5.2 using only data coded with high confidence (a table of these recomputed statistics is available from the author on request). The recomputed statistics yield interpretations quite similar to Table 5.2. Then, we recomputed the Winsorized regression equation (including both effect sizes reported only as nonsignificant and the two Winsorized effect sizes from Bogner and Zielenbach-Coenen). In a hierarchical regression, confidence codes for all variables in the Winsorized regression equation were entered as a block, yielding a nonsignificant multiple R of .20 ($Q_r = 5.33$, df = 8, $p > .50$; $Q_e = 126.65$, df = 50, $p < .005$). Addition of a block of predictors resulted in a significant, well-specified equation (multiple R = .82; $Q_r = 89.12$, df = 21, $p < .005$; $Q_e = 42.86$, df = 37; $.25 > p > .10$). Subtracting the Q_r fit statistics for the first block from those for the second block suggests that the addition of the second block significantly improved prediction ($Q_r = 83.79$, df = 13, $p < .005$). Number of measures (beta = $-.32$) and therapist allegiance ($-.39$) were significant as before; but adult problems no longer was significant. Male therapists were associated with lower effect sizes than female or mixed-gender sets of therapists ($-.23$). Confidence in coding of theoretical orientation was associated with higher effect sizes (.26), and confidence in coding of the state of the presenting problem (i.e., was it behavioral, affective, cognitive, etc.) was associated with lower effect sizes ($-.33$).

Prior to the study, the penultimate draft of the coding manual was submitted to an interrater reliability study (Table 5.3). The present author and a research assistant independently coded all information and one effect size from 30 studies. Reliabilities are generally adequate (Fleiss 1981; Nunnally 1978; Orwin and Cordray 1985). Also, mean percentage agreement for the first 15 studies was 84 percent, but rose to 89 percent for the last 15 studies.[5]

One could use reliability coefficients to deattenuate relationships between variables. Effect size predictor relationships might become stronger

---

[5] One might expect a high correlation between reliability and confidence if guessing leads to lower reliability, but the observed relationships were modest. Between confidence and percentage agreement $r = .50 (p < .01)$; between confidence and kappa $r = .59 (p < .01)$; and the correlation between confidence and Pearson correlations was nonsignificant ($r = .25$). It may be that the underlying relationship between confidence and reliability is weak (Orwin and Cordray 1985) or that restriction of range attenuated coefficients owing to a ceiling effect for confidence ratings, since all these ratings are high.

Table 5.3 Interrater Reliability and Confidence Codings for Variables Coded in the Meta-Analysis

	Percentage Agreement	Kappa	Pearson	Confidence Rating
Publication Characteristics				
Author Profession	87	.70		2.841
Study Category (Marital or Family)	97	.93		2.998
Presenting Problem Characteristics				
Locus of Problem	83	.76		2.963
Problem State	79	.69		2.821
Use of Patient Exclusion Criteria	86	.73		2.980
Methodological Characteristics				
Experimenter Blindedness to Treatment	97		.94	2.485
Source of Clients	86	.76		2.850
Use of University Subjects	97		.94	2.924
Random Assignment	93		.81	2.951
Matching	97	*		2.944
Assignment of Therapist to Conditions	87	.79		2.920
Internal Validity Rating	73		.73	
Treatment Characteristics				
Orientation	80	.71		2.803
Experimenter Allegiance to Therapy Type	83	.67		2.722
Time Focus of Therapy	90		.79	2.922
Use of Communication Training	87	.73		2.827
Adjunct Use of Medication	100	1.00		3.000
Treatment Modality	80	.74		2.880
Treatment Location	70	.55		2.369
Therapist Gender	97	.96		2.756
Therapist Experience	77	.71		2.695
Kind of Control Group	93	.77		2.889
Treatment Standardization	77		.65	2.627
Treatment Implementation	80		.65	2.613
Outcome Characteristics				
Outcome Type	90	.86		2.886
Outcome State	93	.90		2.776
Outcome Mode	90	.86		2.891
Number of Weeks After Treatment That Post-Test Was Taken	100		1.00	2.609

	Percentage Agreement	Kappa	Pearson	Confidence Rating
Blinding of Experimenter to Dependent Variable	70		.59	2.524
Specificity of Dependent Variable	90		.81	2.712
Manipulability of Dependent Variable	80		.87	2.751

Note: This reliability study occurred on the penultimate draft of the coding manual, which was then revised for use in the meta-analysis. Hence some minor differences exist between the variables used in the interrater reliability study and those reported in the rest of this chapter.

*In this case, one rater had no variance, so that computation of any variance based reliability coefficient was impossible. However, agreement was nearly perfect.

(Orwin and Cordray 1985), and previously nonsignificant predictors might become statistically reliable. This would suggest an upper bound for R^2 which is less than 1.0 but which more accurately reflects the fact that one cannot be expected to model and explain unreliability itself. It is worth noting, however, that the reliability of coding of theoretical orientation, although adequate, is one of the lowest in the table. Orientation is notoriously difficult to infer from reading a manuscript. If so, it may be that meta-analyses of psychotherapy need to begin to construct more reliable ways of assessing this variable, such as asking authors to indicate their own orientation.

Using Treatment-Treatment and Treatment-Control Comparisons in the Same Analysis

Previously, we concluded that orientation effects did not persist when one adjusts for other possible confounding variables. But a test of orientation effects exists that may be more powerful and less biased than those reported previously. Specifically, this section explores orientation effects analyzed through treatment-treatment comparisons—direct comparisons of orientation within studies. At the same time we can explore how treatment-treatment and treatment-control comparisons can be combined in the same analysis.

WITHIN-STUDY TREATMENT-TREATMENT COMPARISONS.   When subjects are randomly assigned to two treatments within the same study, the resulting effect size allows inferences that are less confounded by covar-

Table 5.4   Relative Effectiveness of Different Orientations Within the Same Study: Specific Treatment-Treatment Comparisons

Orientation	1	2	3	4	5	6
1 Behavioral	.32*	−.20	.32	−.13	.54*	.60*
	(31)	(5)	(2)	(1)	(8)	(6)
2 Systemic		.07	−.03	−.11	−.09	.21
		(6)	(2)	(1)	(3)	(1)
3 Humanistic			.39	—	—	—
			(2)	(0)	(0)	(0)
4 Psychodynamic				.10	—	−.02
				(1)	(0)	(1)
5 Eclectic					−.19*	.42
					(10)	(3)
6 Other						−.03
						(4)

Note: Positive effect sizes mean that the row orientation produced better post-test effects than the column orientation; negative effect sizes imply the opposite. Numbers in parentheses are the number of study-level sizes on which the estimate is based.
*Significantly different from zero, $p<.05$.

iates than are the treatment-control comparisons used in analyses to this point. For example, comparing a behavioral-control effect size from Study 1 to a systemic-control effect size from Study 2 might well be confounded with the fact that Study 1 was a dissertation and Study 2 a publication, or that Study 1 used more reactive measures than Study 2. Many such covariates *by definition* cannot confound within-study treatment-treatment comparisons. If, for example, Study 3 randomly assigned subjects to systemic or behavioral therapy, the report of results is either a dissertation or a publication but not both. Similarly, since the resulting effect size is computed on a specific measure that is the same for both groups, reactivity of measurement is the same for both therapy orientations. Confounds due to the majority of variables in Table 5.2 are controlled in this fashion.[6]

Table 5.4 presents treatment-treatment effect sizes for specific pairs of orientations such as behavioral versus systemic. Since such specificity leads to very small sample sizes for some pairs, Table 5.5 presents effect sizes for a specific orientation versus all other orientations. Table

[6]Note that these advantages of within-study treatment-treatment comparisons are weakened somewhat (but not completely) when subjects are not randomly assigned to conditions. With random assignment, for example, subject characteristics are (probabilistically) equated over groups, but this is not true in quasi-experiments.

Table 5.5 Relative Effectiveness of Different Orientations: Pooled Comparisons

Orientation	Mean	N
Behavioral vs. All Others	.43*	22
Systemic vs. All Others	−.01	12
Humanistic vs. All Others	−.14	4
Psychodynamic vs. All Others	.06	3
Eclectic vs. All Others	−.25*	14
Other vs. All Others	−.48*	10

*$p<.05$.

5.5 is easier to interpret. Behavioral therapies did significantly better than alternative therapies, and eclectic and "other" therapies did somewhat worse. Table 5.4 provides more specific information. Behavioral therapies did better than eclectic and better than "other" therapies; but behavioral therapies also did better than behavioral therapies, and eclectic therapies did worse than eclectic therapies. These latter two comparisons occurred when some authors compared two versions of the same orientation—for example, behavioral communication training and behavioral contracting, or behavioral training with fading of therapy and such training without fading. (A reviewer of this chapter pointed out that one can use these same-orientation comparisons to compute true error rates for orientations. Treating behavioral-behavioral comparisons as near replications, the same variance estimates the baseline variability within and between studies. One estimates variability by computing the within-study sum of squares and then aggregating over studies for a "pure" error term.)

Three implications are worth noting. First, Tables 5.4 and 5.5 suggest that at least some orientation effects may be present. Second, if different versions of the same therapy can still yield significant results, then some active therapy components may be at a finer level than gross theoretical orientation. For example, the superiority of behavioral therapies to eclectic therapies may not reflect orientation superiority as much as superiority of some component more commonly used in behavioral therapies. Finally, even within-study treatment-treatment comparisons are confounded by some extraneous variables that can covary with treatment, including allegiance, standardization, implementation, differential attrition, therapist experience, and therapy dosage. Sometimes these latter variables are the same over treatments, but there is no logical reason they must be, and often they are not.

COMBINING ANALYSES. In principle, at least, the most accurate estimates of orientation effects would take advantage of all available information from both treatment-control and treatment-treatment effect sizes by combining the data into a single analysis. Most authors do not attempt this, probably because the combining methods are not obvious. We show one way of combining data in this section.

The relationship of treatment-control to treatment-treatment comparisons can be defined by the following equation:[7]

$$d_{AB} = d_{AC} - d_{BC} \tag{1}$$

where d_{AB} is the effect size for a direct comparison between Treatment A and Treatment B, d_{AC} is the effect size for a comparison between Treatment A and a control group, and d_{BC} is the effect size for a comparison between Treatment B and a control group. If the standard error of d_{AC} is se_{AC} and the standard error of d_{BC} is se_{BC}, then under independence the standard error for the difference score on the right side of equation 1, say se^*_{AB}, is approximately equal to $\sqrt{se_{AC}^2 + se_{BC}^2}$. The standard error of d_{AB} is directly computed as se_{AB}, which when multiplied by 1.96 yields 95 percent confidence intervals around d_{AB}. If d^*_{AB} is the result of the subtraction on the right side of equation 1, then one can use se^*_{AB} to compute confidence intervals around d^*_{AB}. If the two confidence intervals overlap, d^*_{AB} and d_{AB} may estimate the same population parameter, and a pooled estimate of the population effect size could be estimated as follows:

$$d^{**}_{AB} = (Q + R)/(S + T) \tag{2}$$

where

$$Q = d^*_{AB}/(se^*_{AB})^2$$
$$R = d_{AB}/(se_{AB})^2$$
$$S = 1/(se^*_{AB})^2, \text{ and}$$
$$T = 1/(se_{AB})^2.$$

Finally, the variance of d^{**}_{AB} would be

$$VAR(d^{**}_{AB}) = 1/(S + T)$$

[7] I am grateful to Larry Hedges and Tom Louis for suggesting the equations used in this section.

The standard error would simply be the square root of this variance, and the product of 1.96 times the standard error would give the 95 percent confidence intervals around d^{**}_{AB}.

This strategy can combine the within study treatment-treatment comparisons in Table 5.4 with the between studies treatment-control comparisons in Table 5.2. Problematically, the number of studies underlying the comparisons in Table 5.4 is often so small that the confidence intervals are impractically large. As an alternative, one can explore this model on a subset of eight studies that are picked because each contained all possible comparisons between a behavioral treatment, a nonbehavioral treatment, and a control group. The behavioral-control $d_+ = 1.10$(se $= .16$), the nonbehavioral-control $d_+ = .93$ (se $= .17$), yielding a $d^* = .17$(se$^* = .23$). The behavioral-nonbehavioral $d_+ = .11$ (se $= .14$), which is not significantly different from d^*. Pooling the two estimates according to equation 2 yields $d^{**} = .13$, with VAR(d^{**}) $= .0143$ and resulting standard error of .12. Since this pooled estimate is not significantly different from zero, we conclude that behavioral treatments do not yield significantly larger effect sizes than nonbehavioral treatments.

However, this model does not fit so neatly if we conduct the same analysis without requiring that the data come from studies that each contain all three contrasts. A total of 36 studies compared a behavioral treatment to a control group with $d_+ = .69$(se $= .07$); 32 studies compared a nonbehavioral treatment to a control group with $d_+ = .72$ (se $= .07$). Some of the 32 nonbehavioral studies are the same as the 36 behavioral studies and some are not. This yields a d^* of $-.03$ (se$^* = .10$), not much like the effect size from 22 studies that compared a behavioral treatment with a nonbehavioral treatment with $d_+ = .43$(se $= .08$). Confidence intervals around these two estimates suggest that they do not overlap and probably should not be pooled. If we nonetheless compute the pooled estimates for the sake of argument, $d^{**} = .25$ with standard error of .06, which is significantly different from zero. Thus some possibility still remains that behavioral treatments outperform nonbehavioral ones, even though it is not clear that the pooling we did is legitimate.

One might interpret in two ways the finding in the last paragraph that results should not be pooled. One could assume that the model in equation 1 does not hold, that effect sizes are not additive over studies. However, this may make less sense than an alternative, that uncontrolled covariates over studies mask additivity. The latter interpretation is supported by the fact that the model fits well on the limited sample of eight studies with all three comparisons; this limited sample should minimize the effects of uncontrolled covariates. If so, an extension of the present model would adjust the effect sizes for available covariates prior to submitting them to this analysis. If the adjusted effect sizes yield a fit to equation 1, then additivity is supported.

Random-Effects Models

The analyses thus far assume a fixed-effects model—that the population of marital and family therapies produce a single true effect size, but we do not observe this one value owing to error from various sources. An alternative model postulates random effects, that *true* effects of marital and family therapies vary so that *observed* effects (sample estimates) vary both from sampling variance and between-studies variance. This section explores the applicability of random-effects models to the present data. However, software for multivariate meta-analytic random-effects models is not fully developed, so that in the latter case we cannot allow within-study random effects on covariates and cannot properly weight effect sizes by the inverse of their variances. Thus the analyses in this section are incomplete, and perhaps inaccurate in some respects. The goal is qualitative, to show that using the random-effects approach "matters" since the results are different in important theoretical ways from those obtained by other approaches and to show how interpretation of results changes under random-effects models.

UNIVARIATE RANDOM-EFFECTS MODELS. A start is to compute variance components—estimates of the variance of the true effect sizes in the population. If a component is not significantly different from zero, the fixed-effects model may be appropriate; if it is different from zero, a random-effects model might apply, and random-effects estimators of effect size should be computed. We did so in the present data using estimation procedures outlined in Hedges and Olkin (1985, chap. 9). The variance component for the overall treatment-control effect size on 58 studies is $\sigma^2(\Delta) = .436$, which is significantly different from zero ($Q = 117.55$, df $= 57$, $p < .005$). The average effect size under the random-effects model over all 58 studies is $\Delta = .73$ with a standard error of .11. The average effect size did not change much from the fixed- to the random-effects model, but the standard error doubled, reflecting the greater inherent variability of effect size under the random-effects model. The 95 percent confidence intervals constructed using this standard error do not contain zero ($.51 < \Delta < .95$), so Δ differs significantly from zero.

In view of this finding, we might conclude that a single population parameter might not underlie observed effect sizes in marital and family therapies. The true population effects of marital and family therapy may be randomly distributed around .73. Assuming that the distribution of population effect sizes is normal (an untestable assumption in the present case), then the square root of the variance component, .66, describes the standard deviation of the distribution. Multiplying .66 by 1.96 yields 95 percent confidence intervals within which the population effect sizes are likely to fall, $-.56 < \Delta < 2.02$. This suggests that marital and family therapy will have mostly positive effects, but about 13 percent of its effects may be zero or negative (an effect size of zero would be 1.11 standard deviations below the mean of .73, and 13.4 percent of the normal curve falls to the left of that point). In fact, about 7 percent of the effect sizes were less than zero in Figure 5.1, although not as far below as $-.56$. This discrepancy might be due to sampling error, to the possibility that the population distribution might be positively skewed like Figure 5.1, or to the possibility that authors might be reluctant to report significantly negative outcomes.

We might also test this random-effects model on other variables to see how they relate to therapy outcome. Toward that end, random-effects statistics are reported in Table 5.2 for the same variables that we previously analyzed using fixed-effects models. Note that 47 of 78 cases had variance components that differed significantly from zero, so that the fixed-effects model may not be appropriate. Table 5.2 also reports the random effects Δ, along with its standard error; the latter can be multiplied by 1.96 to obtain 95 percent confidence intervals around Δ to test if Δ differs significantly from zero. The square root of the variance component can be interpreted as the standard deviation of the population distribution of effect sizes, and if multiplied by 1.96 the result yields an estimate of the boundaries within which most population effect sizes are likely to fall (assuming that population effect sizes are normally distributed).

Take treatment orientation as an example. As with the fixed-effects model, all orientations produced effects that differed significantly from zero under the random-effects model. Recall that the simultaneous confidence intervals reported earlier in this chapter suggested that mean effect sizes did not differ significantly over orientation; this finding also holds for the random-effects analysis of orientation. One is tempted to conclude again that orientation makes no difference. But under a random-effects model, mean differences among categories are not the only important issue. Even though mean effect size may not differ much over orientation, the variance of expected population effect sizes is affected. To judge from these data, we should expect systemic and humanistic treatments to produce fairly consistent results since their variance components are not significantly different from zero. By contrast, these data suggest that we should expect behavior therapies to produce the most variable results of all, producing effect sizes that should be expected to range *in the population* from −.71 to 2.39. Eclectic therapies were nearly as variable.[8]

MULTIVARIATE RANDOM-EFFECTS MODELS.   As was the case with fixed-effects models, univariate random-effects analyses yield results that are almost surely biased by confounding variables. For example, it might be that the greater variability of behavioral treatments is due to the fact that such treatments are highly standardized and use more specific measures, both of which have significant variance components in Table 5.2. As before, one way to assess this possibility is to use multivariate models that partial out potential redundancies. Of course, the same cautions and caveats outlined in the univariate case apply here. However, the estimation of multivariate random-effects models is hindered in practice by lack of computer programs specially adapted to meta-analytic needs or general enough to handle the special need to optimally weight cases. To help initiate exploration of such models with available programs, we present a few very simple and tentative examples in this section. The models make assumptions about the distribution of population effect sizes that cannot be tested in this data; so probability estimates may not be accurate.

A simple model can be run in the computer program BMDP3V (Jennrich and Sampson 1988). The model is a one-way ANOVA in which the single factor is a random-studies effect. In the present case, this factor has 58 levels, one for each study. The model allows a random intercept; that is, an overall random-study outcome; but slopes of predictors are fixed, which implies that the influence of, say, behavioral versus nonbehavioral treatments if fixed. Cell entries are the multiple effect sizes in each study. BMDP3V has no obvious way to weight cases; use of the case-weight command did not result in changes in parameter estimates or significance tests. The model tested with this program had five fixed-effects covariates drawn from the significant predictors of the Winsorized regression equation: number of measures, experimenter al-

---

[8]Table 5.2 also suggests similarities and differences between random Δ and standard error in comparison to fixed-effects models. Random-effects estimates of effect size are about 9 percent larger than fixed-effects WLS estimates and about 7 percent smaller than OLS estimates. The correlation between OLS d and Δ is $r = .92$ ($p < .01$), and between WLS d_+ and Δ was $r = .83$ ($p < .01$). Random-effects standard errors are 69 percent larger than WLS standard errors on average, and 4.6 percent larger than OLS standard errors.

legiance, adult presenting problem, treatment implementation, and behavioral orientation. The between-studies variance component was .399, somewhat less than the .436 component in the univariate random-effects model, presumably because some of the between-studies variation has been explained by covariates; and the within-study (error) component was .443. To test the significance of the variance component for studies, one computes a chi-square difference test for the model estimated with and without that component. Since that chi-square $= 152.049$ (df $= 1$, $p < .001$), treating studies as a random variable significantly improves model fit. Two covariates had significant effects: the number of measures (slope $= -.018$, SD $= .004$, $p < .001$), and treatment implementation (slope $= -.277$, SD $= .138$, $p = .046$), both interpreted as before. Experimenter allegiance did not contribute a significant effect when studies are treated as a random variable, but was significant if studies are a fixed effect. Orientation made no difference.

BMDP5V (Schluchter 1988) computes similar models. An illustration is a model in which studies were again treated as a random effect with four fixed-effects covariates: the number of measures used, experimenter allegiance, behavioral orientation, and treatment implementation. In addition, following Braun's suggestion (1988) of grouping studies with similar precision in estimates of effect size, we included study sample size as a fixed effect by dividing studies into two groups at the median on sample size—useful in this analysis given the inability to weight cases. Results yielded within-study and between-studies variance components of .4219 and .3636, respectively, both statistically significant and similar in magnitude to estimates from BMDP3V. The slope for "number of measures" covariate ($-.02$) was significant, and the grouping factor was not significant. Orientation effects were again not significant.

Over both the BMDP3V and BMDP5V programs, the random-study effect first detected in univariate analyses persisted in the presence of covariates, although its magnitude was slightly less than the .436 study variance component first reported in this section, suggesting that covariates help explain at least some between-studies variation. Thus, the possibility that therapy effects are random effects is still a viable hypothesis. Further exploration of this possibility might also include models that allow both random intercepts for studies and random slopes for covariates (e.g., Braun 1988; Hedges 1988), although the software to support such analyses is less accessible than that used here.

One particular problem that requires further attention concerns decisions about the kinds of generalizations we can make about random effects given the sample of studies we have. Some authors suggest that one must know the entire population of studies, and then randomly sample studies from that population—for example, randomly select such features of studies as random selections of treatment type, or of kinds of presenting problems (Richter and Seay 1987). The problem is that such random selection clearly does not happen, at least not in family and marital psychotherapy studies, and probably not in psychotherapy research generally (and probably not in any literature). Rather, individual researchers often have programs of research yielding highly similar studies over time. Over the population of psychotherapy researchers, study characteristics cluster around therapeutic orientations; and some orientations (behavioral) are probably grossly overrepresented while others (psychoanalytic) are grossly underrepresented. Within orientation clusters, researchers follow currently popular paradigms or currently fundable questions in designing studies. When studies are not randomly sampled, but the researcher interprets results as if they were, inferential tests can be significantly biased. If, for example, studies are systematically chosen only from the center of the study population distribution (e.g., extreme levels of the variable of interest are deliberately ignored), then study variance will be too small relative to the total population of studies. If they are chosen from the tails of the distribution (admittedly unlikely in this case), study variance could be too large. Depending on the design of the meta-analysis, this can seriously inflate Type I or Type II error rates.

Random selection of studies is not the only way to facilitate interpretation of random-effects models, however. The between-studies variance of the sample of studies in the present meta-analysis, for example, arises from choice of approach, patient population, and a host of explicit and "random" choices. Researchers need to know only that there is some process connecting sampled studies to a population, not necessarily the whole story. The point is that in order to interpret the random-effect results, one needs some idea of the reference population. It might be a nonrepresentative population of studies, but as long as this is recognized, inferences can be made. In the present meta-analysis, for example, we do not know how well the present sample of family/marital therapy studies represents the population of all possible family/marital therapy studies. However, especially given the thorough search procedures we used to find our sample of studies, we probably are justified in hazarding the tentative hypothesis that the results from these random-effects models would probably be found for any population of stochastically similar studies.

This problem of interpretation, however, is not an argument that one should rely on fixed-effects models, since they may also be quite biased if the study effect is, in fact, random. Rather, it is an argument that such problems must receive more thorough attention before our understanding of the applicability of random-effects models to psychotherapy meta-analyses will be complete.

Speculative Explorations in Meta-Analytic Causal Models

Is it really any secret that we are interested in causal inferences in meta-analysis? After all, we are exploring orientation effects because we want to know if, say, behavior therapy causes better patient outcome than systemic therapy. Each univariate test is, then, a causal mini-model. Each multivariate analysis is an effort to make those causal mini-models more realistic by trying to partial out possible confounds. Yet for reasons we will see shortly, the multivariate models we typically use in regression are not very realistic themselves. In this section, then, we explore models that may be somewhat more realistic.

Multiple-equation models and latent variable models, in principle at least, can increase the explanatory power of meta-analyses by allowing models that more realistically reflect the processes that may have generated study data. But currently their use is fraught with technical difficulties that preclude confidence in results. Moreover, these models almost certainly cross an ill-defined line between descriptive and causal models (Freedman 1987); and the notion of drawing confident causal inferences from meta-analytic data may make many observers (including the present author) uncomfortable. While we do have causal hypotheses in mind (as we did in analyses prior to this section!), they are of a highly exploratory kind. They are meant to probe possibilities rather than confirm causal connections, to stimulate more realistic thinking about the processes that give rise to study outcomes, and to promote the technical development needed to interpret such models with confidence. These matters are discussed further at the end of this chapter.

MULTIPLE EQUATION MODELS. Previous analyses could be represented with a single regression equation that allows the predictors to affect the dependent variable, but does not allow them to affect each other even when such relationships are plausible. Similarly, it allows predictors to have only *direct* effects on the dependent variable, but some predictors may have *indirect* effects instead of, or in addition to, direct effects, both on the dependent variable and on each other. These problems can be represented using multiple equations (they can also be expressed in matrix algebra form as one equation). To heighten the contrast between single equation models and multiple equation models, consider the single equation model in Figure 5.4. It is conceptually and statistically like the regression models estimated in this chapter, but without predictor covariances. The dependent variable is effect size, and seven predictors are treatment standardization, treatment implementation, number of measures, experimenter allegiance, dependent variable specificity, behavioral orientation, and university setting. Dependent variables have arrows coming into them and error terms caus-

Figure 5.4   Single Equation Model

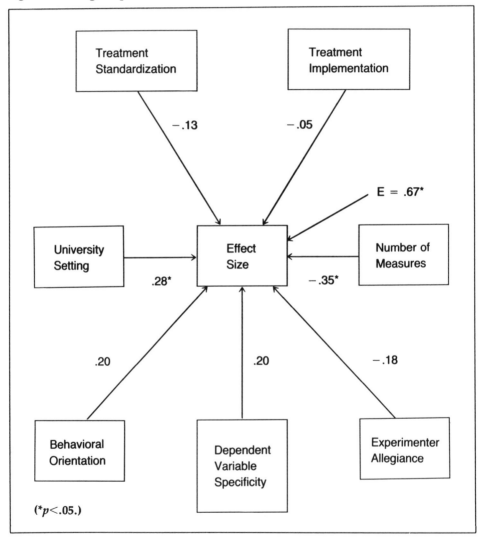

ing them. Predictors have arrows going away from them. The arrows indicate the hypothesized direction of causal influence.

Path coefficients are of primary interest and are generally interpreted as regression coefficients indicating the degree of relationship between two variables. They can be estimated in any regression program. For comparison with the model estimated shortly, they are estimated in EQS (Bentler 1989b) using generalized least squares fit functions rather than maximum likelihood because multivariate normality is mildly violated with some variables in the model (normalized Mardia's coefficient for multivariate kurtosis = 8.06). GLS offers chi-square tests of the overall fit of the model. The covariance matrix was analyzed, but Figure 5.4 reports standardized path coefficients for ease of interpretation.

In Figure 5.4, two variables significantly predict effect size: effect sizes were higher if the study was conducted in a university setting (path coefficient = .28) and were lower in studies with more measures ($-.35$). Behavioral orientation made no difference. This model fits the data reasonably well (chi-square = 26.16, df = 21, $p = .20$, Bentler-Bonett normed fit index = .95). The model is not rejected if the probability is greater than .05; and the normed fit index ranges from 0 to 1, with fits greater than .90 generally taken to be adequate. The fit index assesses how well the model reproduces the observed covariance matrix, something quite different from the model specification tests reported previously in this chapter. Fit indices can yield different results than specification tests, although the conditions under which this might happen are just beginning to be explicated.[9] Also, chi-square is sample size

---

[9]Larry Hedges provides the following example of differences between Hedges-Olkin (HO) model specification tests and analysis of covariance structures (ACS) fit tests. Consider three variables A, B, and D (effect size). Consider the model

$$A \rightarrow B \rightarrow D \leftarrow e,$$

and suppose that it fits *perfectly* (i.e., A influences D only through B). In this case, the ACS fit test should reject the null hypothesis at the nominal level (e.g., 5 percent of the time for $\alpha = .05$). However, note that the disturbance e reflects (or may reflect) not just sampling error about δ, but also a random effect in δ. Hence in Hedges-Olkin notation, $D_i = \Delta + \gamma_i + \epsilon_i$ where Δ is the overall mean population effect, γ_i is the deviation of the ith population effect from Δ, and ϵ_i is the (within study) sampling error. If γ_i is determined entirely by B via

$$\gamma_i = \beta B_i$$

then $e_i = \epsilon_i$ and the two tests will be similar (but not identical because the HO test ignores the AD covariance).

However, suppose that B does not entirely determine γ_i. That is, suppose

$$\gamma_i = \beta B_i + \eta_i$$

with $\eta_i \neq 0$. Then $e_i = \eta_i + \epsilon_i$. If η is uncorrelated with A and B, the ACS fit will be unaffected. However, the HO fit *will* be affected. In fact, by making the variance of η large, dependent, so might overestimate fit given that only $n = 54$ studies contributed to the model; normed fit indices are far less subject to this problem (Bollen 1989a; Bentler 1989a).

Unfortunately, Figure 5.4 is not a very realistic model of the processes that may generate study data. For example, it is not very likely that both treatment standardization and treatment implementation directly and simultaneously cause study outcome. After all, standardization occurs mostly prior to a study being implemented; it probably affects implementation, which in turn influences outcome. Figure 5.5 suggests this kind of a model. The variables are the same as Figure 5.4, but the structure of the model has changed. Five predictors are hypothesized to have a direct influence on effect size. In addition, behavioral orientation is also conceptualized as a proxy for variables reflecting training and methods of doing research, which then indirectly influences outcome through those other variables. For example, behaviorists are more likely to choose reactive dependent variables (Smith et al. 1980). Also, common lore is that behavioral researchers are overrepresented in university settings. Being in a university setting, with pressures to publish and with more frequent exposure to matters of research methods, may make it more likely that the university researcher will attend to methodological matters such as standardizing treatments. Standardized treatments, in turn, may be more likely to be implemented as intended, and effect sizes may increase if the treatment is implemented with fidelity.

This model also fits the data (chi-square = 22.46, df = 19, p = .26, fit index = .95). Lower effect sizes are associated with use of more dependent variables, and the experimenter allegiance effect and the effect for treatment implementation are still nonsignificant. The remaining path coefficients in the model give a different interpretation than Figure 5.4. Behavioral orientations are now directly associated with larger outcomes, as are more specific dependent variables. Further, behaviorists are more likely to be in university settings, which in turn is associated with better treatment standardization and treatment implementation. The path coefficient from implementation to outcome is in the predicted direction but nonsignificant.

The model allows computation of direct effects, indirect effects, and total effects. Direct effects have arrows connecting two boxes. Permis-

---

you can reject HO fit with high probability while keeping *perfect* ACS fit. Less abstractly, random effects that perturb population effect sizes need not (if uncorrelated with other modeled variables) affect ACS fit. Such random effects always hurt HO fit. Thus the two tests do not use the same conception of "fit."

Figure 5.5 A Simultaneous Equation Model

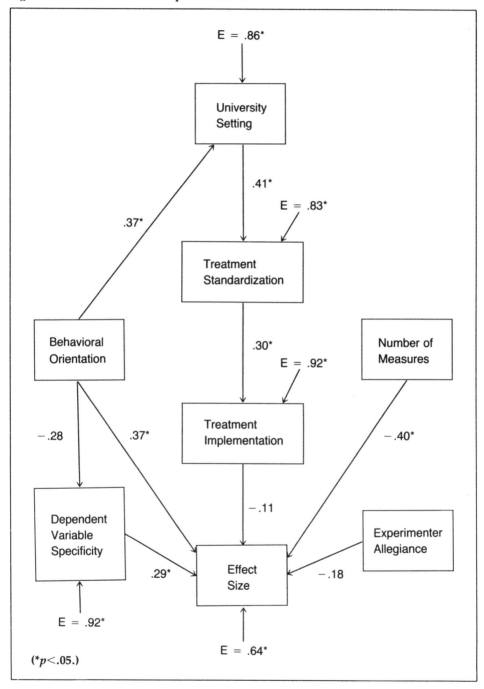

E = .86*

University
Setting

.41*

E = .83*

Treatment
Standardization

.37*

Behavioral
Orientation

.30*

E = .92*

Number of
Measures

−.28

.37*

Treatment
Implementation

−.40*

Dependent
Variable
Specificity

−.11

Experimenter
Allegiance

.29*

Effect
Size

−.18

E = .92*

E = .64*

(*p<.05.)

sible indirect effects are defined by Wright's rules (Loehlin 1987), which for present purposes are mostly limited to cases of indirect effects through mediating variables. Total effects are defined as the sum of the direct and indirect effects. The presence of indirect effects helps explain why single-equation models can yield different results from multiple equation models. The direct effect of behavioral orientation on effect size in Figure 5.5 is .37. Behavioral orientation also has an indirect effect on effect size through dependent variable specificity and through the chain of university setting, treatment standardization, and treatment implementation. Individual indirect effects are the product of the coefficients on each chain, and total indirect effects are the sum of these products. The indirect effect of behavioral orientation through dependent variable specificity is $.29 \times -.28 = -.08$, and through the treatment implementation chain is $-.005$, for a total indirect effect of $-.085$ (since the implementation-effect size link is nonsignificant, one could drop this chain, yielding a total indirect effect of $-.08$). The total effects of behavioral orientation are about $.37 - .085 = .285$. The sum of direct and indirect effects can be *less than* the direct effect alone. So the discrepancy between Figure 5.4 and Figure 5.5 for the effects of behavioral orientation is partly due to the presence of indirect effects. If indirect effects actually contributed to the production of the data (e.g., if behaviorists really are overrepresented in university settings), then models that do not allow for indirect effects produce incorrect estimates of the impact of theoretical orientation (Bollen 1989b).

A key question, of course, is how one knows which underlying processes actually generated the data. Figures 5.4 and 5.5 have similar fits; in fact, a number of models could be generated with good fits, many of which might have very different interpretations from Figure 5.5 (Glymour and Scheines 1986; Stelzl 1986). For instance, I generated several variations of Figure 5.5 to see how sensitive individual parameter estimates are to minor model changes. These variations included adding some interactions terms, adding new variables, or eliminating some variables. A key change to watch concerns the parameter for the direct effect of behavioral theoretical orientation on outcome. Over five such variations, the latter path coefficient ranged from a nonsignificant .21 to a significant .43. Other parameter estimates in the model also changed somewhat; in particular, while they rarely changed direction, they often fell below significance. On the other hand, all the model variations fit well as a whole.

This should caution us not to take individual parameter estimates too seriously, and especially not to take their significance level too seriously, for both of these can change moderately with different model specifications. What should be taken seriously, however, is the demonstration that simple single equation models can be considered a specification of a causal model as in Figure 5.4, but more theoretically realistic models may require estimating multiple equations and can yield quite different results both from single equation models and from each other, depending on how they are specified. All this makes statements about the "true" influence of theoretical orientation on effect size somewhat problematic.

LATENT VARIABLE MODELS.   Researchers should construct good measures of key constructs. In studies of the effects of psychotherapy on depression, for example, outcome measures commonly have 10–30 items, well-explored factor structures, and demonstrated reliability and validity. By contrast, the codings used in most meta-analyses, including the present one, are primitive. The assessment of orientation is an example. In this study each orientation is assessed with one dichotomous item: Either it is behavioral or it is not. The reliability and validity of such an item must be suspect. Indeed, with such poor measurement, it is a wonder that any consistent findings at all have emerged in meta-analyses. With orientation, for instance, we are unlikely to have a firm sense of the constructs involved, or the relationship of those constructs to outcome, until we have more sophisticated, reliable, valid, and differentiated measures to apply to the task. With all the attention to statistical principles in meta-analysis recently, the lack of attention to measurement is all the more glaring.

One illustration of this problem in the present data concerns correlations among items. In response to a request from a reviewer, we submitted four sets of predictors variables (methodology, dependent variable characteristics, therapy inputs, therapy process) to four separate factor analyses to identify patterns of study characteristics. Those results are not presented here partly because they are complex and require more extensive elaboration than space allows, but mostly because the data taken as a whole may be psychometrically inappropriate for factor analysis. Dziuban and Shirkey (1974) suggest three tests of the appropriateness of a correlation matrix for factor analysis that assess whether the variables have much in common. The present data routinely failed two of the three tests badly. One reason for the failures may be that the four sets of variables do not measure the same thing. The solution then would be to isolate more homogeneous subsets to factor; this solution is illustrated shortly. But the failures may also simply reflect poor primary measurement.

To help address this problem, one can use latent variable models rather than observed variable models, at least for the few constructs with multiple items assessing them. Figure 5.5 is an observed variable model, consisting of connections among observed variables. But measurement error, particularly differential error over variables in multiple equation systems, can seriously distort both the magnitude and the direction of path coefficients (Bollen 1989b; Rogosa 1980). This problem can be partly remedied if the meta-analyst has multiple measures of a construct. Figure 5.6 presents such a model, also fit in EQS (chi-square $= 25.89$, df $= 34$, $p = .84$, Bentler-Bonett normed fit index $= .96$); latent variables are in circles and observed variables in squares. The first latent construct, measurement reactivity, is positively related to effect size. Self-report measures (factor loading .78), specific measures (.36), and manipulable measures (.58) significantly relate to overall measurement reactivity; experimenter blindedness to measurement is nonsignificantly related ($-.13$). The second latent construct, completeness of effect size reporting, is also related to higher overall effect sizes. Studies have larger effect sizes when they provide sufficient information to compute effect size exactly (.66) and when they report relatively few effect sizes only as nonsignificant ($-.83$), the latter being coded as zero. Dissertations tend to report effect size information more completely (.31), but this contribution is (barely) nonsignificant. Less intuitively, the number of measures in a study is negatively ($-.31$) but nonsignificantly related to completeness of reporting. It may simply be that the construct is mislabeled; perhaps it should be called "practices that enhance effect size," a bit tautological but consistent with the measures and loadings.

In the presence of these latent variables, behavioral orientation makes no significant contribution to effect size. No modifications of this model yielded a significant effect for behavioral orientation, including retesting the model in Figure 5.5 with measurement specificity replaced by the measurement reactivity construct in Figure 5.6. The implication is that presence of measurement error in Figure 5.5 can lead to biased estimates of path coefficients (Bollen 1989b). To test this further, however, would require a more complete latent specification for all variables in the model, which cannot be done because the present data set lacks multiple measures of most of the other constructs that underly Figure 5.5.

TECHNICAL DIFFICULTIES WITH THESE MODELS.   (The reader who is not interested in methodological matters can skip this section.) The models in Figures 5.4–5.6 suffer from technical difficulties that render their interpretation problematic. First, the primary analytic tool—analysis of

Figure 5.6   A Latent Variable Model

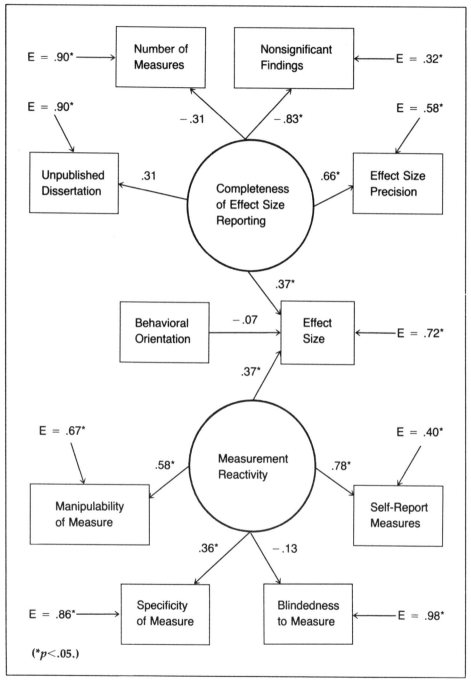

covariance structures (Bentler 1989b; Joreskog and Sorbom 1988)—assumes that observations are a simple random sample of the population covariance matrix. But this is not true since, for example, file-drawer studies are excluded that may have different characteristics than other studies (Shadish, Doherty, and Montgomery 1989). Inferences regarding both parameter estimates and fit tests may be biased as a result. Second, as with multivariate random effects models in BMDP3V and BMDP5V, it is not obvious how to weight effect sizes by the inverse of their variances in available structural modeling programs. This problem may be remediable by using a preprocessor such as PRELIS (Joreskog and Sorbom 1986) to weight data prior to analysis in EQS or related programs like LISREL (Joreskog and Sorbom 1988; see also Shadish and Sweeny, in press). Third, the approach used in Figure 5.5 does not model within-study sources of variation, unlike the generalized least squares model of Becker (this volume; Raudenbush, Becker, and Kalaian 1988); software to remedy this problem in models with latent variables like Figure 5.6 does not exist (Muthen and Satorra 1989).

Becker (this volume) and Premack and Hunter (1988) use an alternative method for constructing meta-analytic path models that suffers less from some of these problems but incurs other problems in the process. The input for all path models requires a matrix of correlations or covariances among variables. But one can construct those correlation matrices in two quite different ways. Premack and Hunter wanted to predict voting to unionize from five variables such as wage level, satisfaction with administration, and intent to unionize; thus their primary input is a 6×6 correlation matrix among these variables. To construct that matrix, Premack and Hunter looked for studies that both measured and reported correlations among any of these variables. They then aggregated those correlations over studies, weighted appropriately, to compute their input matrix. The main problem with this method is that few studies included all or even most of the 15 possible correlations among the six variables, so this procedure yields missing data and correlations based on different numbers of studies. In Premack and Hunter (1988), some aggregate correlations were based on as few as two studies, compared with 54 studies for each correlation used as input to the model in Figure 5.5. We might label this the "study-generated" method of constructing path models. Becker (this volume) also uses this method, but models within-study as well as between-studies covariation, addressing two of the three problems mentioned in the previous paragraph—although as yet this method has no means of including latent variable models.

By contrast, the present chapter used coder-generated data to code all variables except effect size. With this method, one does not have to rely on primary studies either to measure the variables of interest or to report correlations among them. Rather, coders simply do the measurement and the meta-analyst computes the correlations, a procedure that results in very little missing data and in correlations based on a larger and mostly the same sample size. We might label this the "coder-generated" method of constructing path models. The main disadvantage of this latter method is that the resulting correlations are probably less reliable and valid than for study-generated path models.

The two methods are not mutually exclusive. One could use coder-generated correlations as input into the Raudenbush et al. (1988) generalized least squares procedure to yield virtually complete data. Similarly, one could input study-generated correlations into EQS or LISREL, losing the option of modeling within-study covariances but gaining the option to model latent variables. Obviously, the trade-offs among all these options need further study. Unfortunately, one problem is potentially shared by all these path model approaches—that they can all be conceptualized as a form of causal modeling in correlational data. Such models can be severely criticized (see Karlin, Cameron, and Chakraborty 1983; Cloninger et al. 1983; and Wright 1983, for a discussion of some issues involved). In fact, such criticisms may well apply to all meta-analytic data analyses, a topic to which we now turn.

Conclusions

Meta-Analytic Explanation

The theme of this book is that meta-analysis can be explanatory. This chapter suggests two conclusions about this notion. First, a necessary but not sufficient condition for explanation is that it produce results that "fit" the data. Second, different explanations can fit the same data, so certainty about the correctness of an explanation will be elusive.

RESULTS THAT FIT THE DATA. Explanatory models should account for the systematic variance in the data, as assessed by statistics such as Hedges and Olkin's Q tests (1985) or Bentler's normed fit index (1989b). This requirement has a weak and strong rationale. The weak rationale is that explanations that do not fit the data are incomplete and omit some variables. The rationale is weak because incomplete explanations can still be useful if correct. It would be useful, for example, to know that random assignment with matching lowers psychotherapy effect sizes compared with random assignment without matching, even if one knew little else about therapy effectiveness, because randomization with matching could mitigate so-called failures of randomization in small-sample psychotherapy research (Hsu 1989). Table 5.2 supports this explanation, but the explanation does not fit the data to judge from the significant Q_w test. The strong rationale is that explanations that do not fit the data may be simply wrong if, when other explanatory variables are included, the original variable is no longer significant or changes the direction of its contribution. Such changes can rarely be ruled out ahead of time. Table 5.2 contains many variables that were significantly related to effect size, only to be nonsignificantly related once other explanatory variables were included in a regression. Fortunately, sign reversals from significant positive to significant negative influences, or vice versa, are uncommon, but cannot be ruled out.

If so, univariate models may not produce very viable explanations because they implicitly assume that no other variables influence effect size. This is one way of conceptualizing the assumption tested by the Q_w statistic. Every univariate model in Table 5.2 resulted in rejected Q_w tests. Such rejections are the norm in all meta-analyses that conduct the test. Therefore, univariate tests of the relationship between predictor variables and effect sizes must be assumed to yield biased results. Unfortunately, the vast majority of meta-analyses probably still rely entirely on univariate tests. The 1988–1989 issues of *Psychological Bulletin*, for example, contained 13 meta-analyses (Bornstein 1989; Bowers and Clum 1988; Dush et al. 1989; Feingold 1988; Hyde and Linn 1988; Johnson and Eagley 1989; Matthews 1988; Miller and Eisenberg 1988; Parker, Hanson, and Hunsley 1988; Premack and Hunter 1988; Searleman, Porac, and Coren 1989; Swim et al. 1989; Wood, Rhodes, and Whelan 1989). All 13 report categorical analyses such as ANOVAs or the tests reported in Table 5.2. Many also report Hedges and Olkin's Q tests (1985), reject the test, but do not search for additional variables to explain remaining variability. Only four of the 13 use any multivariate procedure. Johnson and Eagley (1989), Parker et al. (1988), and Wood et al. (1989) report single equation regression analyses, and Premack and Hunter (1988) present a multiple equation path analysis. Multivariate explanations do not, of course, guarantee that a model will fit the data; they merely increase the likelihood of doing so.

DIFFERENT MODELS THAT FIT THE SAME DATA.   Even if a model fits the data, it may be wrong. In Figures 5.4 and 5.5, for example, a model without indirect effects fits about as well as a model with them. It is unlikely that both models are correct because this implies that the same indirect effect both does and does not exist. More likely, either one or both models are wrong, but we do not know the true situation. Moreover, they can be wrong even though they fit the data. Thus, the claim that fitting the data is a necessary, and not a sufficient condition for explanation.

The fundamental problem is that all meta-analytic explanatory models share a common flaw that has no obvious solution: They attempt to draw causal inferences from correlational data. In terms of common design classifications, meta-analyses are observational studies rather than, say, experimental studies, random-sample surveys, or other forms of investigation (Louis, Fineberg, and Mosteller 1985). This latter point has been noted before (Cooper 1984), but is worth clarifying because the reader may wonder why the 106 randomized, controlled experiments in this meta-analysis are correlational data. The claim holds for two reasons. First, the conditions under which each study was conducted—for example, use of matching, recruitment of university-based clients, or assessment on general versus specific dependent variables— were not assigned to studies at random. So an inference about the influence of, say, matching on effect size is necessarily quasi-experimental. Most meta-analytic explanations are of this type. Second, the meta-analyst typically does not wish to make a causal inference about the molar treatment package to which subjects were assigned in the original experiment, but rather to make an inference about a subset of qualities associated with that package such as the fact that it was a behavioral treatment. Even in those studies in which subjects were randomly assigned to, say, behavioral versus nonbehavioral treatments, that fact is also not randomly assigned over studies. Experiments with such assignment might be systematically different from other experiments in other ways. Again, the inference is confounded with other potentially active causal agents. Indeed, the only reasonably valid causal inference that is not subject to these confounds is the very first result presented in this chapter: The sample of marital and family psychotherapies in the 106 studies, whatever they may consist of, caused a .70 standardized mean difference between treatment and control groups—marital and family therapies change what people do. Any more specific inference is potentially confounded.

Meta-analysts typically deal with these confounds just as all causal modelers do: They try to adjust them statistically with regression or covariance analysis. This approach results in elaborate causal models of the type in Figure 5.5, or cognate approaches such as selection bias modeling or complete latent variable models (Bollen 1989b; Heckman et al. 1987; Stromsdorfer 1987). If so, then meta-analytic explanation is stymied by the same problem that impedes all causal models: The success of the endeavor depends crucially on correct model specification, and one can rarely know if this condition holds. Correct model specification implies that the researcher includes all the real causes of outcome, correctly identifies causal or correlational relationships among variables, includes multiple measures of latent variables to avoid measurement error bias, correctly models whether the form of a relationship is linear, and knows whether a fixed- or random-effects model holds. Even if all this is done well, and if software packages could do these things simultaneously, meta-analysts face more impediments. For example, causal modeling techniques require large sample sizes, but the number of studies in most meta-analyses is small; and meta-analysts are at the mercy of primary authors who may not report information needed to estimate the model (Orwin and Cordray 1985).

To avoid causal inference problems, some researchers distinguish between descriptive and causal models: The former describe relationships without drawing causal inferences (Freedman 1987). The problem is where to draw the line. Clearly Freedman does not want to prohibit all use of causal language, for he praises Blau and Duncan (1967) with causal-sounding language for their use of "two or three descriptive regression equations and the R^2s as part of a persuasive argument to show that family background *influences* but hardly *determines* educational levels" (Freedman 1987, p. 220; emphasis added). Surely influences involve causes. Rather, he seems to want a prohibition against inferences that the correct population causal structure has been found. So, for example, he criticizes Fox (1987) not for computing regression equations but for reporting *standard errors* of regression coefficients, for this crosses "the boundary separating inferential from descriptive statistics" (p. 208). Freedman is not objecting to the standard errors per se, but to the implied "commitment" to a causal theory which "has to correctly represent the causal relationships being studied" (p. 220). But judging who crossed the line of commitment is much harder than judging who reported standard errors. Figure 5.5 is a good example—reporting significance levels based on standard errors, it may cross the line. But we also recompute them under different model specifications, where sometimes they are significant and sometimes not. Have we crossed the commitment line, or not? Will it help if we endorse healthy skepticism of all particular findings (Shadish 1989a)?

Perhaps it is better to allow discussions of cause, but to foster suspicion of particular causal claims. After all, can any explanatory models, including Premack and Hunter (1988) and Becker (this volume), avoid making some causal hypotheses? More pointedly, can univariate models like those in Table 5.2 avoid being at least partly causally motivated? Not without seriously hampering conceptual discussions of results. For example, Berman et al. (1985), in discussing their univariate findings about therapist allegiance, suggest "that such theoretical allegiances may affect the outcome of a study" (p. 458), a reasonable and interesting causal interpretation. It is hard to think conceptually if one is not at all allowed to use some form of causation. Purely descriptive models skirt dangerously close to being mindless models.

Despite the problems with meta-analytic causal models, therefore, they can accomplish useful things. They force meta-analysts to be explicit about the theoretical models thought to generate study results, help us to lower our expected probabilities for some models that do not fit the data, and suggest hypotheses for future primary studies that test causal links in randomized experiments. Sometimes one obtains the same answer over very different models, increasing one's confidence somewhat, as when the number of measures in a study consistently relates to effect size in every analysis in this chapter. For all these reasons, meta-analytic causal models may be a better option than univariate ANOVAs and *t*-tests. We can agree with Freedman (1987) on this, however: One must constantly remind both the reader and oneself that none of the models presented in this chapter, univariate or multivariate, may capture causal reality very well.

More optimistically, partial remedies to at least some of these problems may exist, and more can be invented. Methodologists interested in causal inference distinguish statistical from design solutions. The latter, such as random assignment of subjects to conditions, help prevent inferential problems from occurring; statistical solutions, such as selection bias modeling, try to remedy a problem after it occurs. The causal modeling represented in Figure 5.5 is of the latter kind, trying to remove bias by proper model specification. In principle, design solutions to causal inference problems in meta-analysis should exist as well. An example is the use of treatment-treatment comparisons within the same, randomized study—like the orientation comparisons in Tables 5.4 and 5.5. Nearly all analyses in this chapter were conducted on between-studies treatment-control comparisons—say, a behavioral-control contrast from one study is compared with a systemic-control contrast from another. Such contrasts are widely used because they are logistically easy to analyze. But they are also the most likely contrasts to be confounded with between-studies differences in how studies are conducted. Direct comparisons of treatments within the same study—say, comparing a behavioral therapy with a systemic therapy in the same study—are subject to fewer confounds. They are likely to use the same therapists and the same measures and be conducted in the same setting. They are somewhat more difficult to analyze, however, especially if a multivariate procedure is used. Nonetheless, especially when treatment-treatment comparisons occur in a randomized study, their results should be preferred to those of between-studies treatment-control comparisons when the two results conflict—as they did several times in this chapter.

One could invent or adapt other design solutions to aid causal inference in meta-analysis. One might use Cook and Campbell's nonequivalent dependent variables design (1979) or their predicted higher-order interactions in meta-analysis. A meta-analytic cohort design might examine sequential studies coming out of the same laboratory to see how variations in the conduct of the study changed results. Coupled with statistical modeling solutions like those in the present chapter, design solutions significantly aid our ability to construct cogent and valid explanations for meta-analytic results.

Effects of Marital and Family Therapies

ORIENTATION EFFECTS.   What have we learned about orientation effects from this meta-analysis? The most common past finding is that orientation makes no difference after potentially confounding variables are partialed out. The present research partly supports this finding, since many simple regression equations suggested no particular orientation effect, and since the latent variable model in Figure 5.6 also found no effect. On the other hand, behavioral therapies were associated with larger effect sizes than some other therapies in the within-study treatment-treatment comparison in Tables 5.4 and 5.5, which may better test the effect than the between-studies treatment-control comparisons on which so much of the literature is based. Also, in the multiple equation path model in Figure 5.5, a nonsignificant behavioral effect in Figure 5.4 became significant when indirect effects were allowed. These findings are not compelling because testing the model in Figure 5.5 on other orientations (e.g., systemic orientations) might have also revealed significant effects for them and because Figure 5.5 is just one of many models that can fit these data (Stelzl 1986). Nonetheless, the findings arose from novel methodological approaches that are probably as appropriate as, or more appropriate than, the traditional meta-analytic approach to the problem. Moreover, the model in Figure 5.5 makes more conceptual sense than the model in Figure 5.4; in statistical jargon, we would have to bet that Figure 5.5 is better specified than either

Figure 5.4 or any of the univariate models in Table 5.2. One could then argue that, at least for causal inference purposes, the orientation effects we found in Figure 5.5 and in Tables 5.4 and 5.5 are probably more accurate estimates than any other in the chapter. Hence we must conclude that orientation effects cannot be ruled out. The causal inferences we are trying to make from observational data, the confounding of orientation with so many other variables, and our inability to know which statistical model specification is correct all make this judgment far more complex than perhaps we first appreciated in meta-analysis.

EXPERIMENTER ALLEGIANCE.   Experimenter allegiance emerges in this and other research as a potentially major explanatory construct. The general conclusion that allegiance is associated with increased effect size makes intuitive sense and is consistent with research both in psychotherapy (Berman et al. 1983; Garfield 1980; Shapiro and Morris 1978) and in other areas. The direction reversal of this finding in some analyses is disturbing, but adds to the need to study allegiance. In addition to the hypotheses suggested earlier, further research should (1) differentiate between therapist and experimenter allegiance, since contact with clients may be a moderating variable; (2) assess allegiance in multiple ways rather than using general, binary allegiance ratings; and (3) rely on other sources than the study for gathering these data, such as surveys of original authors and therapists.

DETERMINING WHAT MAKES A DIFFERENCE IN PSYCHOTHERAPY.   This research paints a more detailed picture about what makes a difference in psychotherapy. Past meta-analyses report that few variables are related to outcome; but many univariate tests in Table 5.2 were significant, perhaps owing to the use of the statistically more appropriate weighted least squares analyses. Subsequent multivariate analyses demonstrated the redundancy of many of these variables; even so, the remaining relationships make intuitive sense and, if replicated, may have practical implications for therapists concerning such matters as treatment implementation.

POTENTIAL RANDOM EFFECTS OF PSYCHOTHERAPY.   Family and marital therapies may yield random effects from time to time, study to study, and client to client. This tantalizing possibility remains largely unexplored, in no small part due to a lack of accessible software. This latter problem will undoubtedly change quickly as statisticians and programmers put their minds to the problem. Meanwhile, the possible viability of random-effects models should make us even more cautious about interpreting the results of single studies—we may be interpreting random variation as much as substantive findings.

FUTURE RESEARCH.   A meta-analysis to explain the effects of marital and family therapies could do much more than was done in this study. Although we had little success with modeling interactions, such interactions are largely unexplored. It might be, for instance, that different therapies work best with different kinds of clients. Micro-mediating therapy process variables of the type presented in Figure 5.5 are also largely unexplored. These latter explorations are hampered by a lack of research strong in both outcome and process assessment. In particular, the best therapy outcome research rarely seems to include process or interactive variables in a way that is systematically cumulative over studies; and the best process research is often devoid of design features that would facilitate causal inference. This need not be the case. Much process research will justifiably be nonexperimental (Gurman et al. 1986). But a call for more process research need not abandon good outcome methodologies. Explanatory meta-analyses could be well served by joint use of the two approaches in which, for example, randomized studies are coupled with extensive measurement of process and interactive variables to allow modeling mediating effects (Alwin and Tessler 1985; Fiske, Kenny, and Taylor 1982; Geiselman, Woodward, and Beatty 1982; Neuberg 1989). Ultimately, explanatory meta-analysis will be only as good as the quality of the primary studies on which it is based.

Clearly, we are just beginning to explore the full richness of data analytic possibilities that could be applied to meta-analysis. The analyses illustrated in this chapter—weighted least squares, random-effects models, multiple equation models, latent variable models—have been available for years and widely used in primary research. Their application to meta-analysis is only now being made, partly because meta-analysis is so new, partly because the analyses are themselves conceptually and logistically difficult, and partly because software technology has not kept pace with meta-analytic needs. Yet application of such models to meta-analysis would facilitate further progress. Light and Pillemer (1984) noted that the real revolution of meta-analysis is that it provides the basis for a scientific methodology for reviewing research. The promise of that revolution cannot be realized until the methodologies for reviewing research are at least as sophisticated as the methodologies for doing research.

Studies Used in this Analysis

Studies indented under another study are multiple reports of the same study and were coded as just one study.

Notes: This research was supported in part by grants to the author from the National Institute of Mental Health (R01MH41097) and from the Russell Sage Foundation. Support was also received from a Centers of Excellence grant awarded to the Department of Psychology, Memphis State University, by the State of Tennessee. The author wishes to thank a number of colleagues and consultants for their help over the years with this project, including Thomas D. Cook, Harris Cooper, David S. Cordray, Alan S. Gurman, Richard Light, William Pinsof, and Richard A. Wells. Thomas Louis and Larry Hedges were exceptionally helpful in providing statistical consultation; many of the procedures used in this chapter were developed by them. Study codings were completed by Linda Montgomery, Paul Wilson, Mary Wilson, Ivey Bright, and Theresa Okwumabua, who also made many other valuable contributions to the project. Correspondence should be sent to the author at the Center for Applied Psychological Research, Department of Psychology, Memphis State University, Memphis TN 38152.

Family Therapy Studies

Alexander, J. F., and B.V. Parsons
    1973   Short-term behavioral intervention with delinquent families: Impact on family process and recidivism. *Journal of Abnormal Psychology* 81:219–225.

Alexander, J. F., and C. Barton
    1976   Behavioral systems therapy for families. In D. H. L. Olson, ed., *Treating Relationships*. Lake Mills, IA: Graphic.

Alexander, J. F.; C. Barton; R. S. Schiavo; and B. V. Parsons
    1976   Systems-behavioral intervention with families of delinquents: therapist characteristics, family behavior, and outcome. *Journal of Consulting and Clinical Psychology* 41:656–664.

Klein, N. C.; J. F. Alexander; and B. V. Parsons
    1977   Impact of family systems intervention on recidivism and sibling delinquency: A model of primary prevention and program evaluation. *Journal of Consulting and Clinical Psychology* 45:469–474.

Malouf, R. E., and J. F. Alexander
    1974   Family crisis intervention: A model and technique of training. In R. E. Hardy and J. G. Cull, eds., *Therapeutic Needs of the Family*. Springfield, IL: Charles C. Thomas.

Parsons, B. V., Jr., and J. F. Alexander
    1973   Short term family intervention: A therapy outcome study. *Journal of Consulting and Clinical Psychology* 41:195–201.

Anderson, C. M.; S. Griffin; A. Rossi; I. Pagonis; D. P. Holder and R. Treiber
1986   A comparative study of the impact of education vs. process groups for families of patients with affective disorders. *Family Process* 25:185–205.

Andreozzi, L. L.
1984   The effects of short-term structural-analytic oriented family therapy on families with a presenting child problem. *Dissertation Abstracts International* 45:3325B. (University Microfilms no. 85-00051)

Anesko, K. M., and S. G. O'Leary
1982   The effectiveness of brief parent training for the management of children's homework problems. *Child and Family Behavior Therapy* 4:113–126.

Bean, A. W., and M. W. Roberts
1981   The effect of time-out release contingencies on changes in child noncompliance. *Journal of Abnormal Child Psychology* 9:95–105.

Bennun, I.
1986   Evaluating family therapy: A comparison of the Milan and problem solving approaches. *Journal of Family Therapy* 8:225–242.

Brunk, M.; S. W. Henggeler; and J. P. Whelan
1987   Comparison of multisystemic therapy and parent training in the brief treatment of child abuse and neglect. *Journal of Consulting and Clinical Psychology* 55:171–178.

Cable, L.
1977   The effectiveness of a leader created agenda parent counseling model in changing parental attitudes of parents of trainable mentally retarded. *Dissertation Abstracts International* 38:1989–90A. (University Microfilms no. 77-21632)

Carns, A. W.
1979   The effectiveness of parent group counseling as compared to individual parent consultation in changing parent attitude and child behavior. *Dissertation Abstracts International* 40:1272A. (University Microfilms no. 79-19717)

Christensen, A.; S. M. Johnson; S. Phillips; and R. E. Glasgow
1980   Cost effectiveness in behavioral family therapy. *Behavior Therapy* 11:208–226.

Dadds, M. R.; S. Schwartz; and M. R. Sanders
1987   Marital discord and treatment outcome in behavioral treatment of child conduct disorders. *Journal of Consulting and Clinical Psychology* 55:396–403.

D'Angelo, R.
1984   Effects coincident with the presence and absence of a one-shot interview directed at families of runaways. *Journal of Social Service Research* 8:71–81.

Delio, A. R.
1982   An investigation of the effectiveness of intervention strategies on ju-
        venile anti-social behaviors. *Dissertation Abstracts International* 43:1466A.
        (University Microfilms no. 82-23594)

DeStefano, T. J.
1981   Family therapy compared to individual play therapy in the treatment
        of young children for behavioral and emotional problems. *Disserta-
        tion Abstracts International* 42:4187B. (University Microfilms no. 82-02698)

Diament, C., and G. Colletti
1978   Evaluation of behavioral group counseling for parents of learning-
        disabled children. *Journal of Abnormal Child Psychology* 6:385–400.

Dorgan, J. P.
1977   A study of the effectiveness of a programmed strategy of behavioral
        intervention for parents of children with learning and behavior dis-
        orders. *Dissertation Abstracts International* 38:1887A. (University Mi-
        crofilms no. 77-21638)

Emshoff, J. G., and C. H. Blakely
1983   The diversion of delinquent youth: Family focused intervention. *Chil-
        dren and Youth Services Review* 5:343–356.

Epstein, G. M.
1977   Evaluating the bereavement process as it is affected by variation in
        the time of intervention. *Dissertation Abstracts International* 38:2362B.
        (University Microfilms no. 77-25096)

Epstein, L; R. R. Wing; R. Koeske; F. Andrasik; and D. J. Ossip
1981   Child and parent weight loss in family-based behavior modification
        programs. *Journal of Consulting and Clinical Psychology* 49:674–686.

Epstein, L. H.; R. R. Wing; R. Koeske; and A. Valoski
1987   Long-term effects of family-based treatment of childhood obesity.
        Journal of Consulting and Clinical Psychology 55:91–95.

Eyberg, S. M., and S. M. Johnson
1974   Multiple assessment of behavior modification with families: Effects
        of contingency contracting and order of treated problems. *Journal of
        Consulting and Clinical Psychology* 42:594–606.

Ezzo, F. R.
1980   A comparative outcome study of family therapy and positive parent-
        ing with court referred adolescents. *Dissertation Abstracts International*
        40:6198. (University Microfilms no. 80-13836)

Falloon, I. R. H.; J. L. Boyd; C. W. McGill; M. Williamson; J. Razani; H. B.
Moss; A. M. Gilderman; and G. M. Simpson
1985   Family management in the prevention of morbidity in schizophrenia.
        Archives of General Psychiatry 42:887–896.

Doane, J. A.; I. R. Falloon; M. J. Goldstein; and J. Mintz
1985   Parental affective style and the treatment of schizophrenia: Pre-

dicting course of illness and social functioning. *Archives of General Psychiatry* 42:34–42.

Falloon, I. R. H.; J. L. Boyd; and C. W. McGill
1984   *Family Care of Schizophrenia.* New York: Guilford Press.

Falloon, I. R. H.; J. L. Boyd; C. W. McGill; J. Razani; H. B. Moss; and A. M. Gilderman
1982   Family management in the prevention of exacerbations of schizophrenia: A controlled study. *New England Journal of Medicine* 306: 1437–1440.

Falloon, I. R. H.; J. L. Boyd; C. W. McGill; J. S. Strang; and H. B. Moss
1981   Family management training in the community care of schizophrenia. In M. Goldstein, ed., *New Directions for Mental Health Services: New Directions in Interventions with Families of Schizophrenics*, no. 12. San Francisco: Jossey-Bass.

Fisher, S. G.
1978   Time-limited brief psychotherapy: An investigation of therapeutic outcome at a child guidance clinic. *Dissertation Abstracts International* 39:377B. (University Microfilms no. 78-11234)

Ford, B. G., and L. W. West
1979   Human relations training for families: A comparative strategy. *Canadian Counsellor* 13:102–107.

Foster, S. L.; F. J. Prinz; and K. D. O'Leary
1983   Impact of problem-solving communciation training and generalization procedures on family conflict. *Child and Family Behavior Therapy* 5:1–23.

Gant, B. L.; J. D. Barnard; F. E. Kuehn; H. H. Jones; and E. R. Christophersen
1981   A behaviorally based approach for improving intrafamilial communication patterns. *Journal of Clinical Child Psychology* 10:102–106.

Garcia-Shelton, L. M.
1980   An evaluation of two treatment programs for families with acting-out adolescents. *Dissertation Abstracts International* 40:5810B. (University Microfilms no. 80-13735)

Garrett, L. D.
1979   Group behavioral therapy versus family behavioral therapy in treatment of obese adolescent females. *Dissertation Abstracts International* 40:1364B. (University Microfilms no. 79-19457) (Contained two different studies which were coded separately.)

Garrigan, J. J., and A. F. Bambrick
1977a   Family therapy for disturbed children: Some experimental results in special education. *Journal of Marriage and Family Counseling* 3:83–93.

Garrigan, J. J., and Bambrick, A. F.
1977b   Introducing novice therapists to "Go-Between" techniques in family therapy. *Family Process* 16:237–246.

1979   New findings in research on go-between process. *International Journal of Family Therapy* 1:76–85.

Ginsberg, B. G.
1971   Parent-adolescent relationship development: A therapeutic and preventative mental health program. *Dissertation Abstracts International* 33:426A. (University Microfilms no. 72-19306)

Glick, I, D.; J. F. Clarkin; J. H. Spencer; A. B. Lewis, J. Peyser; N. DeMane; M. Good-Ellis; E. Harris; and V. Lestelle
1985   A controlled evaluation of inpatient family intervention: I. Preliminary results of the six-month follow-up. *Archives of General Psychiatry* 42:882–886.

Graziano, A. M., and K. C. Mooney
1980   Family self-control instruction for children's nighttime fear reduction. *Journal of Consulting and Clinical Psychology* 48:206–213.

Griest, D. L.; R. Forehand; T. Rogers; J. Breiner; W. Furey; and C. A. Williams
1982   Effects of parent enhancement therapy on the treatment outcome and generalization of a parent training program. *Behavior Research and Therapy* 20:429–436.

Guerney, B.; J. Coufal; and E. Vogelsong
1981   Relationship enhancement versus a traditional approach to therapeutic/preventative/enrichment parent-adolescent programs. *Journal of Consulting and Clinical Psychology* 49:927–939.

Guerney, B.; J. Coufal; and E. Vogelsong
1983   Relationship enhancement versus a tradtional approach to therapeutic/preventative/enrichment parent-adolescent programs. *International Journal of Eclectic Psychotherapy* 2:31–43.

Hampshire, P. A.
1980   Family therapy with lower socioeconomic juvenile offenders: Engagement and outcome. *Dissertation Abstracts International* 41:2761B. (University Microfilms no. 81-02465)

Hannemann, E.
1979   Short-term family therapy with juvenile status offenders and their families. *Dissertation Abstracts International* 40:1894B. (University Microfilms no. 79-22867)

Hardcastle, D. R.
1973   The effects of a family counseling program on parent's family satisfaction, perceived integration, and congruence, and on specific behavior patterns in the family. *Dissertation Abstracts International* 34:2766A. (University Microfilms no. 73-27526)

Harper, M. S.
1963   Changes in concepts of mental health, mental illness, and perceptions of interpersonal relationships as a result of patient, family and

members of family-friend system participation in conjoint family therapy in a hospital. *Dissertation Abstracts International* 25:1395. (University Microfilms no. 64-04248)

Hedberg, A. G., and L. Campbell
1974   A comparison of four behavioral treatments of alcoholism. *Journal of Behavior Therapy and Psychiatry* 5:251–256.

Herold, P. L.
1980   The effects of psychosocial intervention with children who have asthma on children's locus of control and self-esteem scores, and measures of physical status. *Dissertation Abstracts International* 40:5075B. (University Microfilms no. 80-07900)

Hogarty, G.E.; C. M. Anderson; D. J. Reiss; S. J. Kornblith; D. P. Greenwald; C. D. Javna; and M. J. Madonia
1986   Family psychoeducation, social skills training, and maintenance chemotherapy in the aftercare treatment of schizophrenia. *Archives of General Psychiatry* 43:633–642.

Anderson, C. M.
1983   A psychoeducational program for families of patients with schizophrenia. In W. R. McFarlane, ed., *Family Therapy in Schizophrenia*. New York: Guilford Press.

Anderson, C. M.; G. E. Hogarty; and D. J. Reiss
1980   Family treatment of adult schizophrenic patients: A psycho-educational approach. *Schizophrenia Bulletin* 6:490–505.

Anderson, C. M., and D. J. Reiss
1982   Approaches to psychoeducational family therapy. *International Journal of Family Psychiatry* 3:501–517.

Hogarty, G. E., and C. M. Anderson
1986   Medication, family psychoeducation, and social skills training: First year relapse results of a controlled study. *Psychopharmacology Bulletin* 22:860–862.

Jansma, T. J.
1971   Multiple vs. individual family-therapy: Its effects on family concepts. *Dissertation Abstracts International* 33:1288B. (University Microfilms no. 72-22840)

Johnson, J. L.
1976   A time-series analysis of the effects of therapy on family interaction. *Dissertation Abstracts International* 36:5796B. (University Microfilms no. 76-11396)

Johnson, T. M., and H. N. Malony
1977   Effects of short-term family therapy on patterns of verbal interchange in disturbed families. *Family Therapy* 4:207–215.

Katz, A.; M. Krasinski; E. Philip; and C. Wieser
1975   Change in interactions as a measure of effectiveness in short term family therapy. *Family Therapy* 2:31–56.

Kearney, M. S.
1984   A comparative study of multiple family group therapy and individual conjoint family therapy within an outpatient community chemical dependency treatment program. *Dissertation Abstracts International* 45:3945B. (University Microfilms no. 85-03409)

Kent, R. N., and K. D. O'Leary
1976   A controlled evaluation of behavior modification with conduct problem children. *Journal of Consulting and Clinical Psychology* 44:586–596.

Knight, N. A.
1975   The effects of changes in family interpersonal relatioinships on the behavior of eneuretic children and their parents. *Dissertation Abstracts International* 36:783-A. (University Microfilms no. 75-17120)

Kopeikin, H. S.; V. Marshall; and M. J. Goldstein
1983   Stages and impact of crisis-oriented family therapy in the aftercare of acute schizophrenia. In W. R. McFarlane, ed., *Family Therapy in Schizophrenia*. New York: Guilford Press.

Goldstein, M. J., and E. H. Rodnick
1983   *The Interaction of Drug Therapy and Family Treatment of Schizophrenia, Psychopharmacology and Psychotherapy*. New York: Free Press.

Goldstein, M. J., and H. S. Kopeikin
1981   Short- and long-term effects of combining drug and family therapy. In M. Goldstein, ed., *New Directions for Mental Health Services: New Directions in Interventions with Families of Schizophrenics*, no. 12. San Francisco: Jossey-Bass.

Goldstein, M. J.; E. H. Rodnick; J. R. Evans; P. R. A. May; and M. R. Steinberg
1978   Drug and family therapy in the aftercare of acute schizophrenia. *Archives of General Psychiatry* 35: 1169–1177.

King, C. E., and M. J. Goldstein
1979   Therapist ratings of achievement of objectives in psychotherapy with acute schizophrenics. *Schizophrenia Bulletin* 5:118–129.

Langsley, D. G.; K. Flomenhaft; and P. Machotka
1969   Followup evaluation of family crisis therapy. *American Journal of Orthopsychiatry* 39:753–759.

Flomenhaft, K.; D. M. Kaplan; and D. G. Langsley
1969   Avoiding psychiatric hospitalization. *Social Work* 14:38–45.

Langsley, D. G.; P. Machotka; and K. Flomenhaft
1971   Avoiding mental hospital admission: A follow-up study. *American Journal of Psychiatry* 127:1391–1394.

Langsley, D. G.; F. S. Pittman; P. Machotka; and K. Flomenhaft
1968   Family crisis therapy: Results and implications. *Family Process* 7:145–158.

Lea, R. C.
1983   Acting up disorders in children: An evaluation of an intervention. *Dissertation Abstracts International* 44:611B. (University Microfilms no. 83-13733)

Leff, J.; L. Kuipers; R. Berkowitz; R. Eberlein-Fries; and D. Sturgeon
1984   Psychosocial relevance and benefit of neuroleptic maintenance: Experience in the United Kingdom. *Journal of Clinical Psychiatry* 45:43–49.

Berkowitz, R.
1984   Therapeutic intervention with schizophrenic patients and their families: A description of a clinical research project. *Journal of Family Therapy* 6:211–233.

Berkowitz, R.; R. Eberlein-Fries; L. Kuipers; and J. Leff
1984   Educating relatives about schizophrenia. *Schizophrenia Bulletin* 10:418–428.

Berkowitz, R.; L. Kuipers; R. Eberlein-Fries; and J. Leff
1981   Lowering expressed emotion in relatives of schizophrenics. In M. Goldstein, ed., *New Directions for Mental Health Services: New Directions in Interventions with Families of Schizophrenics*, no. 12. San Francisco: Jossey-Bass.

Leff, J.; L. Kuipers; R. Berkowitz; R. Eberlein-Vries; and D. Sturgeon
1982   A controlled trial of social intervention in the families of schizophrenic patients. *British Journal of Psychiatry* 141:121–134.

Liberman, R. P.; I. R. Falloon; and R. A. Aitchison
1984   Multiple family therapy for schizophrenia: A behavioral, problem-solving approach. *Psychosocial Rehabilitation Journal* 7:60–77.

Snyder, K. S., and R. P. Liberman
1981   Family assessment and intervention with schizophrenics at risk for relapse. In M. Goldstein, ed., *New Directions for Mental Health Services: New Directions in Interventions with Families of Schizophrenics*, no. 12. San Francisco: Jossey-Bass.

Lott, E. B.
1981   The effects of pre-therapy training with lower-class families entering family therapy: Attendance and short-term subjective outcome. *Dissertation Abstracts International* 41:3897B. (University Microfilms no. 8107686)

Love, L. R.; J. Kaswan; and D. E. Bungental
1972   Differential effectiveness of three clinical interventions for different socioeconomic groupings. *Journal of Consulting and Clinical Psychology* 39:347–360.

Love, L. R., and J. W. Kaswan
1974 *Troubled Children: Their Families, Schools and Treatments.* New York: Wiley.

Martin, B.
1977 Brief family interventions: Effectiveness and the importance of including father. *Journal of Consulting and Clinical Psychology* 45:1002–1010.

McMahon, R. J.; R. Forehand; and D. L. Griest
1981 Effects of knowledge of social learning principles on enhancing treatment outcome and generalization in a parent training program. *Journal of Consulting and Clinical Psychology* 49:526–532.

McManus, R.
1983 Facilitating family communication and family problem solving abilities through the family reunion game. *Dissertation Abstracts International* 44:612B. (University Microfilms no. 83-12282)

McPherson, S. J.
1980 Family counseling for youthful offenders in the juvenile court setting: A therapy outcome study. *Dissertation Abstracts International* 42:382B. (University Microfilms, no. 81-09550)

Messina, J. J.
1974 A comparative study of parent consultation and conjoint family counseling. *Dissertation Abstracts International* 35:4188B. (University Microfilms no. 74-20014)

Moreno, R.
1985 The effects of strategic-family therapy and client-centered therapy on selected personality variables of juvenile delinquents. *Dissertation Abstracts International,* 46:1195A. (University Microfilms no. 85-15117)

Orchen, M. D.
1983 A treatment efficacy study comparing relaxation training, emg biofeedback, and family therapy among heavy drinkers. *Dissertation Abstracts International* 44:2565B. (University Microfilms no. 83-27825)

Patterson, G. R.; P. L. Chamberlain; and J. B. Reid
1982 A comparative evaluation of a parent-training program. *Behavior Therapy* 13:638–650.

Peed, S.; M. Roberts; and F. Forehand
1977 Evaluation of the effectiveness of a standardized parent training program in altering the interaction of mothers and their noncompliant children. *Behavior Modification* 1:323–350.

Pevsner, R.
1982 Group parent training versus individual family therapy: An outcome study. *Journal of Behavior Therapy and Experimental Psychiatry* 13:119–122.

Phillips, S. B.
    1984   The effect of posttreatment interventions on long-term maintenance
            of behavioral family therapy. *Dissertation Abstracts International* 45:3956B.
            (University Microfilms no. 85-02012)

Raue, J., and S. H. Spence
    1985   Group versus individual applications of reciprocity training for par-
            ent-youth conflict. *Behavior Research and Therapy* 23:177–186.

Regas, S. J.
    1983   A comparison of functional family therapy and peer group therapy
            in the treatment of hyperactive adolescents. *Dissertation Abstracts In-
            ternational* 44:2566B. (University Microfilms no. 83-24051)

Reid, W. J., and A. W. Shyne
    1969   *Brief and Extended Casework.* New York: University Press.

Reiter, G. F., and P. R. Kilmann
    1975   Mothers as family change agents. *Journal of Counseling Psychology* 22:61–
            65.

Ritterman, M. K.
    1978   Outcome study for family therapy, ritalin, and placebo treatments of
            hyperactivity: An open systems approach. *Dissertation Abstracts Inter-
            national* 39:6138B. (University Microfilms no. 79-10077)

Roberts, M. W.; R. J. McMahon; R. Forehand; and L. Humphreys
    1978   The effect of parent instruction-giving on child compliance. *Behavior
            Therapy* 9:793–798.

Robin, A. L.
    1976   Communication training: An approach to problem solving for par-
            ents and adolescents. *Dissertation Abstracts International* 36:5814B.
            (University Microfilms no. 76-11253)
    1981   A controlled evaluation of problem-solving communication training
            with parent-adolescent conflict. *Behavior Therapy* 12:593–609.

Rosenthal, A. J., and S. V. Levine
    1970   Brief psychotherapy with children: A preliminary report. *American
            Journal of Psychiatry* 127:106–111.

Rosenthal, A. J., and S. V. Levine
    1971   Brief psychotherapy with children: Process of therapy. *American
            Journal of Psychiatry* 128:33–38.

Scopetta, M. A.; J. Szapocznik; O. E. King; R. Ladner; C. Alegre; and W. S.
Tillman
    1977   *The Spanish Drug Rehabilitation Research Project: A National Institute on
            Drug Abuse Demonstration Grant, 1974–1977.* Final Report Grant no.
            H81 DA 01696-03. Spanish Family Guidance Clinic—Encuentro, Uni-
            versity of Miami School of Medicine, Department of Psychiatry.

Scopetta, M. A.; O. E. King; J. Szapocznik; and W. Tillman
Undated *Ecological Structural Family Therapy with Cuban Immigrant Families.*
Report of NIDA Grant no. H81 DA 01696. Key Biscayne, Florida: Authors.

Scovern, A. W.; L. H. Bukstel; P. R. Kilmann; R. A. Laval; J. Busemeyer; and V. Smith
1980 Effects of parent counseling on the family system. *Journal of Counseling Psychology* 27:268–275.

Scully, T. M.
1982 Strategic family therapy with conduct disordered children and adolescents: An outcome study. *Dissertation Abstracts International* 43:3042. (University Microfilms no. 82-27817)

Slipp, S., and K. Kressel
1978 Difficulties in family therapy evaluation: I. A comparison of insight vs. problem-solving approaches. II. Design critique and recommendations. *Family Process* 17:409–422.

Stanton, M. D.; T. C. Todd; and Associates
1982 *The Family Therapy of Drug Abuse and Addiction.* New York: Guilford Press.

Stanton, M. D.; I. F. Steier; and T. C. Todd
1982 Paying families for attending sessions: Counter-acting the dropout problem. *Journal of Marital and Family Therapy* 8:371–373.

Steier, F.
1983 Family interaction and properties of self-organizing systems: A study of family therapy with addict families. *Dissertation Abstracts International* 44:863A. (University Microfilms no. 83-16093)

Stover, L., and B. Guerney
1967 The efficacy of training procedures for mothers in filial therapy. *Psychotherapy: Theory, Research, and Practice* 4:110–115.

Stuart, R., B., and T. Tripodi
1973 Experimental evaluation of three time-constrained behavioral treatment for predelinquents and delinquents. In R. Rubin, J. Brady, and J. Henderson, eds., *Advances in Behavior Therapy,* vol. 4. New York: Academic Press.

Stuart, R. B., and L. A. Lott, Jr.
1972 Behavioral contracting with delinquents: A cautionary note. *Journal of Behavior Therapy and Experimental Psychiatry* 3:161–169.

Stuart, R. B., T. Tripodi; S. Jayaratne; and D. Camburn
1976 An experiment in social engineering in serving the families of predelinquents. *Journal of Abnormal Child Psychology* 4:243–261.

Sykes, B. W.
1978   The adaptation of the family training program to an office setting. *Dissertation Abstracts International* 38:3375B. (University Microfilms no. 77-28915)

Szapocznik, J.; W. M. Kurtines; F. H. Foote; A. Perez-Vidal; and O. Hervis
1983   Conjoint versus one-person family therapy: Some evidence for the effectiveness of conducting family therapy through one person. *Journal of Consulting and Clinical Psychology* 51:889–899.
1986   Conjoint versus one-person family therapy: Further evidence for the effectiveness of conducting family therapy through one person with drug-abusing adolescents. *Journal of Consulting and Clinical Psychology* 54:395–397.

Szapocznik, J.; A. Perez-Vidal; A. L. Brickman; F. H. Foote; D. Santisteban; O. Hervis; and W. M. Kurtines
1988   Engaging adolescent drug abusers and their families in treatment: A strategic structural systems approach. *Journal of Consulting and Clinical Psychology* 56:552–557.

Szapocznik, J.; A. Rio; A. Perez-Vidal; W. Kurtines; O. Hervis; and D. Santisteban
1986   Bicultural effectiveness training (BET): An experimental test of an intervention modality for families experiencing intergenerational/intercultural conflict. *Hispanic Journal of Behavioral Sciences* 8:303–330. (Coding of this study included unpublished means and standard deviations sent by its first author.)

Szykula, S. A., and M. J. Fleischman
1985   Reducing out-of-home placements of abused children: Two controlled field studies. *Child Abuse and Neglect* 9:277–283.

Trankina, F. J.
1975   Aggressive and withdrawn children as related to family perception and outcome of different treatment methods. *Dissertation Abstracts International* 36:924B. (University Microfilms no. 75-17824)

Walker, J. M.
1985   A study of the effectiveness of social learning family therapy for reducing aggressive behavior in boys. *Dissertation Abstracts International* 45:3088B. (University Microfilms no. 84-29327)

Walter, H. I., and S. K. Gilmore
1973   Placebo versus social learning effects in parent training procedures designed to alter the behavior of aggressive boys. *Behavior Therapy* 4:361–377.

Wattie, B.
1973   Evaluating short-term casework in a family agency. *Social Casework* 54:609–616.

Webster-Stratton, C.
1984   Randomized trial of two parent-training programs for families with conduct-disordered children. *Journal of Consulting and Clinical Psychology* 52:666–678.

Wellisch, D. K.
1976   A family therapy outcome study in an inpatient setting. *Dissertation Abstracts International* 36:3634B. (University Microfilms no. 76-01246)

Whiteman, M.; D. Fanshel; and J. F. Grundy
1987   Cognitive-behavioral interventions aimed at anger of parents at risk of child abuse. *Social Work* 32:469–474.

Wolfe, D. A.; B. Edwards; I. Manion; and C. Koverola
1988   Early intervention for parents at risk of child abuse and neglect: A preliminary investigation. *Journal of Consulting and Clinical Psychology* 56:40–47.

Zangwill, W. M.
1983   An evaluation of a parent training program. *Child and Family Behavior Therapy* 5:1–16.

Ziegler-Driscoll, G.
1977   Family research study at Eagleville Hospital and Rehabilitation Center. *Family Process* 16:175–189.

Marital Therapy Studies

Azrin, H. H.; V. A. Besalel; R. Bechtel; A. Michalicek; D. Carroll; D. Shuford; and J. Cox
1980   Comparison of reciprocity and discussion-type counseling for marital problems. *American Journal of Family Therapy* 8:21–28.

Barlow, D. H.; G. T. O'Brien; and C. G. Last
1984   Couples treatment of agoraphobia. *Behavior Therapy* 15:41–58.

Baucom, D. H.
1982   A comparison of behavioral contracting and problem-solving/communication training in behavioral marital therapy. *Behavior Therapy* 13:162–174.

Baucom, D. H.
1984   The active ingredients of behavioral marital therapy: The effectiveness of problem-solving/communication training, contingency contracting, and their combination. In K. Hahlweg and N. S. Jacobson, eds., *Marital Interaction: Analysis and Modification*. New York: Guilford Press.

Beach, S. R., and K. D. O'Leary
1986   The treatment of depression occurring in the context of marital discord. *Behavior Therapy* 17:43–49.

Bergner, R. M.
1974   The development and evaluation of a training videotape for the res-
        olution of marital conflict. *Dissertation Abstracts International* 34:3485B.
        (University Microfilms no. 73-32510)

Boelens, W.; P. Emmelkamp; D. MacGillavry; and M. Markvoort
1980   A clinical evaluation of marital treatment: Reciprocity counseling vs.
        system-theoretic counseling. *Behavioral Analysis and Modification* 4:85–
        96.

Emmelkamp, P.; M. Ven Der Helm; D. MacGillavry; and B. Van Zanten
1984   Marital therapy with clinically distressed couples: A comparative
        evaluation of system-theoretic, contingency contracting, and com-
        munication skills approaches. In K. Hahlweg and N. S. Jacobson,
        eds., *Marital Interaction: Analysis and Modification.* New York: Guil-
        ford Press.

Bogner, I., and H. Zielenbach-Coenen
1984   On maintaining change in behavioral marital therapy. In K. Halweg
        and N. S. Jacobson, eds., *Marital Interaction: Analysis and Interaction.*
        New York: Guilford Press.

Brainerd, G. L.
1978   The effect on marital satisfaction of a communication training pro-
        gram for married couples, administered by paraprofessional person-
        nel. *Dissertation Abstracts International* 38:3382B. (University Micro-
        films no. 77-28064)

Carlton, K. A.
1979   An evaluation of the effects of communication skills training on mar-
        ital interaction. *Dissertation Abstracts International* 39:5404A. (Univer-
        sity Microfilms no. 79-05079)

Carrigan, A. E.
1976   A study of interventions in marital relationships using a results ac-
        countability model. *Dissertation Abstracts International* 37:2475B. (Uni-
        versity Microfilms no. 76-25133)

Cassidy, M. J.
1973   Communication training for marital pairs. *Dissertation Abstracts Inter-
        national* 34:3054A. (University Microfilms no. 73-28680)

Christensen, D. N.
1983   Postmastectomy couple counseling: An outcome study of a struc-
        tured treatment protocol. *Journal of Sex and Marital Therapy* 9:266–275.

Cobb, J.; R. McDonald; I. Marks; and R. Stern
1980   Marital versus exposure therapy: Psychological treatment of co-exist-
        ing marital and phobic-obsessive problems. *Behavioral Analysis and
        Modification* 4:3–16.

Cobb, J. P.; A. M. Matthews; A. Childs-Clarke; and C. M. Blowers
1984 The spouse as co-therapist in the treatment of agoraphobia. *British Journal of Psychiatry* 144:282–287.

Collins, J. D.
1971 The effects of the conjugal relationship modification method on marital communication and adjustment. *Dissertation Abstracts International* 32:6674B. (University Microfilms no. 72-13836)

Cookerly, J. R.
1974 The reduction of psychopathology as measured by the MMPI clinical scales in three forms of marriage counseling. *Journal of Marriage and the Family* 36:332–335.

Cookerly, J. R.
1976 Evaluating different approaches to marriage counseling. In D. H. L. Olson, ed., *Treating Relationships*. Lake Mills, IA: Graphic.

Cotton, M. C.
1976 A systems approach to marital training evaluation. *Dissertations Abstracts International* 37:5346B. (University Microfilms no. 77-08742)

Crowe, M. J.
1978 Conjoint marital therapy: A controlled outcome study. *Psychological Medicine* 8:623–636.

Crowe, M. J.
1984 The analysis of therapist intervention in three contrasted approaches to conjoint marital therapy. In K. Hahlweg and N. S. Jacobson, eds., *Marital Interaction: Analysis and Modification*. New York: Guilford Press.

Crowe, M. J.; P. Gillan; and S. Golombok
1981 Form and content in the conjoint treatment of sexual dysfunction: A controlled study. *Behavior Research and Therapy* 19:47–54.

Ely, A. L.; B. G. Guerney, and L. Stover
1973 Efficacy of the training phase of conjugal therapy. *Psychotherapy: Theory, Research, and Practice* 10:201–207.

Emmelkamp, P. M. G., and I. DeLange
1983 Spouse involvement in the treatment of obsessive-compulsive patients. *Behavior Research and Therapy* 21:341–346.

Epstein, N., and E. Jackson
1978 An outcome study of short-term communication training with married couples. *Journal of Consulting and Clinical Psychology* 46:207–212.

Everaerd, W.
1977 Comparative studies of short-term treatment methods for sexual inadequacies. In R. Gemme and C. G. Wheeler, eds., *Progress in Sexology*. New York: Plenum Press. (Reports three different studies which were coded separately.)

Everaerd, W., and J. Dekker
   1981   A comparison of sex therapy and communication therapy: Couples complaining of orgasmic dysfunction. *Journal of Sex and Marital Therapy* 7:278–289.

Friedman, A. S.
   1975   Interaction of drug therapy with marital therapy in depressive patients. *Archives of General Psychiatry* 32:619–637.

Glisson, D. H.
   1977   A comparison of reciprocity counseling and communication training in a treatment marital discord. *Dissertation Abstracts Internatioinal* 37:7973A-7974A. (University Microfilms no. 77-12462)

Graham, J. A.
   1968   The effect of the use of counselor positive responses to positive perceptions of mate in marriage counseling. *Dissertation Abstracts International* 28:3504A. (University Microfilms no. 68-01649)

Hafner, R. J.; A. Bradenock; J. Fisher; and H. Swift
   1983   Spouse-aided versus individual therapy in persisting psychiatric disorders: A systematic comparison. *Family Process* 22:385–399.

Hahlweg, K.; D. Revenstorf; and L. Schindler
   1982   Treatment of marital distress: Comparing formats and modalities. *Advances in Behavior Research and Therapy* 4:57–74.

Hahlweg, K.; D. Revenstorf; and L. Schindler
   1984   Effects of behavioral marital therapy on couples' communication and problem-solving skills. *Journal of Consulting and Clinical Psychology* 52:553–566.

Hahlweg, K.; L. Schindler; D. Revenstorf; and J. C. Brengelmann
   1984   The Munich marital therapy study. In K. Hahlweg and N. S. Jacobson, eds., *Marital Interaction: Analysis and Modification*. New York: Guilford Press.

Schindler, L.; K. Hahlweg; and D. Revenstorf
   1983   Short- and long-term effectiveness of two communication training modalities with distressed couples. *American Journal of Family Therapy* 11:54–64.

Harrell, J. E.
   1974   Efficacy of a behavioral-exchange program for teaching conflict negotiation skills to enhance marital relationship. *Dissertation Abstracts International* 36:4063A. (University Microfilms no. 75-27618)

Hartman, L. M.
   1983   Effects of sex and marital therapy on sexual interaction and marital happiness. *Journal of Sex and Marital Therapy* 9:137–151.

Hartman, L. M., and E. M. Daly
   1983   Relationship factors in the treatment of sexual dysfunction. *Behavior Research and Therapy* 21:153–160.

Hefner, C. W., and J. O. Prochaska
1984   Concurrent vs. conjoint marital therapy. *Social Work* 29:287–291.

Henry, N. R.
1977   The effects of a communication group for married couples: An outcome study. *Dissertation Abstracts International* 37:4683B. (University Microfilms no. 77-05095)

Huber, C. H., and B. Milstein
1985   Cognitive restructuring and a collaborative set in couples' work. *American Journal of Family Therapy* 13:17–27.

Hudson, J. A.
1978   Development and evaluation of problem-solving treatment packages for marital therapy. *Dissertation Abstracts International* 39:5070B. (University Microfilms no. 79-06832)

Jacobson, N. S.
1977   Problem-solving and contingency in the treatment of marital discord. *Journal of Consulting and Clinical Psychology* 45:92-100.
1978   Specific and nonspecific factors in the effectiveness of a behavioral approach to the treatment of marital discord. *Journal of Consulting and Clinical Psychology* 46:442–452.
1984   A component analysis of behavioral marital therapy: The relative effectiveness of behavior exchange and communication/problem-solving. *Journal of Consulting and Clinical Psychology* 52:295–305.

Jacobson, N. S., et al.
1985   A component analysis of behavioral marital therapy: 1-year follow-up. *Behavior Research and Therapy* 23:549–555.

Jacobson, N. S., and W. C. Follette
1985   Clinical significance of improvement resulting from two behavioral marital therapy components. *Behavior Therapy* 16:249–262.

Jacobson, N. S.; K. B. Schmaling; and A. Holtzworth-Munroe
1987   Component analysis of behavioral marital therapy: 2-year follow-up and prediction of relapse. *Journal of Marital and Family Therapy* 13:187–195.

Johnson, S. M.; and L. S. Greenberg
1985   Differential effects of experiential and problem-solving interventions in resolving marital conflict. *Journal of Consulting and Clinical Psychology* 53: 175–184.

Kind, J.
1968   The relationship of communication efficiency to marital happiness and an evaluation of short-term training in interpersonal communication with married couples. *Dissertation Abstracts International* 29:1173B. (University Microfilms no. 68-11956)

Libman, E.; C. S. Fichten; W. Brender; R. Burstein; J. Cohen; and Y. M. Binik
1984   A comparison of three therapeutic formats in the treatment of sec-

ondary orgasmic dysfunction. *Journal of Sex and Marital Therapy* 10:147–159.

Margolin, G., and R. L. Weiss
1978   Comparative evaluation of therapeutic components associated with behavioral marital treatments. *Journal of Consulting and Clinical Psychology* 46:1476–1486.

Matanovich, J. P.
1970   The effects of short-term counseling upon positive perceptions of mate in marital counseling. *Dissertation Abstracts International* 31:2688A. (University Microfilms no. 70-24405)

Mathews, A.; J. Bancroft; A. Whitehead; A. Hackman; D. Julier; J. Bancroft; D. Gath; and P. Shaw
1976   The behavioral treatment of sexual inadequacy: A comparative study. *Behavior Research and Therapy* 14:427–436.

McCrady, B.S.; N. E. Noel; D. B. Abrams; R. L. Stout; H. F. Nelson; and W. M. Hay
1986   Comparative effectiveness of three types of spouse involvement in outpatient behavioral alcoholism treatment. *Journal of Studies on Alcohol* 47:459–467.

McCrady, B. S.; T. J. Paolino; R. Longabough; and J. Rossi
1979   Effects of joint hospital admission and couples treatment for hospitalized alcoholics: A pilot study. *Addictive Behaviors* 4:155–165.

McCrady, B. S.; J. Moreau; T. J. Paolino; and R. Longabaugh
1982   Joint hospitalization and couples therapy for alcoholism: A four-year follow-up. *Journal of Studies on Alcohol* 43:1244–1250.

McLean, P. D.; K. Ogston; and L. Grauer
1973   A behavioral approach to the treatment of depression. *Journal of Behavior Therapy and Experimental Psychiatry* 4:323–342.

McLean, P. D., and K. D. Craig
1975   Evaluating treatment effectiveness by monitoring changes in problematic behaviors. *Journal of Consulting and Clinical Psychology* 43:105.

Mehlman, S. K.; D. H. Baucom; and D. Anderson
1983   Effectiveness of cotherapists versus single therapists and immediate versus delayed treatment in behavioral marital therapy. *Journal of Consulting and Clinical Psychology* 51:258–266.

Miaoulis, C. N.
1976   A study of the innovative use of time and planned short-term treatment in conjoint marital counseling. *Dissertation Abstracts International* 37:1993–1994A. (University Microfilms no. 76-23022)

O'Farrell, T. J.; H. S. G. Cutter; and F. J. Floyd
1985   Evaluating behavioral marital therapy for male alcoholics: Effects on

marital adjustment and communication from before to after treatment. *Behavior Therapy* 16:147–167.

O'Farrell, T. J., and H. S. G. Cutter
  1984 Behavioral marital thepy for male alcoholics: Clinical procedures from a treatment outcome study in progress. *American Journal of Family Therapy* 12:33–46.

O'Leary, K. D., and H. Turkewitz
  1981 A comparative outcome study of behavioral marital therapy and communication therapy. *Journal of Marital and Family Therapy* 7:153–169.

O'Leary, K. D., and H. Turkewitz
  1978 Marital therapy from a behavioral perspective. In T. J. Paolino and B. S. McCrady, eds., *Marriage and Marital Therapy: Psychoanalytic, Behavioral, and Systems Theory Perspectives.* New York: Brunner/Mazel.

Padgett, V. R.
  1983 Videotape replay in marital therapy. *Psychotherapy: Theory, Research, and Practice* 20:232–242

Price, M. G., and S. N. Haynes
  1980 The effects of participant monitoring and feedback on marital interaction and satisfaction. *Behavior Therapy* 11:134–139.

Ross, E. R.; S. B. Baker; and B. G. Guerney
  1985 Effectiveness of relationship enhancement therapy versus therapist's preferred therapy. *American Journal of Family Therapy* 13:11–21.

Swan, R. W.
  1972 Differential counseling approaches to conflict reduction in the marital dyad. *Dissertation Abstracts International* 32:6629B. (University Microfilms no. 72-13009)

Teismann, M. W., and B. Rodgers
  1982 A comparison of a traditional and a marital approach to rehabilitation counseling. *Journal of Marital and Family Therapy* 8:91–93.

Vancamp, J. V.
  1971 Modification of adult aggressive behavior by using modeling films. *Dissertation Abstracts International* 32:7327B. (University Microfilms no. 72-15870)

Wells, R. A.; J. A. Figurel; and P. McNamee
  1977 Communication training vs. conjoint marital therapy. *Social Work Research and Abstracts* 13:31–39.

Welsh, E. C.
  1977 Counseling and self-help: The effect of three treatment modes on interpersonal perception and marital adjustment. *Dissertation Abstracts International* 38:643A. (University Microfilms no. 77-16878)

Wieman, R. J.
1973   Conjugal relationship modification and reciprocal reinforcement: A comparison of treatments for marital discord. *Dissertation Abstracts International* 35:493B. (University Microfilms no. 74-16097)

Ziegler, J. S.
1973   A comparison of the effect of two forms of group psychotherapy on the treatment of marital discord. *Dissertation Abstracts International* 34:143–144A. (University Microfilms no. 73-13262)

6

Models of Science Achievement: Forces Affecting Male and Female Performance in School Science

with Christine M. Schram, Lin Chang,
Mary M. Kino, and Maria Quintieri

For the past decade scientists and science educators around the world have regarded the small numbers of women in scientific careers as a critical problem (e.g., Bruer 1983; Science Council of Canada 1981; National Research Council 1983). Although women's levels of participation in scientific enterprises have increased over the past two decades (see, e.g., National Research Council 1983, p. 16), the increases have not occurred uniformly throughout all of science. In the United States, for instance, the numbers of women pursuing education in the physical sciences have been very small, whereas in biology and other life sciences almost 30 percent of Ph.D.s went to women as long ago as 1980 (again, see National Research Council 1983). And though increases have occurred in many areas, concern continues that equity has not yet been achieved in either participation or the norms and methods of science (e.g., Schiebinger 1987).

The extent to which the lack of women in science is viewed as a problem in different countries depends not only on the actual numbers of women scientists but also on perceived needs for scientists. Specific "manpower" needs have motivated many of the studies and funding initiatives concerning gender issues in American science education (e.g., Bruer 1983; McMillen 1987; Subcommittee on Science Research and

Note: Studies used in this analysis are indicated by a †.

209

Technology 1982). Educators and policymakers alike state their interest in increasing the numbers and proportions of women pursuing careers in science. Many efforts have been made to that end worldwide (see, e.g., Kelly, Whyte, and Smail 1984; Kreinberg 1982).

At a symposium entitled "Women in Scientific Research," Vetter questioned, "How can we affect the choices of young women towards appropriate precollegiate studies or science careers if we do not understand how those choices are made, which factors are influential, or whether boys and girls, men and women, utilize the factors differently to reach career decisions?" (Vetter quoted in Bruer 1984, p. 5). Are the processes the same by which males and females decide to persevere in science? A first step in the process of achieving in the world of science involves persistence and achievement in school science. This review attempts to examine factors which predict achievement and persistence in school science[1] for males and for females.

The bulk of the literature on science education and gender issues has focused on differences between males and females in such characteristics as liking for science, interest in scientific careers, science achievement, and aptitude for science. A smaller yet still significant portion of the literature, dealing with associations among these characteristics for males and females, is the focus of the current synthesis.

Goals and Rationale

Model-Driven Research Synthesis

A primary goal of this research synthesis is to examine a set of models showing the role of gender in science achievement and to use these models for the prediction of science-achievement behaviors to direct the research-synthesis process. Problem formulation, literature-searching strategies, data evaluation and analysis all can be organized to focus on a particular model or models (see Becker 1989b for details).

One motivation for attempting this novel approach in this synthesis is the growing number of primary research studies which have attempted to study particular models of achievement in science for males and females. The problem of how to combine results from studies of multivariate models of science achievement with results based on simpler research designs is one of the challenging aspects of this synthesis.

Organizing the synthesis around a particular model or set of models

---

[1]Generally, school science includes biological (life) and physical sciences, earth science, and geology, but excludes mathematics, engineering, and computer sciences.

**Figure 6.1   Hypothetical Models of Prediction of Science Achievement
for Males and Females**

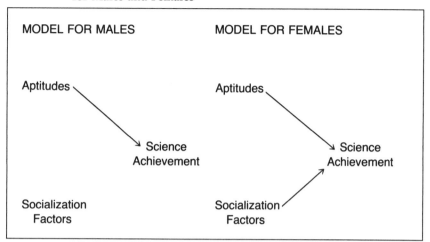

provides a context for determining what is known and what is not known about gender and science achievement. Gaps in our knowledge about factors posited to have different influences on science achievement for males and females can be easily seen in relation to such models; information on these gaps may be useful in developing focused agendas for future research. In this synthesis models are also used to guide the analysis of extant study results.

A concrete product of a model-driven synthesis could be a model (or set of models) synthesized from a collection of diverse studies. Figure 6.1 shows a hypothetical pair of models of science achievement for males and females.

Models can comprise the same or different numbers of components and paths between components. The models in Figure 6.1 contain the same components but different paths. Specifically, socialization relates to achievement for females but not for males. Also (though it is not indicated in this figure), the strength of existing relationships may differ by sex. These two hypothetical models also suggest that different and possibly manipulable factors may be critical for the science achievement of males and females.

What Is Known and What Is Not

As mentioned above, part of the rationale for this synthesis is to develop an understanding of what is and is not known about the associations among science-related variables. One goal is simply to describe the interrelationships that have been studied for males and females. Some relationships, perhaps those which are simpler or more basic, will have been more thoroughly investigated than others. This review provides a "catalog" of the interrelationships among science-achieve-ment behaviors and their predictors which have been studied for males and females separately.

The use of explicit models of interrelationships also enables us to assess, within the context of particular models, what is *not* known. For example, suppose that a model posits that males' and females' assess-ments of task value mediate the influence of parental encouragement on science achievement. If empirical studies have all examined the di-rect relationship of parental encouragement to achievement (for males and females), or if the only studies of the mediating influence of task value have not reported results by gender, then we have no informa-tion on the mediating effect of task value proposed in the particular model for the two sexes. That part of the model may be a good area for further research.

The analyses in this synthesis attempt to address several questions about relationships among science variables. What are the most impor-tant predictors of science achievement for males? Are different predic-tors important for females? More broadly, this work concerns gender differences in the nature of interrelationships among science-related variables. Measures of science achievement, attitudes, and science course-taking have received much attention as outcomes. Persistence in sci-ence course-taking, science-related aptitudes and attitudes, and social-izers' influences on male and female students are among the factors which have been hypothesized to relate to these outcomes. The general question addressed by this review is whether, on the basis of the exist-ing literature, each of these factors appears equally important for males and females.

Modeling Dependent Study Results

The conceptualization of this synthesis as a study of models of systems of variables also suggests an important consideration for data analysis. The results to be analyzed should be considered as a single set of in-terrelated (dependent) indices, rather than several sets of results for a collection of bivariate relationships. That is, dependencies should be modeled among multiple study outcomes that arise from individual samples.

Most reviews of relationships have proceeded by studying only a few relationships. Indices representing each bivariate relationship are typically analyzed separately. When individual studies contribute indices measuring several of the relationships under study, unmodeled dependencies may cause mild to severe deviations in error rates for the analyses. By explicitly modeling within-sample dependencies in study results the analysis should have more accurate error rates.

Also modeling dependencies among results will ensure that samples which contribute many indices to the review do not unduly influence the character of the overall results. If dependencies in results are modeled, a sample which contributes three highly interrelated results will likely have less influence than three results from independent samples.

Examining an Assumption of Gender-Differences Research

A consequence of studying interrelationships among science variables will be the examination of a key assumption of many studies of science gender differences. The assumption is that the same set of variables (i.e., the same model) is appropriate for predicting the achievement behaviors of males and females and that each predictor is equally important for both sexes.

The premise underlies many studies of gender differences in science achievement and possible predictors. The following exemplary quotation is from Kavrell and Petersen (1984).

> As conceptualized by Crandall, Katkovsky, and Preston (1960), there are three factors which can be used to predict achievement behaviors in different areas: (a) attainment value, . . . (b) achievement standards . . . and (c) achievement expectancy. . . . *The existence of sex differences in these aspects would certainly have implications for the choice of and performance in science and mathematics courses.* (1984, p. 3, emphasis added)

Each of these three factors must be assumed to relate to achievement in the same way for males and females, otherwise the finding of a sex difference on any of the factors might *not* lead to a sex difference in the outcome. Figure 6.2 shows two cases in which a predictor *(X)* is linearly related to an outcome, say science achievement, called *Y*. In part A the magnitude of the relationship of *X* to *Y* is the same for boys and girls, and a difference in *X* means corresponds to a difference in *Y* means. However, in part B, though the means on *X* differ to the same extent as those in part A, the variables *X* and *Y* have different linear relationships for males and females. Thus the mean gender difference on *X* does not imply a corresponding average difference on *Y*.

This assumption is implicit in many studies of gender differences in

Figure 6.2   Two Possible Relationships Between X and Y
                for Males (M) and Females (F)

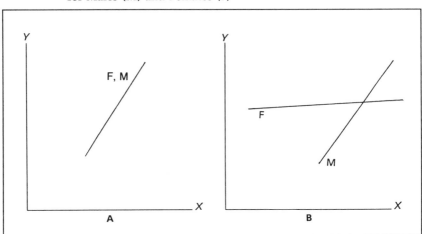

average performance on science variables. It also underlies most studies which examine achievement outcomes via multiple regression. The literature examined for this review identified 38 studies which had included gender of subjects as a predictor in a single (common) multiple-regression model with a science outcome. Only five studies, however, included any coefficients to represent the *interaction* of gender with other predictors. Five similar studies had included gender in path analyses, discriminant analyses, and other correlational analyses, without any interaction terms involving gender.

Although many of these authors have described their results as providing predictive "models" for males and females, none had investigated whether *different* models could apply for the two sexes. This synthesis examines that part of the research literature which directly addresses the question of whether *different* models, or collections of variables, can be used to predict the achievement and persistence of males and females in science.

Deriving Practical Knowledge

A final goal of the synthesis is to derive, from the existing knowledge base, practical knowledge to inform the process of science education for males and females. If different correlates of science performance are found for males and females, perhaps the prediction of science achievement and the process of educational decision-making should be based on different considerations for the two sexes.

For example, if high levels of mechanical aptitude relate to high physics performance for boys but not for girls, then the fact that a girl has low mechanical aptitude should not strongly influence her decision to study physics. Knowledge of different correlates of science performance for males and females may enable teachers to design instruction that is more effective for each sex. For instance, †Smail and Kelly (1984b) have found that for girls the perception of science as a male domain was significantly related to lower scores on a science-knowledge scale (with social class and general ability level controlled).[2] However, for boys these two variables were unrelated. Do some girls fail to develop knowledge about stereotypically male topics because they avoid studying such topics? A concerted effort to portray science as a gender-neutral (or even female-oriented) topic may draw in those females who see science as a male domain.

Using Models in Research Synthesis

One approach to the synthesis of studies predicting achievement and persistence in science might be to simply amass all available studies with those variables as outcomes and to synthesize the existing results for each of the predictor-outcome relationships found in the literature. The synthesis would be "empirically guided" because the relationships to be examined would arise from the collected literature.

A second approach, used here, is to guide the synthesis by the use of conceptual and empirical models. The models were drawn from the literature on science achievement and the literature on social and psychological influences on the development of general achievement behaviors. One model is based on the work of Eccles and her colleagues (e.g., Meece et al. 1982). A simpler model derived from an earlier review of gender differences in science achievement (Steinkamp and Maehr 1983) is also examined. Other models (e.g., from Dunteman, Wisenbaker, and Taylor 1979 and Keeves 1975) may be examined in future analyses.

The models from Eccles and Steinkamp and Maehr were incorporated as comprehensively as possible into the conduct of the research

---

[2]This one finding is taken from a more complex set of results, which contains several correlations with interpretations that differ slightly from those described here.

synthesis. Specific ways in which the models were used are described in later sections of this report. The two models are described in this section.

Simple Model, after Steinkamp and Maehr (1983)

The simplest of the models examined in this synthesis was empirically derived. That model was presented in an earlier review of correlational studies of affect, ability, and achievement in science (Steinkamp and Maehr 1983). The augmented version of the model, used in this synthesis, is shown in Figure 6.3.

The correlations shown in Figure 6.3 are from Steinkamp and Maehr's synthesis. These unweighted averages are based on correlations from a relatively few studies conducted in 1979 and earlier. Steinkamp and Maehr concluded that *"in pedagogical situations in which achievement in science is the immediate goal, cognitive ability is more important than is positive affect"* (1983, p. 388, italicized in original).

The primary problem with the uncritical acceptance of Steinkamp

Figure 6.3   Simple Model, with Average Correlations from Steinkamp and Maehr (1983)

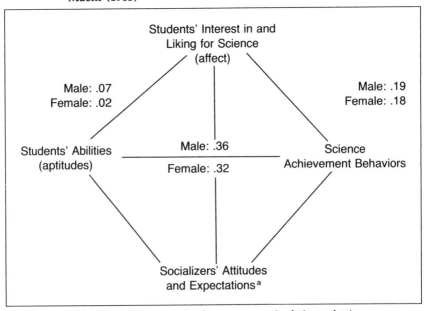

[a]Steinkamp and Maehr did not consider this component in their synthesis.

and Maehr's conclusions is that their analyses separately examined the interrelationships of ability, affect, and achievement. Other predictors or correlates of achievement were ignored (whether they were studied or not). Many kinds of ability, achievement, and affective measures were considered equivalent.[3] Nonetheless, the model in Figure 6.3 has an appealing simplicity.

Several questions must be asked about Steinkamp and Maehr's model. Does a model with only these broad components adequately describe the constructs involved in science achievement? Is it necessary to distinguish among the several varieties of aptitude, affect, and achievement behaviors that appear in the research literature in order to understand associations among them? Similarly, does a single model (regardless of the number of its components) describe the interrelationships for males and females? Essentially, is the conceptualization of relationships shown in their model a good one?

The three components in Steinkamp and Maehr's model represent achievement and two broad *psychological* factors. One way of elaborating the model slightly is to acknowledge the importance of aspects of the social context of achievement. The addition of the fourth component, dealing with the influence of socializers and social factors on achievement, is supported by research on the roles of the home environment (e.g., Bridgeman, Oliver, and Simpson 1985), social climate (e.g., †Anderson 1969a), and teacher characteristics (e.g., Rothman, Welch, and Walberg 1969). An even more complicated model is described below.

A critical evaluation of Steinkamp and Maehr's empirically derived model should also consider the quality of their data and how they were analyzed. Since the publication of Steinkamp and Maehr's work more relevant studies have been published[4] and some questions have been raised about their initial set of studies (Becker 1989a). One question to ask is whether this simple model still appears adequate in light of the data available today.

Another question about Steinkamp and Maehr's results concerns their

[3]Steinkamp and Maehr's statistical tests indicated the measures could be considered equivalent. However, the tests they used are problematic (see, e.g., Hedges 1986), calling their results into question.
[4]The fact that more research has been produced is not a criticism of the work of Steinkamp and Maehr. However, since their reviews were done schools have implemented explicit efforts toward gender equity (e.g., Pennsylvania State Department of Education 1984), and the perceived importance of science has grown for all students. Researchers can now ask whether the interrelationships under consideration appear to be, or actually are, different than they were in the past.

analyses. Three sets of correlations, pertaining to the three bivariate relationships among ability, affect, and achievement, were analyzed. Dependencies arising when several correlations came from individual studies were ignored. The review's conclusions were based on a simplification of a complex multivariate system of interrelationships. From Steinkamp and Maehr's analyses it is impossible to evaluate whether this simplification is justifiable.

Eccles' Model of Social and Psychological Factors in Achievement Behaviors

A much more complex idealization of the process by which individuals achieve (across domains) has been proposed and studied by Eccles and her colleagues (e.g., Eccles et al. 1983; Meece et al. 1982). This research has focused primarily on the use of the model of academic choice to describe the development of achievement behaviors in mathematics. It has not been applied to the study of science achievement for males and females separately. Its form, however, is general enough that it can be easily applied to science achievement.

Meece et al. (1982) described both the model, shown in Figure 6.4, and research on sex effects in mathematics achievement which is relevant to the model. Most of the research asked whether gender differences existed for the various components in the model. Thus the justification for using the model to examine gender differences in mathematics achievement appears to require the assumption of equivalent relationships across gender detailed above. With few exceptions (see Meece et al. 1982, pp. 326–327, 330) the researchers assumed the presence of common relationships or paths between model components for males and females.

Similarly, other studies of the model (e.g., Eccles et al. 1983) show analyses in which sex functions as a predictor in causal-analysis models. Eccles and her coworkers have, however, also looked for sex differences in correlations in their analyses.

Several important contrasts exist between Eccles' model and the simpler model derived from Steinkamp and Maehr (1983). Eccles' model is theoretical and is derived from a different research literature than that under review. Steinkamp and Maehr empirically derived their model from studies of science and gender; thus it is more likely that evidence exists about the Steinkamp and Maehr model than about the Eccles model.

Also, Steinkamp and Maehr's model is simpler. It may appear to be better understood (i.e., better studied) at the cost of being an oversim-

Figure 6.4 General Model of Academic Choice

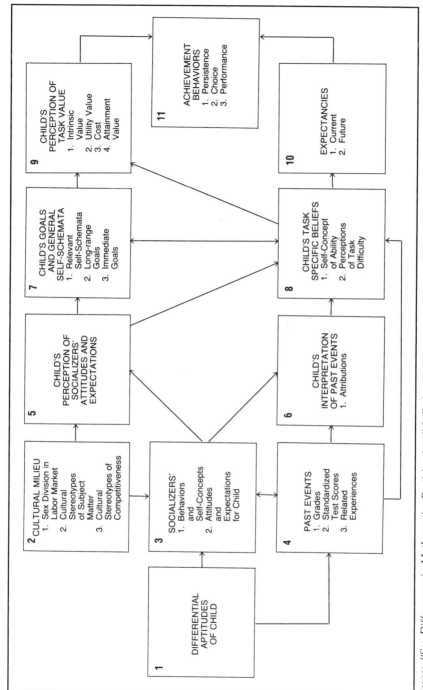

Source: "Sex Differences in Mathematics Participation," J. Eccles (Parsons). In Steinkamp, W., and M. L. Maehr, eds. *Advances in Achievement and Motivation, Vol. 2.* Greenwich, CT: JAI Press, 1984, pp. 93–137.

plification. One obvious omission from Steinkamp and Maehr's original model was any representation of the context of science achievement. The augmented simple model contains a broad category for social factors, but Eccles' model identifies specific social influences on achievement.

One benefit of this added complexity is that it suggests specific avenues of intervention or influence on students' achievement. For example, one facet of Eccles' second "cultural milieu" component is the nature of stereotypes of subject matter (here, science or scientists). If science stereotypes play a role in student achievement, teachers and parents can perhaps work to change negative stereotypes that exist.

The model also provides "hypotheses" about the complex mechanisms underlying the development of achievement behaviors. Proposed roles of intervening variables suggest that simple cause-effect relationships are less than likely. For instance, the "child's perceptions of task value" component moderates the effects of both the general goals of the student and the student's task-specific beliefs on achievement. This suggests that even if a student's goals involve science achievement and his or her science self-concept is positive, if there is too high a cost or insufficient value in achieving in science, the student may choose not to strive for success. A program that changes only goals, or science self-concept, might still not produce achievement if this model is accurate.

Finally, the paths in Steinkamp and Maehr's model connect all components to one another.[5] However, in the model from Eccles' work the connections are more often indirect. For example, the relationship between aptitude and achievement is indirect in Eccles' model. The nature of the paths has consequences for the assessment of evidence about the models, as will be seen below.

Relationships Between the Two Models

The model proposed by Steinkamp and Maehr (1983) included three components labeled aptitudes, affect, and achievement. In Eccles' model there are many more components. The literature does not relate these models to one another, thus leaving many possible ways of conceiving of their interrelationship.

One idea is to think of the simple model derived from Steinkamp and Maehr's work as a "submodel" of the individual components from

---

[5]The issue of directionality in these paths is complex in meta-analysis. Temporal precedence cannot usually be assumed for the different components, either within or across studies. See Becker (1989b) for more discussion of this issue.

Eccles' model. The question then becomes one of determining which four components from Eccles' model represent the components in the simple model. The aptitudes and achievement components are clear, but the choice of single components to represent affective factors and socialization effects is not as obvious. For analyses based on this view of how the models are related, the components labeled "child's perception of task value" and "socializers' behaviors and attitudes" were used as the affective and social components, respectively, of the simple model.

A second conception of the simple model views it as a distillation of the more complex model. In this framework the components of the simpler model may relate to either single components from the more complex model or to collections of those components. Analyses based on this view of the relationship between the models represented the affective factors (in the simple model) as components 7–9 from Eccles' model: the child's goals and self-schemata, task-specific beliefs, and perception of task value. The social factors were considered to be components 2 and 3: cultural milieu and socializers' behaviors and attitudes. Aptitudes and achievement were taken singly, as above.

This view of the simple model is shown in Figure 6.5. Several components from Eccles' model do not appear in this model. Those for past

Figure 6.5   The Simple Model as a Distillation of Eccles' Model

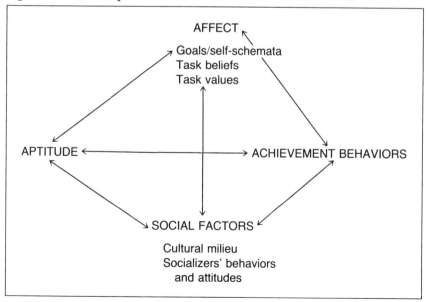

events, child's interpretation of past events, and demographics do not appear, thus implicitly are considered irrelevant.

Overview

In the current synthesis of science-related variables the simple model and Eccles' model of academic choice were used to inform the synthesis process and analyses. Data-collection strategies were designed to consider components in the two models, as discussed below. Many studies identified through these model-driven data-collection procedures examined only gender differences (in average performance) on model components. Studies of the paths or relationships among model components were the focus of the synthesis.

Data-evaluation procedures (described below) also focused on the two models. Each variable studied in the collected literature was examined, and almost all were classified into or associated with a component in the Eccles' model.

Although they were developed independently, the models are not completely unrelated. Eccles' model might be considered a richer elaboration of the simple model from Steinkamp and Maehr. Grouping or eliminating the components in Eccles' model can produce a smaller collection of components similar to those in the simple model. Analyses which compare the models are provided below.

Methods

Data Collection

TARGET STUDIES.   The target collection of studies was conceptualized to include studies of relationships among all components of the Steinkamp and Maehr and Eccles models. Studies were required to have measured at least one variable that was related to "school science" and that appeared in the Eccles or Steinkamp and Maehr models. School science was defined to include the general areas of physical and life sciences studied in most primary and secondary curricula (general science, biology, chemistry, and physics), as well as common variations of and specializations within those fields (e.g., physical sciences, electrostatics, earth science, biochemistry). Fields such as mathematics, computer science, and social science were excluded, as were all areas of engineering.

Each study must have provided data for males and for females on at least one interrelationship involving a science-related variable. (Not all variables in the two models are limited to the science context.) That is, the study need not have considered science *achievement* as an outcome. Other science-related variables such as science attitudes or parental encouragement to pursue science could have been examined. Results had to be presented separately for males and for females, regardless of what statistical analyses were used.

Both achievement as measured by grades or achievement tests (for example) and persistence were considered school-science achievement behaviors. Persistence was measured as continued course-taking or as choice of a science college major. Persistence was considered an achievement behavior since a student cannot achieve or learn in science classes if he or she does not first enroll in a science course (see, e.g., DeBoer 1984).

The search was also limited on several specific study features, including publication date and school level of subjects, as described below.

PUBLICATION DATE. Studies published prior to 1965 were not included in the synthesis. Steinkamp and Maehr (1983, 1984) used 1965 as the cut-off date in their earlier reviews of research on gender differences in science achievement. Since it appears that the bulk of studies on gender and science have been conducted within the past decade, it is unlikely that many additional studies would be found prior to 1965.

SCHOOL LEVEL. The school level of subjects was also used to delineate the target collection of studies. Studies of preschool, graduate, and postgraduate students and studies of science teachers were excluded. Studies of preschool science are quite rare and often have focused primarily on "science-related" aptitude measures, such as Piagetian tasks (e.g., Nelson 1976). Measures of achievement at graduate and postgraduate levels seem qualitatively different from the more typical (and more structured) achievement tests used at primary and secondary levels. Studies of college science courses were included because most involve the more structured settings and outcomes common to lower grades and because most college students have not made the kind of decision to study science that is required of graduate students in science. Gender differences in level and prediction of science achievement are likely to be quite different for graduate and postgraduate science, if only because of the restricted range of performance observed.

SEARCH PROCEDURES. Most of the documents were obtained through three computerized database searches. Additional sources included past reviews on related topics and the scanning of relevant journals, dissertation catalogs (University Microfilms International 1987), and reference lists. Information was also obtained from the Center for the Advancement of Science, Engineering, and Technology (CASET) and the American Association for the Advancement of Science (AAAS). (A more detailed discussion of the search procedures is available from the author upon request.)

Table 6.1 summarizes information concerning the strategies used to identify the literature on gender and science. The Educational Resources Information Center (ERIC) database, the *Psychological Abstracts* (PSYC) database, and the *Dissertation Abstracts* (DISS) database provided 346 documents, nearly three-fourths of the obtained documents.

From earlier syntheses of studies on gender differences in science achievement, 110 documents had been assembled (Steinkamp and Maehr 1983, 1984). Over 100 other sources were identified by scanning journals and dissertation catalogs and by examining the reference lists of sources in the collected literature.

More than 70 documents initially thought to be relevant to the issue of gender and school science were eliminated from the collection. Many studies either did not examine school science outcomes or did not use students as subjects (e.g., Bernard 1979). Several studies focused on science career achievement rather than school achievement. Several dissertations did not report data separately by gender, and other documents were not empirical.

STUDIES OF GENDER AND SCIENCE. The initial 522 documents relevant to gender effects in school science included 64 dissertations, 24 books,

Table 6.1 Numbers of Documents Identified by Different Search Procedures

Source	Number	
	Identified	Obtained
Computerized Databases		
Educational Resources Information Center	626	177
Psychological Abstracts	534	156
Dissertation Abstracts	25	13
Past Reviews	—	110
Journals	53	35
Dissertation Catalogs	—	11
Reference Lists	117	54
Other Sources	—	39

Table 6.2   Types of Documents in Literature on Science and Gender

Type	Number
Policy and Program Documents	60
Other Related Documents (e.g., books, discussions of models)	69
Empirical Studies	
Studies of models (by gender)	41
Correlational studies (by gender)	32
Studies of gender differences	320
Total	522

and 434 articles, book chapters, and unpublished documents. Most of these sources, however, did not bear directly on the question of gender differences in relationships among science variables.

To identify those studies which provided empirical data on interrelationships by sex, the collected documents were read or skimmed by at least two people. Documents were classified into five categories, as shown in Table 6.2.

Documents discussing policies regarding gender equity in the sciences (including textbook bias), women employed in science, and other nonempirical documents were included in the initial collection of studies. A total of 393 empirical documents with the potential of providing information on gender and school science remained after all other sources were set aside.

The empirical studies of interest in the present synthesis are from the two categories labeled studies of models and correlational studies. Studies of models usually examined regression models for male and female students, though some more complex analyses were also presented.

STUDIES OF GENDER DIFFERENCES IN RELATIONSHIPS.   The 73 studies of interrelationships by gender included 15 studies which had examined interrelationships for only one gender (e.g., Astin 1968). Although it will eventually be interesting to try to incorporate those studies into this synthesis, these 15 studies were not included in this initial set of analyses. Thus the initial set of documents which had presented data on relationships separately for both sexes included 58 documents.

Six of the 58 documents were eliminated from the analysis because the outcomes studied did not conform to our definition of school science. Brewer and Blum (1979), Dunteman, Wisenbaker, and Taylor (1979)

Matyas (1985, 1986), and Ware, Steckler, and Leserman (1985) had examined enrollment in "science" courses or choice of "science" major, but both science and *math* courses (or majors) were considered to be "science." Kelly and Weinreich-Haste (1979) had examined student interest on a continuum labeled "arts-sciences." This did not seem to be a clearly scientific outcome, thus the study was also omitted. One study by Cline, Richards, and Needham (1963) that was included by Steinkamp and Maehr (1983) had been published prior to 1965.

The 51 relevant remaining sources are listed at the end of this chapter. Several authors had contributed more than one document to the set of 51 relevant sources. Inspection of abstracts and sample descriptions for these documents indicated that in all but one case the same samples were featured in the multiple reports. The documents for each of these seven cases were considered one "study" for the purpose of analyses. Altogether, 19 documents were interdependent, reducing the total number of independent sources from 51 to 39. These 39 sources were coded and considered for analysis.

Coding

DEVELOPMENT OF CODING SHEET AND CODEBOOK. The coding sheet and codebook for this analysis were extensions of the coding sheet from a previous, smaller scale meta-analysis on gender differences in science achievement (Becker 1989a). The final versions of the coding sheet and codebook contained 14 categories of variables, which are listed in Table 6.3. The coding sheet, codebook, and a description of training procedures are available from the author.

VARIABLES CODED. Although the complete coding included large amounts of information extracted from studies, a subset of variables was used in this analysis. Those variables included characteristics of the document or source itself, characteristics of the author(s) and subjects, characteristics of outcomes and measures, and results.

Each document was read and coded by two separate coders. Differences were identified and resolved by a third coder. In all, five persons served as coders. Reliabilities of coded variables ranged from a low of 59.4 percent (for author status) to 100 percent for source of article (e.g., journal versus book) and author gender. Reliabilities of variables representing information about tests and measure characteristics ranged from 78 to 99 percent.

Table 6.3   Types of Variables Coded in the Synthesis

Article Characteristics
Author Characteristics
Subject Characteristics
Time of Study
School/Classroom Characteristics
Special Characteristics of Design/Study
Teacher Characteristics
Family/Parent Influence/Socializers
Outcomes/Measures
Type of Study
Results
Quantitative Assessments
Notes
Abstract

CODING OF RESULTS.   Results of the 39 studies were coded first by the principal investigator, who recorded any information presented regarding the interrelationships of interest. Although many studies had reported Pearson's zero-order product-moment correlation coefficient *(r)*, some studies presented partial correlations (e.g., †Smail and Kelly 1984b), while others presented multiple correlations and other regression-related results.

Seven studies which used more complex correlational analyses (e.g., discriminant analysis, multiple regression) and which did not present sufficient information to allow the computation of zero-order correlation coefficients were withheld from the initial analyses reported here. Thus results of 32 studies were eventually included in the data analysis.

Some sources contained correlations which were excluded from the analysis. Some of the correlations did not involve science-related variables and thus were irrelevant. This occurred especially when complete correlation matrices were reported (e.g., †Smith 1966). Two studies (†Gilmartin et al. 1976; †Haladyna, Olsen, and Shaughnessy 1982) presented very large numbers of relevant correlations. In each case the correlations related many different predictors to a single outcome. For these and the studies of †Peng and Jaffe (1979) and †Van Harlingen (1981), a reduced set of correlations was included in the analyses. The sets were formed by selecting total scores rather than subtest scores (as in †Van Harlingen) or by selecting a representative set of variables (e.g., SES was represented by four different variables in †Peng and Jaffe). Theoretically all relevant correlations, even if somewhat redundant, could be included in the synthesis.

A second coder examined the 32 studies and recorded relevant correlations. All correlations, including those omitted from the initial analyses, were recorded. Agreement of values across the 223 pairs of correlations recorded by the principal investigator was 98.7 percent.

MODEL-RELATED CODING.   Evaluation of the two models discussed above required the determination of which correlations were relevant to each path in each model. This was accomplished by categorizing each of the 192 measures used in the 32 studies into one of 12 classes. The classes corresponded to the (numbered) components in Eccles' model in Figure 6.4, and one additional group of demographic measures.

The classification required the coders to make a "forced choice" among the available categories. That is, each measure was placed into the category that best matched its definition. For example, most science-achievement tests and counts of science courses taken by students were considered science-achievement behaviors (component 11). Science self-concept measures (e.g., †Welch, Rakow, and Harris 1984) were considered task-specific beliefs (component 8), whereas general academic self-concept measures (e.g., †Handley and Morse 1984) were categorized as general self-schemata (component 7).

There are two serious drawbacks to this procedure. One is that many studies did not provide enough information about the measures to ensure a good basis for such judgments. Another is that some of the 192 measures may not actually fit into Eccles' model (which does not claim to be inclusive of *all* important predictors). Thus alternative categorization schemes must be considered carefully as assessment of these (and other) models continues.[6]

Judgments about the component measured by each of the variables were made by pairs of coders, and all discrepancies were resolved. The classification was done in two steps. A first set of 156 variables was coded, and the pairs of coders agreed on 81 percent of classifications (126 of 156 decisions) before discrepancies were resolved. Additionally, 9 of the 30 discrepancies arose because of a single decision (about how to classify a test used by †Linn and Pulos 1983). Thus the interrater agreement on classifications was reasonably high. A set of rules was

---

[6]The task of determining which studies actually examine paths in particular models may best be informed through consultation with researchers who study those models.

developed which assigned model-component codes according to the type of variable that had been studied. Later an additional 36 variables were classified using these same rules and were independently judged as well. All classifications and independent judgments agreed for these 36 variables.

Table 6.4 shows the number of variables judged to be relevant to each of the components in Eccles' model. Additionally the topics examined by measures of the different components are shown.

All but one of the demographic measures tapped socioeconomic or occupational status, and one measure related to socializers' influence (from †Bridgham 1969) was an indicator of whether the subjects' fathers were employed in scientific careers (and thus might serve as career role models). The cultural context measures were all scales which probed subjects' views of scientists, including several scales regarding male and female roles in science (†Handley and Morse 1984; †Welch, Rakow, and Harris 1984).

Aptitude measures most often tapped inquiry and general reasoning skills, but verbal and number abilities also were measured (e.g., †Grobman's Differential Aptitude Test (1965)). Very few instruments were classified as measures of science aptitude. Exceptions include †Gilmartin and his colleagues' Scientific Potential Index (1976), †Jensen's Iowa Placement Examination in Chemistry (1966), and †Linn and Pulos' Piagetian science reasoning measure, Predicting Displaced Volume (1983).

Achievement measures were classified as either "past events" or "achievement behaviors," depending on the time at which the measures were administered. One exception was a set of largely nonscientific achievement scales, administered by †Roberts (1965) to a sample of National Merit Scholars. These and the corresponding "criterion" measures were all administered at one time, but the nonscience measures were considered "past events," thus as potential predictors of science achievement.

Measures of the other five components in Eccles' model covered a variety of topics. †Ormerod's three scales for liking of science teachers (1975) were classified as "perception of socializers." Measures of masculinity and femininity (†Handley and Morse 1984), of career preferences and objectives (e.g., †Erb and Smith 1984), and various measures of life goals from †Peng and Jaffe (1979) were subsumed under "goals and general self-schemata."

Science-related beliefs, interests, and values are included in the components labeled "perception of task value" and "task-specific beliefs." Many science attitude scales fell into the prior category, including the

Table 6.4 Topics Examined by Measures of Components in Eccles' Model

Component	Affect	SES/Occupation	Age	Father in Science Occupation	Views of Scientists
Demographics		5	1		
Culture				1	10
Socializers	1			1	

Component	Art/Music	Science Topics	General	Inquiry/Reasoning	Math	General/Science	Social/Interpersonal	Verbal
Abilities	3	6	6	35	6	1	1	10
Past Events		5	5		3	5		3
Achievement Behaviors	2	13		2		17	1	2

Component	Affect	Art/Music	Science Topics	Education	Inquiry/Reasoning	Math	Non-academic	General/Science	Social/Interpersonal	Scientists	Verbal
Perception of Socializers			3					1			
Self-Schemata	2			2	3				2	4	
Task-Specific Beliefs	1			2				2			
Task Value	2	2	8		1	2	3	14	2	1	1
Expectancies			1								

Note: "Science topics" measures include those related to biology, chemistry, physics, and other specific topics while "general/science" measures include nonspecified science measures and measures labeled general science.

scientific scale of the Kuder General Interest Survey (†Cohen 1979) and the Kuder Preference Record (†Jensen 1966). Many of the science attitude measures were experimenter-made (as detailed below); examples include †Erb and Smith's Image of Science and Scientists scales (1984) and †Neale, Gill, and Tismer's semantic-differential scale for science attitudes (1970).

Several measures of attitudes and interest in nonscientific areas were also classified under these two labels. An alternative approach would have been to include them as part of the more general "goals and general self-schemata." However, because we considered that component to be broad rather than specific we classified all specific interest and attitude measures under the more limited task-value label. Analysis of the correlational results may suggest whether this was a reasonable approach.

Finally, only one study of expectancies was included in our review, that by †Weimer (1985), which asked students about their expected future academic success in science.

Data Analysis

The correlations from the 32 studies which had examined interrelationships for both male and female subjects were analyzed using two approaches, an analysis of individual paths and a generalized least squares regression analysis. Both approaches required that certain assumptions be made regarding the role and structure of interrelationships among study results.

Neither approach is optimal because both require the acceptance of assumptions that are or appear to be untenable. However, the optimal hierarchical analysis is computationally complex and could not be accomplished using currently available computing software.

HIERARCHICAL ANALYSIS.   The structure of the data suggests that their analysis should account for several sources of sampling and parameter variation, both between and within studies. A hierarchical analysis, such as that proposed by Raudenbush, Bryk, and others (e.g., Raudenbush 1988; Raudenbush and Bryk 1985), would be appropriate.

Between-studies variation, due to differences in study design, methodology, and other study-level features, is of first importance in explaining differences in sample (and population) correlation coefficients. Two sources of within-study variation and covariation are critical in this synthesis. The first source of variation arises because different relationships have been studied. If a study has measured three or more relevant variables, two or more correlations may be available for each sex. Differences in the values of those correlations may relate to characteristics of the variables studied. A second source of within-study variation is the gender effect. The samples of males and females within each study will be similar in many ways, because of common characteristics as well as common study methodology.

Current software for hierarchical data analysis requires that data follow a strict multivariate structure. Each study must have the same number of correlations, and all studies must have examined the same interrelationships. For example, if all studies had examined only the correlations between spatial aptitude and achievement and between verbal aptitude and achievement, the conditions would be met. However, neither of these conditions obtain in this data set, thus the hierarchical analysis can not be completed at this time.

ANALYSIS OF INDIVIDUAL PATHS.   One approach used to analyze the correlational data in this synthesis was to conduct a number of univariate analyses of correlations representing the individual paths in the models, assuming independence among all the outcomes. This assumption is incorrect, because of dependencies among correlations both between and within paths.

Twenty-four of 32 studies contributed more than one relevant correlation per sample. In many cases the correlations were somewhat redundant, as mentioned above. Intercorrelations among rs for similar relationships (i.e., for rs from the same paths) tend to be relatively high. For example, †Peng and Jaffe (1979) related a series of aptitude measures (and a set of measures of life goals) to science course-taking. Correlations among the rs relating aptitudes to course-taking ranged from .39 to .64 for males (median = .52) and .42 to .67 for females (median = .58).

Intercorrelations among rs across all studies ranged from small negative values to large positive intercorrelations among rs for similar relationships (e.g., $r = .67$ between the correlations of physical science achievement with English and with math achievement in †Marjoribanks 1976). Even rs representing different paths may be correlated when they arise from the same sample. Thus though some studies may have correlations that are essentially independent, others have highly correlated outcomes, which may heavily influence the review conclusions.

Although it is less than optimal because these dependencies are ignored, the analysis of individual "univariate" outcomes has been used in many meta-analyses (e.g., Giaconia and Hedges 1982; Hyde 1981). Premack and Hunter (1988) used this approach in examining research on the unionization process, in a synthesis reflecting on a theoretical model of that process.

The advantage of this approach is that it does not require any data beyond the correlations to be synthesized. All of the 446 zero-order correlation coefficients retrieved from the 32 studies were included in this analysis.

GENERALIZED LEAST SQUARES ANALYSIS.   One alternative approach which accounts for *within-study* covariation is the generalized least squares (GLS) analysis. This approach explicitly models the within-study dependencies which arise because of the interrelated correlations. However, covariation between results for males and females which may arise across studies is ignored with this approach.

An analysis based on the approach suggested by Raudenbush, Becker, and Kalaian (1988) was used to examine the results from 23 of the 32 studies. This approach requires the estimation of within-study covariances for each study's results. The normal approximation to the distribution of a vector of correlations (Olkin and Siotani 1976) was used to estimate the variance-covariance matrix for the correlations from each study.

GLS Model for Correlations.   Notation is needed to describe the model used to analyze the correlations via GLS regression. Let r_{ij} be the jth correlation from study i, with a corresponding population correlation ρ_{ij}. Let k be the number of studies and m_i be the number of correlations in study i. Then $\mathbf{r}$ and $\boldsymbol{\rho}$ represent column vectors of these sample and population values; that is,

$$\mathbf{r}' = (r_{11}\ r_{12}\ .\ .\ .\ r_{1m_1}\ .\ .\ .\ r_{k1}\ .\ .\ .\ r_{km_k})$$

and

$$\boldsymbol{\rho}' = (\rho_{11}\ \rho_{12}\ .\ .\ .\ \rho_{km_k}).$$

The total number of correlations is

$$m = \sum_{i=1}^{k} m_i.$$

The model for the GLS analysis is

$$\rho = X\beta + \epsilon,$$

where **X** is a $m \times p$ matrix of predictors such as study characteristics and features of the correlations and the measures they interrelate, β is a vector of p regression coefficients, and ϵ is an $m \times 1$ vector of errors.

The matrix **X** is very important in the GLS analysis. Through **X** a multitude of different possible models can be designed to "explain" or account for variation in the values of the population correlations. In the present analyses **X** will be used to specify regression models that are analogous to the theoretical and empirical models presented by Eccles and by Steinkamp and Maehr. Other predictors can be added in order to account for variability that remains after features of the models themselves have been examined.

This brief illustration shows how the predictor matrix **X** can be formulated to represent the components in a very simple model with two paths. Each correlation in the data set represents a link between two of the components in Eccles' model (or between demographic information and a component).

Suppose that we are interested in examining a model which suggests that two factors are critical in predicting science achievement: aptitudes and any affective factors (attitudes, self-concept, and so on). The model suggests that there are three kinds of relationships with achievement: aptitudes-achievement, affect-achievement, and other (unimportant) relationships. Consequently there are three kinds of correlation.

Table 6.5 shows that one column (X_1) represents a grand mean, while X_2 and X_3 are dummy variables which take on the value 1 when a correlation represents the specific relationship. (Only two dummy variables are needed to specify three groups.) A regression model based on X_1 through X_3 would provide estimates of an average aptitude-

Table 6.5 Construction of X for Two-Path Model

Relationship (Path) Represented by r	Values of Elements of the X Matrix for						
	Grand Mean and Paths			Interactions with Sex			
				Males		Females	
	X_1	X_2	X_3	X_4	X_5	X_4	X_5
Aptitude-Achievement	1	1	0	0	0	1	0
Affect-Achievement	1	0	1	0	0	0	1
Other	1	0	0	0	0	0	0

achievement correlation, an average affect-achievement correlation, and the average of all remaining correlations.

In order to model gender differences in these interrelationships it is necessary to include interaction terms. The variables X_4 and X_5 are computed as the product of a dummy variable for gender (coded 0 for males and 1 for females) multiplied by each of X_2 and X_3, respectively. If the regression coefficient b_4 (for example) differs significantly from zero, the average correlation of aptitude with achievement is likely to be different for males and females. Estimates of the correlations representing the two paths in the model can be obtained as predicted values ($\hat{\rho}_{ij}s$) from the regression model. Separate overall means for males and females are not included in this model, though it would be possible to do so.

Approximate Distribution for Correlation Coefficients.   Correlations within each study were assumed to be independent if they arose from separate samples, but dependencies among multiple correlations from single samples were modeled explicitly. The set of correlations from each study was considered to be a vector, and the distribution of each study's correlation vector was estimated using results from Olkin and Siotani (1976). Covariances among dependent correlations were computed using the PlanPerfect spreadsheet and were then used to weight the correlational results in the GLS regression analysis, which was computed using Statistical Analysis System (SAS Institute 1979) Proc Matrix.

The computation of covariances among correlations requires not only the correlations which are related, but also other correlations from the study's correlation matrix. In the simplest case, when two correlations from one study share one index, their asymptotic covariance is

$$\text{Cov}(r_{st}, r_{su}) = [0.5(2\rho_{tu} - \rho_{st}\,\rho_{su}) \times (1 - \rho_{st}^2 - \rho_{su}^2 - \rho_{tu}^2) + \rho_{tu}^3]/n,$$

where n is the sample size (Olkin and Siotani 1976, p. 238). Thus one needs an estimate of ρ_{tu} to compute the covariance between r_{st} and r_{su}. If the related correlations do not share an index the covariance formula is even more complex.

These complicated formulas lead to the biggest difficulty in applying the GLS approach: Data are often missing. Eleven studies of the 32 in this collection had not reported the data needed to estimate covariances (by sex) among the correlations of interest. These studies were omitted from the initial GLS analyses, with two exceptions ([†]Kaminski and Erickson 1979; [†]Schock 1973),which were included to illustrate how one might include studies with missing data.

Ideally, one might estimate or impute values for the missing correlations and then compute covariances using the imputed values. Although in theory the idea is relatively straightforward, it is difficult to put into practice. The underlying question is what values are "reasonable" values to impute for the missing correlations. Many sophisticated methods exist for data imputation (e.g., Rubin 1987) which could be applied in this situation. Similarly, it may be possible to rely on other sources of information, such as test manuals or other research studies, to develop "reasonable" values.

A principle behind most imputation is to obtain the imputed values from cases (here, studies) which are as similar as possible to the study with missing data. In meta-analysis, however, some studies will be unique. That is, it may be very difficult to find even one other study presenting a correlation between the same two variables measured by the missing correlation.

One such study in this synthesis was by †Haladyna, Olsen, and Shaughnessy (1982). Science attitude was correlated with 39 different variables, characterizing the teacher, student, and learning environment for a group of seventh through ninth graders. Only four of the 39 relationships were included in this synthesis. Science attitude, the importance of science, and teacher support for the individual were not studied in any of the other sources in our collection. SES and science self-concept were studied elsewhere. Each predictor had been conceptualized with considerable specificity. This suggested that the substitution of a correlation representing an overall "environment-attitude" relationship would not acknowledge differences between these variables. A search of other work by these authors did not provide additional data (and the authors indicated that their raw data are no longer available).

Another possible approach would be to look for an estimate of the interrelationship for a combined population in other research on science attitudes, but research which does not examine gender differences. This approach was used to estimate the covariances among the correlations from †Kaminski and Erickson (1979). The imputation of values for †Kaminski and Erickson (1979) and for †Schock (1973) is described in Appendix 6.A.

Results

Description of Studies

SOURCES.   The 32 studies in the analysis were published or completed between 1965 and 1985, with more than two-fifths of the studies and more than one-third of the correlations from the 1980s. These documents were primarily published articles (20, or 63 percent), with 6 dissertations and 6 ERIC documents constituting the remainder. All authors were male for 18 studies (56 percent), while all authors were female for only 6 studies.

Thirty of the 32 studies were judged to have a major focus on gender, which is not surprising. Purposeful presentation of detailed regression or correlational analyses by gender most likely indicates an interest that guided the authors' analyses from the start.

SAMPLES.   Correlations for 38 independent samples were extracted from the 32 sources. Thirty-three samples (87 percent) involved North American subjects; the remaining five were of British students. Three samples were of volunteers, while 24 (or 63 percent) were required to participate or did so unknowingly. This information was unavailable for 11 other samples.

The bulk of the studies (21, or 66 percent) had used convenience samples. The probability samples were often from large-scale surveys such as Project Talent (e.g., †Gilmartin et al. 1976). The average age of the subjects was slightly more than 14 years, and the average grade was ninth grade. Correspondingly, 27 samples were of junior high or high school students. Table 6.6 shows the counts of samples at four

Table 6.6   Characteristics of Samples

Characteristic	Frequency
School Level	
Elementary	5
Middle/junior high	12
Secondary	15
College	6
Presence of Attrition or Selection	
Attrition	12
Selection	4
Attrition and selection	4
Neither	3
No information	15

different school levels, as well as information on sample selection. Over half of the samples showed attrition or selection. The total number of subjects from the 38 samples included 19,785 males and 18,770 females.

MEASURES.   Table 6.7 describes the types of measures used in the 32 studies. Of the 192 different measures used, 93 were tests, 86 were self-reports, and 13 were based on transcripts (e.g., GPAs, course-taking). Aptitude and achievement measures constituted over half (52 percent) of the measure types.

The majority of the measures (87 percent) were either standardized (36 percent), experimenter-made (33 percent), or research-based (18 percent). (Measures developed for the purpose of other research were classed as research-based instruments.) The mean number of items (NITEMS) per measure was about 39 (based on the 136 measures for which NITEMS was reported). This varied by type of measure, however, as is shown in Table 6.7.

Table 6.8 shows the methods of construction for the achievement, aptitude, and attitude and interest measures. Methods of construction varied for the different types of measures. Most aptitude measures were standardized, while most attitude and interest measures and other measures were research instruments. Achievement was measured with standardized and research instruments and teacher-made tests and grades.

The 38 achievement measures were mainly tests, with 3 self-reports and 7 measures based on transcripts. About one-third of the achievement tests were labeled only as science measures, and most of the rest

Table 6.7   Measures of Science-Related Variables

Type of Measure	Frequency	Percent	Median Numbers of Items
Aptitude	61	32	29
Achievement	38	20	45
Science Course-Taking	5	3	1
Attitude	24	13	8
Interest	25	13	33
Self-Concept	7	4	7
SES	6	3	1
Other	26	14	—

Note: "Other" includes such variables as activities, goals, number of math courses, and test anxiety. These variables occurred no more than four times each. No median number of items is reported across these measure types.

Table 6.8 **Methods of Construction of Different Measure Types**

	Frequencies			
Method of Construction	Achievement	Aptitude	Attitude/ Interest	Other
Research Measures				
Experimenter-made	6	12	19	26
Research-based	4	7	15	9
Standardized Measures				
Standardized	12	36	15	6
Curriculum-based	4	1	—	—
School Measures				
Teacher-made	9	—	—	—
School records	3	5	—	3
Total	38	61	49	44

were of achievement in specific science topics (biology, chemistry, general science, and physics). Table 6.9 shows the topics examined by each measure type.

The 61 aptitude measures were nearly all tests, with five measures based on student records. Twenty-two of these measures assessed in-

Table 6.9 **Topics Investigated with Different Measure Types**

	Frequencies		
Topics	Achievement	Aptitude	Attitude/Interest
Affect	—	—	1
Biology	4	1	2
Chemistry	3	2	2
General Intelligence	5	6	—
General Science	1	—	—
Inquiry	2	8	2
Math	2	6	—
Nonacademic Interests	—	—	5
Physics	6	—	4
Reasoning	—	14	2
Science	12	1	10
Scientists	—	—	15
Spatial	—	13	—
Verbal	3	10	1

Note: This table does not present a complete list of topics.

quiry and reasoning aptitude, while 29 others measured numerical, spatial, and verbal abilities.

The 49 attitude and interest measures were all self-reports from the subjects. The majority of these self-reports assessed interest and attitude toward science and scientists (25) or specific science topics such as biology, chemistry, and physics (8).

The remaining 44 measures included assessments of background characteristics (e.g., SES and age), numbers of science and math courses taken, test anxiety, self-concept, teacher support, and importance of science. Most of these measures were constructed by researchers or drawn from past research.

CORRELATIONS.   Four hundred forty-six correlations were retrieved from the 38 samples of males and females. In all but two studies the correlations represented the same interrelationships for males and females. Only †Baker (1981) and †Ignatz (1982) had presented different correlations for the sexes. Thus 221 relationships were measured by pairs of correlation values and four other interrelationships were documented by Baker and Ignatz.

Data Analysis

The analyses in this section examine the two models discussed in the first section of the chapter. Many other analyses could be conducted to examine different models or to examine other different subgroups of the correlations in relation to these two models.

Much can be learned about what is and is not known about the interrelationships among science-related variables by examining the numbers of correlations which bear upon the components in the two models. Because each measure had been categorized as representing one of the components in Eccles' model (plus demographics), each correlation represented a possible relationship among those components.

Table 6.10 presents counts of each of the correlations relevant to the different possible paths between 12 categories of variables. The table is in the form of a matrix, and the row and column labels represent the 11 components of Eccles' model plus demographics.

The matrix shows dramatically the information in these 32 studies. Most of the correlations in the matrix represent relationships involving measures of aptitude and of science achievement. Over half of the correlations (53 percent) involve an achievement measure. Science attitudes ("perception of task value") also appear to be thoroughly studied, with 160 correlations (80 pairs) representing paths between attitudes and other components.

Table 6.10 Number of Correlations for Paths in Simple Model

First Component	Second Component											
	Dem	Apt	Cul	Soc	Pas	Per	Int	Goa	Sel	Per	Exp	Ach
Demographics	0	4			2					6		6
Aptitudes		54		○	12			3		(26)		(100)
Culture			0		2			4	2	20		18
Socializers				0						(6)		(2)
Past Events					2			8		2		28
Perception of Socializers						0				16		
Interpretation of Past Events							0					
Goals/Self-Schemata								0				5
Self-Concept									0	14	2	12
Perception of Task Value										26		(44)
Expectancies											0	
Achievement												20

241

Those components dealing with socialization and the more psychological aspects of achievement, such as the student's attributions (component 6), goals and overall self-concept (component 7), and expectancies (component 10), are not well investigated in this literature.

More specific information could be obtained by further categorizing each of the sets of correlations. For instance, though achievement appears to be well studied, achievement as measured by course-taking is represented in only 7 of the 235 correlations involving achievement.

SIMPLE MODEL WITH PATHS BETWEEN INDIVIDUAL COMPONENTS.   The circles drawn around the counts in the matrix of Table 6.10 indicate paths in the simple model viewed as a submodel of Eccles' model, as discussed above. The simpler model has six possible paths, and Table 6.10 shows that five of those six paths among individual components have been studied. The paths connected to the "socializers" component (which was added to the original three-component model) have been studied less extensively; only eight correlations are on those paths.[7] The three other paths in the model all had more than ten pairs of correlations from these studies. The aptitudes-socializers relation did not appear in any study in this collection.

SIMPLE MODEL WITH COMPOSITE COMPONENTS.   Table 6.11 shows the counts of correlations between the composite components of the "distilled" version of the simple model. (Counts for paths not in the model are not shown.)

All of the paths that have been studied at all for this version of the model are represented by at least ten pairs of results. However, the role of social factors in the process still appears least well studied. The lowest numbers of correlations to achievement behaviors are for social factors, and relationships between social factors and aptitudes have not been studied.

ECCLES' MODEL.   Table 6.12 shows 18 circles for the paths outlined in the model in Figure 6.4. Only four of those paths are represented by studies in this collection. The complex model describing the whole process of boys' and girls' development of achievement behaviors *may* be more realistic than the simple model. However, it is hard to evaluate

---

[7]The inclusion of the additional relevant correlation coefficients from the four studies that produced these *rs*, especially from †Haladyna et al. (1982), would change this picture slightly. Several other "learning climate" variables may fit into this component of Eccles' model.

**Table 6.11 Number of Correlations for Paths in Simple Model
with Composite Components**

First Component	Aptitudes	Social Factors	Affect	Achievement Behaviors
		Second Component		
Aptitudes	54	0	29	100
Social Factors		0	32	20
Affect			40	61
Achievement Behaviors				20

whether that is the case because most paths in the model for science achievement have not been studied separately by sex. Because the model posits many indirect paths (between, for instance, aptitude and achievement) it appears not to be well understood. Below, the *results* of the existing research are investigated.

Analysis of Individual Paths

The analysis of the full set of 446 correlation values followed procedures outlined by Hedges and Olkin (1985). Correlations were transformed using Fisher's Z transformation, and averages were weighted by the inverses of within-sample error variances (i.e., $w = (n-3)$). Homogeneity tests and average correlations were computed for the groups of correlations representing the 24 possible paths enumerated in Table 6.10. Results of this analysis are shown only for Eccles' model and the simple model involving individual components (i.e., the simple model viewed as a subset of Eccles' model).

SIMPLE MODEL WITH PATHS BETWEEN INDIVIDUAL COMPONENTS. Table 6.13 shows the weighted average correlations and values of the homogeneity test H_T (Hedges and Olkin 1985) for males and females for five paths in this model. Under the null hypothesis of equal population ρ values, H_T is distributed as a chi-square statistic with degrees of freedom equal to one less than the number of correlations being combined. All of the tests displayed are significant, indicating considerable variation in the values of the correlations on each of the paths. The weighted mean r values must be considered as *average* rather than common correlation values.

Table 6.12 Number of Correlations for Paths in Eccles' Model

First Component							Second Component					
	Dem	Apt	Cul	Soc	Pas	Per	Int	Goa	Sel	Per	Exp	Ach
Demographics	0	4			2					6		6
Aptitudes		54	0	○	⑫			3	2	26		100
Culture				○	2	○		4		20		18
Socializers				0	○	○	○		○	6		2
Past Events					2		○	8		2		28
Perception of Socializers						0		○	○	16		
Interpretation of Past Events							0					
Goals/Self-Schemata								0	○	○		5
Self-Concept									0	⑭	②	12
Perception of Task Value										26		㊹
Expectancies											0	○
Achievement												20

244

Table 6.13   Analysis of Correlations in Simple Model

Path in Simple Model	Number of Pairs of Correlations	Males		Females	
		Mean Correlation	H_T	Mean Correlation	H_T
Aptitude–Task Value	13	.39	47.13	.28	23.68
Aptitude–Achievement	50	.33	725.06	.32	578.73
Task Value–Achievement	22	.16	187.60	.12	98.09
Socializers–Task Value	3	.28	14.62	.34	14.53
Socializers–Achievement	1	.18	0	.66	0

Table 6.14 displays stem-and-leaf diagrams, by gender, for three of the paths. The diagrams show the variation in correlations indicated by the H_T values and overall show great similarity between the sexes.

Aptitude–Task Value.   One exception to that similarity appears in the diagram for aptitude–task value correlations. Both the average values and median values reflect a stronger influence of aptitude on perception of task value for the males. The 95 percent confidence intervals for the weighted average correlations for females and males are from .24 to .31 and .36 to .42, respectively. The fact that these do not overlap indicates a significant gender difference.

The diagram in Table 6.14 shows two very low values for females. Even if these low values are "trimmed" the new median *r* for females is still only .30. A somewhat simplified interpretation of these results is that males who do well in science also like science and see its value, whereas this is not necessarily true for females.

Aptitude–Achievement.   The distributions of correlations between aptitude and achievement look similar for both sexes, and both cover a wide range. One indication of how similar the results are for the two sexes is the correlation between the *r*s themselves. The Pearson product-moment correlation for the 50 pairs of male and female *r*s for the aptitude-achievement relationship is .79. Although the 50 pairs of correlations are not completely independent (several studies contributed more than one pair of *r*s), this gives an approximate indication that the patterns of study results are fairly similar for the two sexes.

It is likely, however, that particular kinds of aptitudes relate differently to science achievement. When the correlations in this collection were grouped according to the eight kinds of aptitudes that were measured, some differences appeared. The highest correlations were be-

Table 6.14 Stem-and-Leaf Diagrams of Correlations
for Paths in Simple Model

Aptitude and Task Value

Males ($k = 13$)		Females ($k = 13$)	
.5	4 9	.5	
.4	0 4 5 6 6	.4	8
.3	7	.3	0 3 6 6 8
.2	3 5 6 7 7	.2	0 1 4 5 9
.1		.1	
+.0		+.0	8 9

median = .40   median = .29

Aptitude and Achievement

Males ($k = 50$)		Females ($k = 50$)	
.7	0 1 4	.7	
.6	3 3 4	.6	0 3 3 5 6 6 8 9
.5	1 1 7	.5	0 0 1 1
.4	1 1 2 2 3 3 7 7 8 8 9	.4	1 1 3 3 4 5 7 7 8
.3	0 1 2 3 4 6 8 9	.3	0 0 2 2 4 5 5 5 6 6 9 9
.2	3 4 5 7 8 8 8 9	.2	0 1 1 3 3 5 6 9
.1	0 2 2 2 4 4 6 7	.1	3 4 5 6 7
+.0	6 7 7 8 9	+.0	7 8
−.0	7	−.0	5 6
−.1		−.1	

median = .32   median = .36

Task Value and Achievement

Males ($k = 22$)		Females ($k = 22$)	
.6		.6	
.5	1 9	.5	
.4	0 6	.4	0 0 1
.3	0 2 3	.3	3 9
.2	0 2 4 8	.2	0 2 4 7
.1	1 8	.1	0 1 2 9
+.0	4 4 7 9	+.0	4 4 7 9
−.0	2 3 7	−.0	1 2 5 5 6
−.1	0 0	−.1	
−.2		−.2	

median = .19   median = .12

tween general intelligence measures and science achievement, with average correlations of .56 for both sexes (based on six results). Verbal aptitude and reasoning abilities (which are often highly verbal) showed moderate average correlations, ranging between .35 and .39. Mathematical and spatial aptitudes surprisingly showed even weaker relationships to achievement. The average correlation with science achievement of math-aptitude measures was only .24 for both sexes, while the average correlation of spatial aptitude with achievement was .22 for males and .28 for females.

All of these subsets of correlations, however, were quite inconsistent. The exception was the set of four correlations of achievement with chemistry aptitude, which was consistent and had average rs of .39 for males and .32 for females. All of these results suggest that the undifferentiated category "aptitude" is too broad to explain much variation in correlation values.

Task Value–Achievement.   An equally broad range of values is covered by the distributions of rs for task value and achievement. The distribution for females is less diverse than for males and is again centered about a slightly lower value. In this case, however, the 95 percent confidence intervals for mean rs overlap, indicating no gender difference in the average strength of the task value–achievement relationship. Again, however, the correlations for each sex are so variable that further analysis is warranted.

SUMMARY OF SIMPLE MODEL.   Figure 6.6 shows the 95 percent confidence intervals for the five sets of correlations relevant to the simple model. Although the sets of values are heterogeneous, the confidence intervals are still quite narrow, mainly because of the large samples within studies. As discussed above, the relationships on several paths appear slightly stronger on average for males than females, though the only pair of means which differs significantly is for the aptitude–task value relationship. Further study of results from other affective components of Eccles' model may suggest whether this gender difference applies to all aptitude-affect relationships.

ECCLES' MODEL.   Table 6.15 shows the mean correlations and fit tests for the paths studied in Eccles' model, and Figure 6.7 shows the 95 percent confidence intervals for the mean correlations. Not only have relatively few paths in this model been studied, but (with the exception of task value–achievement) those that have been studied have not been studied extensively. Only one study (†Gilmartin et al. 1976) looked at

Figure 6.6 Ninety-five Percent Confidence Intervals for Average Correlations for Paths in Simple Model

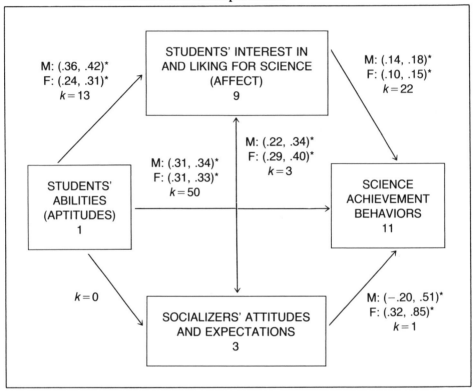

Note: Asterisks represent sets of heterogeneous correlations. The number of correlations for each sex is denoted as *k*.

Table 6.15 Analysis of Correlations in Eccles' Model

Path in Eccles' Model	Number of Pairs of Correlations	Males Mean Correlation	H_T	Females Mean Correlation	H_T
Aptitude–Past Event	6	.29	52.22	.23	24.01
Task Belief–Task Value	7	.15	59.54	.14	39.63
Task Belief–Expectancies	1	.52	0	.09	0
Task Value–Achievement	22	.16	187.60	.12	98.09

Figure 6.7 Ninety-five Percent Confidence Intervals for Paths in Eccles' Model

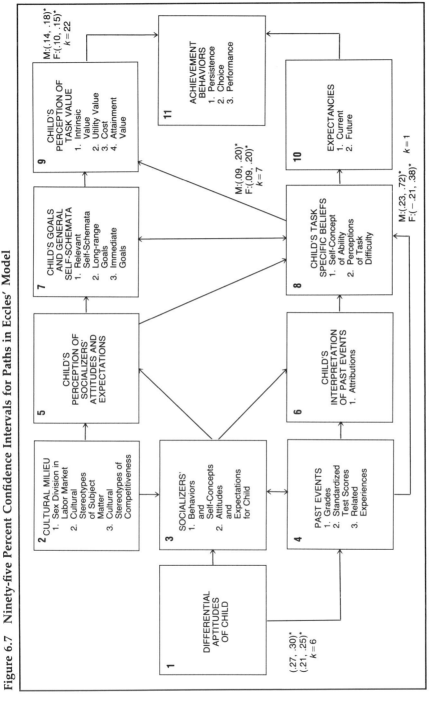

Note: Asterisks represent sets of heterogeneous correlations. The number of correlations for each sex is denoted as *k*.

249

the aptitude–past events relationship, and only †Weimer (1985) studied task beliefs and expectancies. The task value–achievement relationship is represented here by the same correlations discussed above (and listed in Table 6.12) for the simple model. The average correlation for the males is slightly, but not significantly, higher than that for the females.

OTHER PATHS.   Detailed results are presented for only one of the 12 other interconnections between model components that were examined in the 29 studies. In most other cases there were very few correlations and the correlations were highly inconsistent, suggesting that other explanations of interstudy and intrastudy differences (e.g., subject matter, measure type, or age differences) in the correlation values should be sought. Appendix 6.B shows a table of estimated mean correlations between all components of the model examined by these studies.

Past Events–Achievement.   The past events–achievement relationship is the only path not appearing in either model that was measured by more than ten pairs of correlations in the sample. Table 6.16 shows the stem-and-leaf diagrams of past events–achievement correlations for males and females. Since many of the measures of "past events" were achievement tests or grades (and some in science), these correlations show more high values than those for some of the other paths. Both distributions are highly negatively skewed, and the median *r*s of .61 and .60 for males and females, respectively, are better measures of the typical correlations than means (.46 for males and .47 for females) would be. These results depict the common finding that past success is often

Table 6.16   Stem-and-Leaf of Correlations Between Past Events and Achievement

Males ($k = 14$)		Females ($k = 14$)	
.7	2 3 4 4 4	.7	0 1 1 4 7
.6	4 5	.6	2 7
.5	5 8	.5	1 9
.4	3 9	.4	5
.3	1	.3	1 5
.2		.2	9
.1	2	.1	
+.0	8	+.0	7
	median = .61		median = .60

the best predictor of future success. The results for males and females are again quite similar within this collection of values, with a correlation of .82 between males' and females' *r*s.

COMMENTS ON ANALYSIS OF INDIVIDUAL PATHS. The analyses of correlations representing individual paths in the two models revealed some interesting results. Of the paths that have been studied more than once, the strongest (average) relationships seem to be for general-intelligence measures and past activities with science achievement. Other kinds of aptitudes, including numerical, spatial, and verbal abilities, showed small to moderate correlations with achievement. This analysis suggested that the strength of most of these relationships was the same for the two sexes. One exception was the relationship of aptitudes to task value, which was significantly stronger for males than for females.

The homogeneity tests for the collections of correlations for each path indicated great variation even within very specific subsets of correlations. For instance, even when the aptitude-achievement correlations were categorized according to the type of aptitude measured, considerable variation remained.

It may be necessary to classify the correlations according to very specific measure characteristics. Alternately some other explanatory variables may be needed to understand the differences in the *r*- values. Further clustering of sets of correlations, such as in the second version of the simple model, is not warranted on the basis of these analyses.

LIMITATIONS. The analyses of individual paths in the two models were conducted under the assumption that the correlations for each path were independent. This assumption was clearly violated for many paths, however, since several studies have contributed more than one *r* per sample to those paths. Additionally, the high positive correlations between the *r*s for males and females across studies indicated that serious within-study covariation (due to similarities in samples, measures, and study design) was ignored by this analysis. These findings suggest that modeling the within-study correlation between male and female results (as discussed in the methods section) should be an explicit objective of future methodologies for synthesizing correlational results in which gender differences are an issue.

Generalized Least Squares Analysis

In conducting a model-driven meta-analysis the question arises of how to incorporate models into an analysis of existing research results. The question for the GLS analysis concerns which paths should be "modeled" in the regressions and consequently which paths should be represented in the **X** matrix for the analyses.

The initial part of this GLS regression analysis focuses on the paths that have been investigated between individual components in Eccles' model. A second part of the analysis considers collections of several related components and the paths between those composite components. First, however, the samples from which the data for the GLS analyses were drawn are described.

DESCRIPTION OF SAMPLES IN THE GLS ANALYSIS. Twenty-four samples had sufficient data for the computation of covariances for the GLS analysis. The GLS analysis used 244 correlations for which variance-covariance matrices could be obtained. The studies with covariances (the "GLS studies") differed from the studies without covariances in several ways. The median year of publication for the GLS studies was 1977, while for the others the median was 1983. More of the GLS studies used random or probability samples (21 percent) than did the excluded studies (7 percent). Sixty-three percent of the samples from GLS studies had either attrition or selection, or both. Information about attrition and selection was available for 75 percent of the GLS studies, while only 36 percent of the excluded studies contained this information.

More of the GLS studies were written solely by male authors (58 percent versus 36 percent), and fewer had a major focus on gender (58 percent versus 71 percent). The excluded studies generally used larger numbers of subjects who were slightly younger (mean age 13.8 years) than those in the GLS studies (mean age 14.7 years).

Several of the articles which did not contain sufficient data to compute covariances had more than one sample or had multiple variables correlated with a science outcome (e.g., †Haladyna et al. 1982; †Linn and Pulos 1983). All six dissertations contained data with which to compute covariances. These two facts suggest that space limitations may have prohibited the publication of all available data, the fact that the GLS studies were somewhat older also raises the possibility that publication practices or the nature of the research being conducted may have changed over time. If more recent studies tend to be bigger and more complex they may be less well reported in journals because more information had to be omitted.

Table 6.17 displays the paths investigated by the studies in the GLS analysis. A comparison between Table 6.17 and Table 6.10 or 6.12 shows that ten paths examined by the excluded studies were not examined by those in the GLS analysis.

Table 6.17 Number of Correlations for Individual Paths in 23 Generalized Least Squares Studies

First Component	Second Component											
	Dem	Apt	Cul	Soc	Pas	Per	Int	Goa	Sel	Per	Exp	Ach
Demographics	0	2										6
Aptitudes		0			<u>10</u>			3		<u>22</u>		<u>88</u>
Culture			0									2
Socializers				0								
Past Events					2			8		2		<u>28</u>
Perception of Socializers						0						
Interpretation of Past Events							0					
Goals/Self-Schemata								0				1
Self-Concept									0		2	4
Perception of Task Value										4		<u>40</u>
Expectancies											0	
Achievement												20

Note: Underlined numbers denote paths which were assigned predictor variables in the GLS regression.

253

OVERALL REGRESSION ANALYSIS. The overall homogeneity test value across all 244 correlation coefficients in this analysis was $H_T = 6691.6$ (df = 243, $p < .001$). The correlations were highly variable, thus they most likely did not arise from a common population. The weighted average correlation across all samples and types of relationships was .11, which though modest in magnitude still differed significantly from zero.

It is informative to compare the H and mean r values from the full multivariate analysis to the values obtained from an analysis in which all covariances are ignored. That is, the latter analysis represents the results that would have been obtained if all correlations were assumed to be independent. The homogeneity value obtained from this "univariate" analysis of the 244 correlations was $H_T = 8536.8$, slightly larger than the H_T from the GLS regression. Also, the average correlation obtained under the assumption of independence was .28, which is considerably larger than the value of .11 from the GLS analyses.

Both of these findings indicate the effects of using the GLS approach. The average correlation from the GLS analyses was much lower than the average from analyses which assumed independence. Some studies have apparently contributed a number of relatively large and highly intercorrelated r- values. Furthermore the results appeared more variable when the dependencies were not modeled than would be expected under the null hypothesis (of a common ρ).

Overall significance tests for each regression model are shown in Table 6.18. The regression model test is analogous to the usual F test for the significance of predictors in a regression model. The model specification test indicates whether the variability remaining among the correlations exceeds what would be expected because of sampling error. A significant model specification test means that significant amounts of variability remain and suggests that other important predictors may need to be added to the equation. Also given are values of the statistic which represent the change in explanatory power (i.e., in the regression model statistics) due to the added variables. These statistics, in the last column of Table 6.18, are also distributed as chi-square statistics, with degrees of freedom equal to the number of added predictors.

The initial regression model, positing distinct correlations for males and females (and with an overall grand mean), produced a regression model statistic of 55.3 (a chi-square with 1 degree of freedom, $p < .001$). Although this was significant and indicated a stronger average correlation for males (a difference of .04 across all studies), much variability remained among the correlations. The model specification test shown in Table 6.18 for this model (number 1: sex effect) was highly significant.

Table 6.18 Regression Model Tests for the Generalized Least Squares Analysis

Description of Model	Regression Model Test, H_R	(df)	Model Specification Test, H_E	(df)	Change Due to Added Predictors	(df)
1. Sex Effect	55.3	(1)	6636.3	(242)		
2. Sample and Study Features	3316.4	(7)	3375.2	(236)	3261.1	(6)
3. Model 2 and five Paths	3809.9	(12)	2881.7	(231)	493.5	(5)
4. Model 3 and five Sex-by-Path Interactions	3836.9	(17)	2854.7	(226)	27.0	(5)
5. Model 2 and four Paths for Composite Components	3861.2	(11)	2830.3	(232)	544.8	(4)
6. Model 5 and four Sex-by-Composite-Path Interactions	3886.2	(15)	2805.4	(228)	25.0	(4)

Note: All tests are significant at the .05 level or beyond.

Several sets of predictors were added to the initial regression model. The addition of publication year, sample grade, and nationality whether a volunteer sample was used, sample type (whether a convenience sample was used), and whether the study had a major focus on gender differences changed the regression fit significantly ($\chi^2 - 3261.1$, df $= 6$, $p < .001$), and all coefficients were significant. Thus seven predictors were retained, and additional coefficients representing paths between model components were added for analyses relevant to Eccles' model and the two views of the simple model.

None of the analyses below contains predictors for all of the possible paths among components that have been studied. A model containing all of these could be trivial, if each correlation were uniquely specified by the coefficients for the paths together with other predictors (e.g., year, nationality). The question to be addressed is how much of the variation among the correlations can be accounted for by an equation containing a limited number of these predictors.

PATHS AMONG INDIVIDUAL COMPONENTS.   The X matrix for the analyses of paths among individual components contained predictors representing article and sample characteristics, as well as specific paths between model components. Publication date, gender as a focus of the study, nationality, sex, and grade level of subjects, and the two sampling features mentioned above were included in the analyses.[8] The columns of X relating to specific paths were constructed after inspecting the matrix shown in Table 6.17. Five paths, between components which had been studied by ten or more pairs of correlations, were each assigned a dummy variable (taking value 1 for correlations investigating that path and 0 otherwise).

The paths between individual components represented in the X matrix were for aptitudes with past events, task value, and achievement behaviors (components 1 and 4, 1 and 9, and 1 and 11) and for both past events and task value with achievement behaviors (i.e., between 4 and 11, and 9 and 11). Five columns of X represented these paths, two of which appear in Eccles' model and three of which do not. Four of the six paths in the simple model are represented in the predictor matrix. Additionally, interaction terms were computed for gender with each of the five paths, thus making it possible to test for differences

---

[8] Additional predictors such as percentage of male authors, the presence of attrition and selection, and the nature of the publication (e.g., published versus unpublished), as well as additional unprocessed measure and sample characteristics, remain available for future analyses.

between males and females in the strengths of specific relationships or paths.

MODEL WITH PATHS BETWEEN INDIVIDUAL COMPONENTS.   Adding predictors to represent the aptitude–past events and task value–achievement paths (i.e., the two paths in Eccles' model which had predictor indicators) significantly improved the regression statistic, though the aptitude–past events predictor was not significant. This model is shown as number 3 in Table 6.18. When predictors representing the aptitude–achievement, past events–achievement, and aptitude–task value paths were added, the fit of the regression improved further. However, the aptitude–achievement path predictor did not reach significance in this model. The model is shown as number 3 in Table 6.18.

The addition to model 3 of interaction terms for sex with each path in the model further reduced the regression specification statistic ($\chi^2 = 27.0$, df = 5, $p < .001$). This model is shown as number 4 in Table 6.18, and the coefficients from this model are given in Table 6.19. The model leaves much variation in the magnitudes of correlations unexplained, accounting for 57.3 percent of the variation in these study outcomes. Terms representing the interaction of sex with grade and year were also added, but they did not improve regression fit significantly.

Addition of predictors for the individual paths and their interactions with sex caused the effect of publication date to become nonsignificant. Additionally, some of the path predictors and their interactions were not statistically significant. The differences between correlations on these paths may have been accounted for by the sample characteristics.

The significant negative slope coefficient for the aptitude-task value coefficient suggests that this relationship is weaker than those for the other paths in the model. Additionally, the significant interaction term for this path suggests that the relationship is weaker for females than for males, with all other factors held constant. This finding is consonant with the results of the univariate analysis and the correlations for this path shown in Table 6.14.

Another significant coefficient was found for the past events-achievement relationship. The positive slope for this path indicates that past events show a stronger than average relationship to achievement. This finding also agrees with those of the univariate analysis. Finally, the interaction of the task value-achievement path with sex indicates that this relationship is stronger for females, with all other factors held constant. This result does not agree with the overall findings of the

Table 6.19 Predictors in Regression with Individual Paths (Model 4)

Predictor	Regression Coefficient	Standard Error	z Test
Grand Mean	−0.017	0.0682	−0.24 ns
Sex	−0.030	0.0096	−3.19
Year	−0.001	0.0094	−0.74 ns
Grade	0.030	0.0029	10.54
Nationality	0.522	0.0225	23.22
Focus on Gender	−0.034	0.0108	−3.12
Voluntariness	−0.256	0.0140	−18.37
Sample Type	0.117	0.0152	7.72
Aptitude–Past Events Path	0.010	0.0143	0.71 ns
Aptitude–Task Value Path	−0.063	0.0267	−2.36
Aptitude–Achievement Path	−0.005	0.0128	−0.42 ns
Past Events–Achievement Path	0.060	0.0134	4.52
Task Value–Achievement Path	−0.226	0.0146	−15.51
Sex × Aptitude–Past Events Path	−0.014	0.0211	−0.66 ns
Sex × Aptitude–Task Value Path	−0.092	0.0416	−2.22
Sex × Aptitude–Achievement Path	0.027	0.0149	1.79 ns
Sex × Past Events–Achievement Path	0.027	0.0169	1.61 ns
Sex × Task Value–Achievement Path	0.043	0.0117	3.68

univariate analysis, however, the set of correlations examined here differs slightly from that in the univariate analysis.

Table 6.20 lists the actual and predicted values of the correlations for six samples from the data set. Values of four demographic variables and names of the paths these samples have studied are also listed. Each pair of results from a study had measured correlations between the same two variables.

The correlations for the †Kaminski and Erickson samples are predicted to be higher than the others since they are on the aptitude-achievement path. (No past events–achievement results are shown.) Although the aptitude-achievement correlations shown are predicted fairly well by this model, the same is not true for the other results, especially those from †Bridgham's study (1969). The path for "socializers"–achievement relationships was not included in the model because Bridgham's was the only study which had measured that path. Consequently, these correlations are not well predicted by the model.

The regression model can also be used to predict results for situations not manifest in actual studies. However, if the predictor values are out of the range of values for which the equations were estimated, extrapolation may be problematic. For example, this model predicts that the correlation for a past events–achievement relationship for a convenience sample of twelfth grade males from outside the United States would be 1.04.

Although this regression model had the smallest model specification test of all models involving individual paths (and explained 57 percent of the variance among the rs), significant amounts of variability remained among the results. It is possible that additional demographic variables or study characteristics could help explain more of the variation among these study results.

PATHS AMONG COMPOSITE COMPONENTS.   For the second part of the GLS analysis the initial model was "collapsed" by combining several subsets of components. A second X matrix was created representing clusters of components. Four clusters were formed which represented aptitude (component 1 alone), social factors (components 2 and 3), affective factors (components 7, 8, and 9), and achievement behaviors (component 11), as described above. The components representing demographics, past events, student's interpretations of past events, student's perceptions of socializers, and expectancies were not included in these categories because they were represented in very few or no studies. Consequently, this omission should not have a great effect since few studies have been overlooked.

Table 6.20 Results and Predicted Correlations from Six Samples

Study	Sex (Value)	Year	Grade	Focus on Gender (Value)	Path	Correlations		
						Actual	Model 4	Model 6
Kaminski and	M (0)	79	12	Major (1)	Apt-Ach	.42	.25	.26
Erickson	F (1)	79	12	Major (1)	Apt-Ach	.30	.25	.25
Cohen	M (0)	79	8	Minor (0)	Task-Ach	.09	-.05	-.04
	F (1)	79	8	Minor (0)	Task-Ach	.19	-.04	-.03
Bridgham	M (0)	69	3	Minor (0)	Soc-Ach	.18	.03	.24
	F (1)	69	3	Minor (0)	Soc-Ach	.66	.00	.46

Note: Labels for the paths are: Ach = Achievement, Apt = Aptitude, Soc = Socializers' behaviors and attitudes, Task = Task value. All studies shown examined U.S. students (Nationality = 0) and did not use volunteer samples (Voluntariness = 0).

Table 6.21   Numbers of Correlations for Composite Paths in 23 Generalized Least Squares Studies

First Component	Second Component			
	Aptitudes	Social Factors	Affect	Achievement Behaviors
Aptitudes	0	0	25	88
Social Factors		0	0	2
Affect			4	45
Achievement Behaviors				20

Dummy variables were created for each of the four paths that had been studied between composite components. Table 6.21 shows the numbers of correlations for each of those paths. For instance, the dummy variable for the social factors–achievement path had the value 1 if the correlation was *either* for a relation between cultural milieu and achievement (original path 2 to 11) or between socializers' behaviors and achievement (path 3 to 11). The X matrix based on these dummy variables was used to investigate the predictive or explanatory power of

Table 6.22   Predictors in Regression for Composite Components (Model 6)

Predictor	Regression Coefficient	Standard Error	z Test
Grand Mean	0.062	0.0685	0.91 ns
Sex	−0.025	0.0081	−3.06
Year	−0.001	0.0009	−0.55 ns
Grade	0.028	0.0030	9.32
Nationality	0.499	0.0230	21.74
Focus on Gender	−0.035	0.0106	−3.28
Voluntariness	−0.314	0.0121	−25.94
Sample Type	0.117	0.0151	7.78
Aptitude–Affect Path	−0.123	0.0250	−4.92
Aptitude–Achievement Path	−0.062	0.0101	−6.13
Affect–Achievement Path	−0.287	0.0131	−21.98
Social Factors–Achievement Path	0.128	0.1681	0.76 ns
Sex × Aptitude–Affect Path	−0.105	0.0404	−2.60
Sex × Aptitude–Achievement Path	0.013	0.0128	1.03 ns
Sex × Affect–Achievement Path	0.037	0.0105	3.50
Sex × Social Factors–Achievement Path	0.243	0.2020	1.20 ns

the second conceptualization of the simple model, described above. Again, interaction terms were computed for gender with each of the paths between the composite components, thus making it possible to test for gender differences on particular paths.

ANALYSIS OF PATHS BETWEEN COMPOSITE COMPONENTS. The second part of the GLS regression analysis is based on equations constructed with the demographic variables and predictors representing paths among the *composite* components.

Model 5 in Table 6.18 included the same demographic variables as in model 2 plus predictors representing paths for aptitudes, affect, and social factors with achievement, and for aptitudes with affect. This model had the lowest model specification value of models without interaction terms, and explained 57.7 percent of the variation in the correlation values. All slope coefficients for the composite paths in this model were significant.

Model 6 in Table 6.18 includes all composite paths and the corresponding interaction terms. The addition of the interactions improved the fit of the model ($\chi^2 = 25.0$, df $= 4$, $p < .001$), though some of the interaction terms were not significant. Table 6.22 shows the coefficients. All but one of the composite-path slopes remained significant in this model, which explained 58.1 percent of the variation in the correlations. Terms representing the interaction of sex with year and grade were also added to this model, but did not change the model's fit significantly.

The slopes for the aptitude-affect and affect-achievement paths and their interaction terms in this model are very similar to those for the corresponding paths in the model for individual components. Unlike the model for individual components, however, this model shows that correlations for the aptitude-achievement relationship are significantly lower on average than the other correlations. This has occurred in part because the past events-achievement relationship is not represented in model 6. The correlations of past events with achievement were the strongest in the synthesis.

Comparisons of the coefficients for the aptitude-achievement and affect-achievement paths and interactions together show that, for both sexes, the affect-achievement correlations are predicted to be smaller than the aptitude-achievement correlations. The remaining correlations (such as past events-achievement correlations) are predicted to be larger on average than any of those with dummy variables representing their paths.

Finally, the slope and interaction for the social factors-achievement path were not significant in this model. However, it is clear from the predicted values for the †Bridgham (1969) samples shown in Table 6.20 that the addition of these terms has enabled the few correlations on this path to be predicted better by model 6 than by model 4. Two correlations from the samples in †Bridgham (1969) represent this path which has a large positive slope and interaction in this model, and thus have large predicted values.

COMPARISON OF REGRESSION MODELS.   Because the "univariate" analysis of paths between the composite components was not computed, it cannot be compared with these results. The regression analysis can, though, be compared with the regression analysis based on individual paths. First, the GLS model with coefficients representing the composite components (model 5) explained more variation than any of the equations specifying paths between individual components. When the criterion of simplicity is used, the model based on composite components again has the advantage. It contains fewer predictors, and without interaction terms explains more variation than the individual paths analysis in model 4.

However, even this analysis should be viewed somewhat tentatively because much unexplained variability remains among the sample correlations. Additionally, none of the analyses reported here were able to model all sources of within-study variation, and the intercorrelations between the correlations for males and females suggest that this covariance may be considerable. Future analyses must control for such implicit dependencies.

FURTHER GLS ANALYSES.   Additional variables can be incorporated into the GLS analysis of these correlations. Indicators of subject matter content of achievement tests and topic or focus of affective measures are two kinds of important indicators that will be considered in further analyses, as well as subject characteristics such as sex of the students' teachers, sex composition of schools, and so on.

Conclusions

This synthesis informs us in several areas. First, it provides a new perspective on models for the development of science achievement and the factors that are important in science-achievement behaviors for both males and females. Also the synthesis provides insights into the process of, and problems associated with, conducting a model-driven synthesis. Finally, the synthesis provides a base from which to draw implications for science-education policy and practice, and recommendations for future research.

Process of Achieving in Science

Although much information is available about a few relationships among science variables, little or nothing is known about many others. Unfortunately, relationships involving most of the variables that might be thought of as alterable or manipulable fall into that latter category.

In particular, little is known about whether socializers' attitudes and behaviors are more influential for the development of boys' or girls' attitudes toward and achievement in science. This lack of knowledge is particularly surprising in light of the amount of attention paid to "environmental" explanations in other areas of achievement (e.g., Grieb and Easley 1984). Even less is known about the mediating role that male and female students' perceptions of their experiences and their socializers' beliefs and behaviors might play in that development.

Mean correlations representing the cultural milieu–task value and cultural milieu–achievement paths suggested that cultural context plays a limited role in the development of students' attitudes and achievements. Mean correlations between cultural milieu measures and task value were .30 for males and .26 for females. Correlations of cultural milieu with achievement were more modest, .23 and .25 on average for males and females, respectively. (These results are presented in Appendix 6.B.)

One complicating factor in the interpretation of these correlations involving cultural milieu is the possibility that some of the measures might be expected to relate differently to the criterion measures than others. For instance, †Handley and Morse (1984) used two measures: One examined the appropriateness of science careers for males and the other examined the appropriateness of science careers for females. If male and female career roles differ, measures of the appropriateness of the two roles may relate differently (even in opposite ways) to science outcomes for males and females.[9]

The role of past experiences in science achievement is well studied. Past successes appear to relate positively to later achievement for both sexes. This suggests that one obvious strategy for the encouragement and retention of students in science education is to ensure that they have the opportunity to experience some initial achievements in science.

[9] The correlation of Handley and Morse's measure of "appropriateness of science careers for males" with science attitudes was positive and significant for males but essentially zero for females, and the difference was significant ($p<.01$). However, correlations of "appropriateness of science careers for males" with science achievement and "appropriateness of science careers for females" with both attitudes and achievement did not show gender differences. Thus there is mixed evidence about this concern.

A second well-studied linkage is the aptitude-achievement relationship. The present analyses indicate that prior-achievement measures are stronger predictors of later achievement than are aptitude measures. Similarities in instrumentation between the past-events and achievement measures may contribute to this finding, but that has not been explicitly investigated in this analysis. Aptitude measures had some predictive value, though aptitudes did not seem more important for girls than for boys.

Given the modest size of the relationships between specific aptitudes and science achievement, interventions designed to increase or simply bolster girls' science-related cognitive skills (number and spatial skills) may be only moderately effective. This may be particularly true because the efficacy of some skill training (e.g., spatial-skill training) in altering target abilities is still controversial.

Attitudes toward science are widely believed to play a mediating role in the development of academic achievement behaviors. In this synthesis their role as direct predictors of achievement appeared to be subordinate to those of both aptitudes and past achievements. However, the GLS analyses suggested that attitudes may be more important for females than for males.

On the other hand, aptitudes did appear to relate moderately strongly to perceptions of task value, as indicated by both analyses of individual paths and the GLS analysis. Task value may mediate the role of aptitudes in the prediction of achievement (an issue not investigated in this analysis). The role of gender in the aptitude–task value relationship is still controversial, though our analyses suggested the link was weaker for females than for males.

Further information about aptitudes, attitudes, and past achievement is available in several studies which were not included in the above data analyses. Cannon and Simpson (1985) concluded that while seventh-grade girls scored lower than boys on measures of attitudes toward (and motivation in) science, for both sexes attitudes were less important in predicting achievement outcomes than were prior abilities. DeBoer (1984) showed that high school science grade point average (GPA) was the most important predictor of college science GPAs for both men and women. For both sexes high school GPA was more important than either Verbal or Mathematical Scholastic Aptitude Test scores or math grades in either high school or college.

Finally, Thomas's analysis (1981) of National Longitudinal Survey data on college science course-taking also emphasized the importance of past achievement for both sexes. A measure of general verbal and mathematics skills and intention to pursue a "hard" or technical college major (as stated during high school) were the best predictors of number of science courses taken in college for both males and females. These were more important than educational expectations, family socioeconomic status, race, or high school rank in predicting science course enrollment. However, attitudinal variables were not examined in Thomas's analysis.

Conducting a Model-Driven Meta-Analysis

Conducting a model-driven synthesis is difficult. While the substantive and statistical problems that arise are similar to those that arise in traditional meta-analyses, the consequences may differ in the context of the model-driven synthesis.

SUBSTANTIVE ISSUES.   Substantive problems may arise in the selection and interpretation of models (here, of the process of science achievement). Are the models selected the most informative ones to examine? Have they been studied by enough primary researchers to support meaningful syntheses? In this synthesis the model derived from Eccles' work in mathematics seemed conceptually sound, but little empirical evidence was available about its applicability to science achievement for the two sexes. Because the selected models define the focus of a model-driven synthesis, a reviewer can likely portray a distinctive picture of the literature in a research domain by questionable (or idiosyncratic) selection of models to guide the synthesis.

The selection of studies for the synthesis also has consequences for the quality of the synthesis. The limits on the process of examining models through meta-analysis are set in large part by the constraints of the data. Certain relationships may never have been studied. Even a model-driven synthesis cannot provide evidence where none exists. A model-driven synthesis can provide information on constellations of variables never studied *together*, but if there are no data at all, the synthesis is stymied.

A third issue concerns the connections between different models and between the models and primary research. Judgments about similarities between models rest upon conceptualizations of constructs in the models. The degree to which constructs are delineated also determines whether it is possible to match operationalizations of variables used in primary research with those constructs. Such connections are critical to a model-driven synthesis, otherwise judgments about models or comparisons between them will not be well founded. This synthesis examined two different views of how Eccles' model and a simpler model might relate to each other, but other views might also be interesting.

A last practical problem with a substantive aspect concerns limiting the selections of studies relating to paths in the models. If the reviewer does not set some criteria for limiting the collection of studies to be synthesized, he or she risks including many tangentially relevant studies.

Eccles' model contains, for example, a link between student aptitudes and past achievement. This relationship has undoubtedly been examined for a number of subject matter areas. The model itself does not limit the context to science achievement. However, in this synthesis all studies were required to have measured at least one variable related to school science, which in this illustration could include either science aptitudes or past science achievement.

STATISTICAL ISSUES.   Numerous statistical problems arise in conducting a model-driven synthesis. In this synthesis most studies provided results based on dissimilar collections of variables. Some primary-research studies reported complex data analyses while others presented simple bivariate results. Additionally many studies did not fully report all analyses (e.g., complete correlation matrices), thus causing the problem of missing data. Varied reporting conventions added to problems with retrieval of data, especially for the generalized least squares analysis.

The generalized least squares analysis used here required that full within-study correlation matrices (by sex) be presented so that covariance matrices for the correlations of interest could be estimated. Many studies did not report full matrices because of page limits (set by journals) and other practical constraints of reporting. Another challenge for model-driven synthesis is to explore more methods of data imputation in order to incorporate all available evidence into the analyses of the models.

Finally, a complete model-driven synthesis should account for both between-studies and within-study variation and covariation. This was not possible with currently available software, even though the statistical theory for the analysis is straightforward. Achieving such complete analyses should be an eventual goal for model-driven syntheses.

ADVANTAGES OF MODEL-DRIVEN SYNTHESES.   Although it is not easy to conduct a model-driven research synthesis, there are benefits to the approach. Most social phenomena are complex multicomponent processes. Theoretical models can acknowledge the complexity of such processes while helping to focus the attention of the reviewer on a particular view (or views) of the process. Thus while this model-driven synthesis acknowledged the many potential influences on science achievement behaviors, it also focused the analysis of those influences rather narrowly.

Model-driven syntheses can provide information about relationships or differences in relationships (i.e., interactions). In this synthesis the focus was on the role of gender in determining patterns of relationships among science variables. Regression predictors were used to identify the "path" that each correlation represented within each model. These variables indicated whether (across all subjects) the magnitudes of interrelationships differed, and the patterns of relationships and differences helped identify opportunities to impact the process of developing achievement behaviors.

The interactions of sex with each path are more pertinent to gender-differences research. Interaction variables indicated whether the relationships on particular paths differed for males and females. Other effects which could be investigated in a similar fashion are cross-national effects or interactive effects of gender and nationality.

Model-driven research syntheses also provide a lens through which a reviewer can identify what is *not* known about a process and about particular views of a process. Models propose specific paths or connections among variables. In this study Eccles' model proposed a series of indirect links leading to the development of achievement behaviors. But few researchers had studied most of the paths proposed in the model for the two sexes. Current knowledge about that view of the development of science-achievement behaviors is limited.

A model-driven synthesis can reveal what is not known about a process or phenomenon. Conventional meta-analysts would view a topic that was not well studied (or not studied at all) as a bad choice for quantitative review *because* of the lack of data. Yet clearly such information can be valuable for the planning of future research as well as for other purposes. For instance, Cordray has noted how the policy question for an evaluation synthesis "often revolves around the question of whether there is any evidence that could be brought to bear on a policy issue" (1986, p. 5).

Models that do not appear to be well studied should not be considered poor or inadequate models, however. In this synthesis the most thoroughly studied relationships were arguably the most fundamental (e.g., the aptitudes-achievement relationship). But they also provide the least vision into how science achievement can be modified. Some of the less-studied paths may not have been studied because they involve unstable or modifiable variables.

Such variables may serve to generate excellent ideas for interventions that seek to purposefully change the process, if that is desired. For example, if teacher (or counselor) expectations influence student attitudes and achievement, programs could be designed to both inform teachers of that fact *and* try to instill in them positive expectations for all students.

Implications for Policy and Practice

Models of achievement have been most frequently applied in the study of achievement in mathematics (e.g., Eccles et al. 1983; Ethington and Wolfle 1988). Extensive research on Eccles' model has outlined factors important in predicting student intentions to persist in math, their math course-taking, and their actual achievement in math. These factors are often invoked as potential predictors of (and solutions to) findings of gender differences in science achievement.

Much of the research on gender differences in achievement has focused on mathematics exclusively. However, summaries of that literature often blur the distinction between math and science. Recommendations for policies and interventions for science and mathematics instruction often look similar or identical. This synthesis strongly suggests that although there are some similarities between these results and others for the prediction of math outcomes, considering math and science to be identical is *not* warranted.

In some cases there are similarities. For instance, Eccles and her colleagues (1983) found that math achievement for females was predicted by prior achievement and self-concept in math, whereas for males task value was an additional important predictor. This model for math parallels, in part, the present analysis for science, showing the importance of prior achievement for both sexes. However, the role of science self-concept as a strong predictor of science achievement was not supported by this synthesis.

Many programs designed to address issues of sex equity and women's participation in math and science have considered math, engineering, and the sciences together (e.g., Malcom 1984). Since some scientific careers require a level of quantitative sophistication, interventions aimed at the various scientific fields have often been linked with math. However, this synthesis suggests that achieving in science requires much beyond a background in and appreciation for mathematics.

Furthermore, many of the factors that have been well investigated in the domain of mathematics (e.g., the relationship of students' self-concept in the subject matter to parents' or teachers' perceptions of their abilities) have not been studied at all (by gender) for science outcomes. The evidence is still inconclusive about whether the many social factors identified as important in the mathematics literature are also implicated in science participation and achievement.

Focusing on science outcomes, this synthesis suggests that on average, the same variables are important in the prediction of science for males and females. Thus the same kinds of interventions should be effective at increasing levels of participation and achievement for both sexes. Specially designed or exclusive programs aimed at one sex or the other can be avoided. This is congruent with the thrust of Title IX legislation, which discourages sex-segregated situations and programs in the public schools. On the basis of the present evidence, the same kinds of characteristics and abilities seem to lead to positive science outcomes for all students.

STATE-LEVEL POLICY.   Jacobs and Wigfield (1989) have argued that most of the impact of sex-equity policies has been at either the state or district and school levels. They outlined three primary policy concerns at the state level, including teacher licensure, testing programs, and course (curriculum) requirements.

This synthesis speaks in part to each of these concerns. Regarding teacher licensure, Jacobs and Wigfield suggested that courses in sex equity might be required of all teachers in training. This research suggests that teachers must not only be informed of the importance of their support and instruction for the science achievement and persistence of their students, but be convinced that the *same* rather than different factors are important for both boys and girls.

Another state-level responsibility mentioned by Jacobs and Wigfield is the design and administration of achievement-testing programs. The assessment issue raised by this synthesis is not that of sex bias in tests of science achievement (though that should not be ignored). Rather this research brings to mind the use of tests to monitor skill levels and even to guide the curriculum in ways that are seen as desirable. Much controversy exists about whether and how tests ought to be used to influence curriculum and instruction, but little doubt exists that testing can have such influences (Resnick 1980). If states wish to assure that all students have the prerequisite mathematics and verbal skills for later science achievement, one way of moving in that direction would be to institute programs to assess students' levels on those prerequisite skills.

Similarly, this synthesis pointed out the importance of prior science achievement in later science achievement. A statewide science testing program could provide a method of monitoring the progress of individuals and identifying students in need of attention in the early grades, before the student reaches choice points at which he or she might decide to opt out of science on the basis of poor past performance.

States also set course and graduation requirements. Jacobs and Wigfield suggested that "course choices may be introduced at the wrong time for females, leading them to choose stereotypically female courses, thus limiting their later career possibilities" (1989, p. 45). This synthesis supports the strategy of increasing course requirements in the sciences at the high school (or earlier) levels; additionally, it indicates that such a strategy would likely benefit both males and females (because of the strong importance of past achievement for both sexes).

DISTRICT- AND SCHOOL-LEVEL POLICY.   The district and the school itself are the settings for the most direct impacts of policy on teachers and students. In-service programs are often organized at the district level, and Jacobs and Wigfield (1989) suggested the appropriateness of in-service sex-equity training, possibly in combination with district-level Title IX activities. In the area of science teaching, district-level in-service activities could present strategies for teachers to use to identify and deal with students whose prerequisite skill levels are inadequate for further science instruction. Teaching teachers how to make reasonable assessments of *relevant* abilities for both males and females could be another theme of such in-service training.

Districts or schools could organize cooperative programs involving science teachers and teachers from other relevant subject matter areas in team efforts (to teach students, for instance, the terminology and language of the sciences). Additional focus points for in-service instruction would be the roles of nonintellectual variables such as student attitudes and, more important, teacher attitudes and support for the achievement of all students.

Schools provide the setting for direct interventions with students. Matyas (1988) has described many programs aimed at increasing the participation of precollege females in mathematics and the sciences. Her review noted that many programs have targeted either a general audience or minority students. The present synthesis showed the similarity of influences for both males and females, suggesting that common programs should be effective at increasing participation for both sexes. (An assumption here is that "common" programs would be designed to be relevant to both sexes. For example, both male and female role models or mentors would be included in a role model/mentoring program.)

An important aspect of continued participation is making students aware of the value and opportunities in science. School career days, for instance, may serve to spark student interest, but "are *not* designed to *sustain* those interests" (Matyas 1988, p. 4). Sustained efforts should involve parents and counselors as well as teachers. The evidence (in this synthesis) on the involvement of counselors and teachers is not as strong as that for parental involvement or for some of the more direct influences of achievement mentioned above. However, if any of these individuals can induce more positive student attitudes toward science, this synthesis suggests that they can likely promote persistence for both sexes.

Much of the above discussion emphasizes the similarities in results for males and females. What are similar are the important predictors of achievement, not the levels of performance of boys and girls on those variables. Other research has suggested that gender differences do exist on science achievement (Becker 1989a) and on variables related to science achievement (e.g., cognitive variables: Linn and Hyde 1989; science attitudes: Steinkamp and Maehr 1983, 1984). Although similar interventions may produce positive changes in persistence and achievement for males and females, different levels of effort may need to be directed at males and females to get the same levels of final performance for the sexes. This synthesis, however, did not examine gender differences in performance on science-related variables and thus cannot inform this issue.

Areas for Future Research

Several areas for future research are suggested by this review. The role of socializers and socializers' attitudes on student achievement and attitudes is an excellent area for further study. Because it is possible to intervene with students' families as well as their teachers, it would be valuable to know what effects parents and siblings can have on students' achievement in science. Additional research might be aimed at determining if teachers' behaviors differentially affect their male and female students.

Additional research is also needed on the effects of cultural or societal attitudes on students' science attitudes and achievement. All of the correlations bearing upon this relationship arose from a single study by †Handley and Morse (1984). Although Handley and Morse measured several aspects of the view of science in U.S. culture, generalizability would be improved if the same (or additional) variables were observed for other samples.

Perhaps the most needed research, assuming that Eccles' model has theoretical validity, concerns the mediating role that students' perceptions play in the development of their attitudes and achievement behaviors. If students have high aptitudes for science, why do they not hold positive attitudes toward science? Attribution theory (e.g., Deaux 1976; Frieze et al. 1982) may provide some explanations. Do some students' expectations for failure in science intervene when their aptitudes are high? Only †Weimer (1985) had studied the role of students' expectations in science, and she did not investigate their relationship to either aptitudes or achievement.

Compared with the vast amount of information that is available about the magnitudes of gender differences in science, little is known about the role of gender in interrelationships in science. Additional research might also include attention to the role of other demographic variables in the development of science attitudes and achievement behaviors. Few studies had information about the socioeconomic or ethnic backgrounds of their subjects. It would be difficult to argue that all primary researchers should provide separate analyses for the socioeconomic, ethnic, and gender subgroups of their samples. However, in cases where researchers report information by gender they could be requested to present all (or most) relevant descriptive statistics by gender.

Because it has become a critical "manpower" issue, women's participation in science has been a focus of numerous funded programs and interventions in recent years. However, the amount of information that has *not* been gathered from those programs is disconcerting. Even if a program involves girls alone, information can be gained about interrelationships among pertinent variables for the female participants. And programs that involve families as well as students may provide information that is otherwise very difficult to obtain, about attitudes and influences from home. Thus a final suggestion for future research is to couple it with ongoing and new program initiatives so that longitudinal information is gained about the process of the development of science attitudes and achievement, as well as how to impact that process.

Appendix 6.A Imputation in Covariance Matrices

Kaminski and Erickson

†Kaminski and Erickson (1979) reported the correlations of IQ, socio-economic status (SES), grade point average (GPA) and past science grades with achievement separately for senior high school boys and girls. However, only a common (pooled) correlation matrix showed the intercorrelations among IQ, SES, GPA, and grades. These common values were used to compute the covariances for both girls and boys.

We were concerned that the consequence of using the pooled r values would be that the intercorrelations among the rs (based on the estimated covariance matrix) could contain "impossible" values, for example, correlations larger than 1.0. That is, when the covariance between two correlations of interest is divided by the standard deviations for the two rs, the resultant intercorrelation could be larger than 1.0.

Therefore, the covariance matrix was also computed using both the upper and lower limits of 95 percent confidence intervals constructed about the common correlation values. However, all of these sets of values also produced reasonable correlations, thus the original pooled values were retained.

Several studies had unusual patterns of correlations among rs, suggesting that there may be hidden (unreported) inconsistencies in even the available r values. For example, pairwise deletion of missing data in the computation of correlations may produce *matrices* of values which could not have been obtained from the full sample. In some cases not enough information was reported so that an explanation of the peculiarities could be deduced.

Further work to investigate the effects of imputing common values (when separate-sex data are missing) may be fruitful. Also, reports of primary research ought always to include explicit descriptions of the nature of the reported results, so that problems such as these can at least be expected.

Schock

†Schock (1973) examined correlations among the pre-test and post-test scores of students on the Nelson Biology Test (NBT) and A Scientific Literacy Test (ASLT). Pre-test–post-test correlations were reported for each test separately, and the ASLT and Nelson scores were also correlated at both pre-test and post-test. Missing values were for correlations between NBT and ASLT different tests at two different times, as shown in Table 6.A.1.

The reported NBT-ASLT correlations provided upper limits for the missing NBT-ASLT correlations over time. It would be unlikely that two different tests given at different times would correlate more highly than two different tests administered together. Table 6.A.1 shows that for males the correlation between NBT and ASLT pre-tests was $r = .28$, while the NBT-ASLT post-tests correlated with $r = .22$. A slightly smaller "round" number is the value .20. The value $r = .20$ was substituted for the missing r for males.

The smaller of the NBT-ASLT correlations at the two time points was $r = .19$ for the females. The value of $r = .15$ was imputed for females, following the logic described above. The covariances among the correlations of interest could thus be computed with the "complete" correlation matrices for the two sexes.

The vector of correlations included in the synthesis was

$$r' = (.28 \ .22 \ .19 \ .27).$$

The estimated covariance matrix for the Schock correlation vector, based on the real and imputed correlation values, was

$$
\begin{bmatrix}
0.004799 & 0.002148 & 0 & 0 \\
0.002148 & 0.005121 & 0 & 0 \\
0 & 0 & 0.004489 & -0.00006 \\
0 & 0 & -0.00006 & 0.004163
\end{bmatrix}
$$

Table 6.A.1   Correlations from †Schock (1973)

	Pre-test		Post-test	
	NBT	ASLT	NBT	ASLT
Pre-test NBT	1.0	**.28**	**.86**	r_M
Pre-test ASLT	.19	1.0	r_M	**.52**
Post-test NBT	.69	r_F	1.0	**.22**
Post-test ASLT	r_F	.58	.27	1.0

Note: Males' correlations are above the diagonal (in boldface) and females' correlations are below. The values $r_M = 0.20$ and $r_F = 0.15$ were substituted for the missing values. The numbers of males and females were 177 and 206, respectively.

Appendix 6.B

Table 6.B.1 Average Correlations Among All Components in Eccles' Model

First Component						Second Component						
Component	Dem	Apt	Cul	Soc	Pas	Per	Int	Goa	Sel	Per	Exp	Ach
Demographics		.40			.06					.32		.23
Aptitudes	.44				.29			.33		.39		.33
Culture					.14			.23		.30		.23
Socializers									.15	.28		.18
Past Events	.02	.23	.25					.10		.08		.46
Perceptions of Socializers										.49		
Interpretation of Past Events												
Goals/Self-Schemata		.17	.21		.07							.18
Self-concept			.18							.15	.52	.25
Perception of Task Value	.31	.28	.26	.34	.18	.45			.14			
Expectancies									.09			.16
Achievement	.25	.32	.25	.66	.47			.17	.23	.12		

Note: Values for males are above the diagonal; those for females are below the diagonal. Underlined values do *not* differ from zero at the .05 level of significance. Table 6.10 shows the numbers of correlations averaged for each path.

Studies Used for This Review

Studies in Analysis and Dependent Studies

Anderson, G. J.
1969a Effects of classroom social climate on individual learning. Unpublished doctoral dissertation, Harvard University.
1969b Effects of classroom social climate on individual learning. Paper presented at the annual meeting of the American Educational Research Association, Washington, DC. (ED 045 530)
1970 Effects of classroom social climate on individual learning. *American Educational Research Journal* 7:135–152.

Baker, D. R.
1981 The differences among science and humanities males and females. Paper presented at the annual meeting of the National Association for Research in Science Teaching, Ellenville, NY, April. (ED 204 143)

Bodner, G. M.; T. L. B. McMillen; T. J. Greenbowe; and E. D. McDaniel
1983 Verbal, numerical and perceptual skills related to chemistry achievement. Paper presented at the annual convention of the American Psychological Association, Anaheim, CA, August. (ED 238 349)

Bridgham, R. G.
1969 Classification, seriation, and the learning of electrostatics. *Journal of Research in Science Teaching* 6:118–127.

Cohen, M. P.
1979 Scientific interest and verbal problem solving: Are they related? *School Science and Mathematics* 79:404–408.

Dunlop, D. L., and F. Fazio
1977 Piagetian theory and abstract preferences of secondary science students. *School Science and Mathematics* 77:21–25.

Erb, T. O., and W. S. Smith
1984 Validation of the attitude toward women in science scale for early adolescents. *Journal of Research in Science Teaching* 21:391–397.

Gilmartin, K. J.; D. H. McLaughlin; L. L. Wise; and R. J. Rossi
1976 *Development of Scientific Careers: The High School Years.* Palo Alto, CA: American Institutes for Research in the Behavioral Sciences. (ED 129 607)

Grobman, H.
1965 Identifying the "slow learner" in BSCS high school biology. *Journal of Research in Science Teaching* 3:3–11.

Haladyna, T.; R. Olsen; and J. Shaughnessy
1982 Relations of student, teacher, and learning environment variables to attitudes toward science. *Science Education* 66:671–687.

Handley, H. M., and L. W. Morse
1984 Two-year study relating adolescents' self-concept and gender role perceptions to achievement and attitudes toward science. *Journal of Research in Science Teaching* 21:599–607.

Hardester, L. M.
  1976   An analysis of immediate and delayed verbal cognition information processing, mental operation levels and course performance of college freshmen by sex. Unpublished doctoral dissertation, University of Pittsburgh.

Ignatz, M.
  1982   Sex differences in predictive ability of tests of Structure-of-Intellect factors relative to a criterion examination of high school physics achievement. *Educational and Psychological Measurement* 42:353–360.

Jensen, J. A.
  1966   An analysis by class size and sex of orthogonalized interest and aptitude predictors in relation to high school chemistry achievement criteria. Unpublished doctoral dissertation, University of Rochester.

Jones, W. P.
  1970   Sex differences in academic prediction. *Measurement and Evaluation in Guidance* 3:88–91.

Kaminski, D. M.
  1978   Entry into science: The effect of parental evaluations on sons and daughters. Unpublished doctoral dissertation, Western Michigan University.

————, and E. Erickson
  1979   The magnitude of sex role influence on entry into science careers. Paper presented at the meeting of the American Sociological Association, Boston. (ED 184 855)

Kelly, A., and B. Smail
  1986   Sex stereotypes and attitudes to science among eleven-year-old children. *British Journal of Educational Psychology* 56:158–168.

Linn, M. C., and S. Pulos
  1983   Male-female differences in predicting displaced volume: Strategy usage, aptitude relationships, and experience influences. *Journal of Educational Psychology* 75:86–96.

Marjoribanks, K.
  1976   School attitudes, cognitive ability, and academic achievement. *Journal of Educational Psychology* 68:653–660.

  1978   The relation between students' convergent and divergent abilities, their academic performance, and school-related affective characteristics. *Journal of Research in Science Teaching* 15:197–207.

Morse, L. W., and H. M. Handley
  1982   *Relationship of Significant Others, Parental and Teacher Influences to the Development of Self Concept, Science Attitudes and Achievement Among Adolescent Girls.* Washington, DC: National Institution of Education. (ED 238 902)

Neale, D. C.; N. Gill; and W. Tismer
  1970   Relationship between attitudes toward school subjects and school achievement. *Journal of Educational Research* 63:232–237.

Ormerod, M. B.
  1973   Social and subject factors in attitudes to science. *School Science Review* 54:645–660.
  1975   Single sex and co-education: An analysis of pupils' science preference and choice and their attitudes to other aspects of science under these two systems. Conference at the Centre for Science Education, Chelsea College, London, England.
  1981a   The effects of single sex and coeducation on science subject preferences and choices at 14 + . *School Science Review* 62:553–555.
  1981b   The social implications of science and science choices at 14 + . *School Science Review* 63:164–167.
  ———; M. Bottomley; W. P. Keys; and C. Wood
  1979   Girls and physics education. *Physics Education* 14:271–277.

Payne, B. D.; J. E. Smith; and D. A. Payne
  1983   Sex and ethnic differences in relationships of test anxiety to performance in science examinations by fourth and eighth grade students: Implications for valid interpretations of achievement test scores. *Educational and Psychological Measurement* 43:267–271.

Pell, A. W.
  1985   Enjoyment and attainment in secondary school physics. *British Educational Research Journal* 11:123–132.

Peng, S. S., and J. Jaffe
  1979   Women who enter male-dominated fields of study in higher education. *American Educational Research Journal* 16:285–293.

Rakow, S. J.
  1985   Prediction of the science inquiry skill of seventeen-year-olds: A test of the model of educational productivity. *Journal of Research in Science Teaching* 22:289–302.

Roberts, R. J.
  1965   Prediction of college performance of superior students. *National Merit Scholarship Research Reports* 1:11–19.

Schock, N. H.
  1973   An analysis of the relationship which exists between cognitive and affective educational objectives. *Journal of Research in Science Teaching* 10:299–315.

Smail, B., and A. Kelly
  1984a   Sex differences in science and technology among 11-year-old schoolchildren: I—Cognitive. *Research in Science and Technological Education* 1:61–76.

1984b   Sex differences in science and technology among 11-year-old school-children: II—Affective. *Research in Science and Technological Education* 2:87–106.

Smith, I. R.
1966   Factors in chemistry achievement among eleventh-grade girls and boys. Unpublished doctoral dissertation, Catholic University of America.

Storey, A. G., and Desson, G. H.
1966   Achievement as a function of assigned grades. *Alberta Journal of Educational Research* 12:269–274.

Van Harlingen, D. L.
1981   Cognitive factors and gender related differences as predictors of performance in an introductory level college physics course. Unpublished doctoral dissertation, Rutgers University.

Weimer, L. J.
1985   Sex differences in achievement beliefs of bright children. Unpublished doctoral dissertation, University of Washington.

Welch, W. W.; S. J. Rakow; and L. J. Harris
1984   Women in science: perceptions of secondary school students. Paper presented at the annual meeting of the National Association for Research in Science Teaching, New Orleans, April.

Studies Withheld from Analysis and Dependent Studies

Cannon, R. K., Jr., and R. D. Simpson
1985   Relationships among attitude, motivation, and achievement of ability grouped, seventh-grade, life science students. *Science Education* 69:121–138.

deBenedictis, T.; K. Delucchi; A. Harris; M. Linn; and E. Stage
1982   Sex differences in science: "I don't know." Paper presented at the annual meeting of the American Educational Research Association, New York, March. (ED 216 868)

Deboer, G. E.
1984   A study of gender effects in the science and mathematics course-taking behavior of a group of students who graduated from college in the late 1970s. *Journal of Research in Science Teaching* 21:95–103.

Emmeluth, D. E.
1979   *An Assessment of Selected Variables Affecting Success in Community College Introductory Biology.* Johnstown, NY: Fulton-Montgomery Community College. (ED 174 298)

Khan, S. B.
  1973   Sex differences in predictability of academic achievement. *Measurement and Evaluation in Guidance* 6:88–92.

Linn, M. C.; T. de Benedictis; K. Delucchi; A. Harris; and E. Stage
  1987   Gender differences in National Assessment of Educational Progress science items: What does "I don't know" really mean? *Journal of Research in Science Teaching* 24:267–278.

Szabo, M., and J. F. Feldhusen
  1970   Personality and intellective predictors and academic success in an independent study science course at the college level. *Psychological Reports* 26:493–494.

Thomas, G. E.
  1981   *Choosing a College Major in the Hard and Technical Sciences and the Professions: A Causal Explanation.* (Report No. 313) Baltimore: John Hopkins University, Center for Social Organization of Schools. (ED 206 829)

7

Some Generic Issues
and Problems for Meta-Analysis

The studies included in this volume reveal that meta-analysis, like any other research endeavor, has its own internal logic. As described in Chapter 1, the organizational structure can be seen in Cooper's model (1989) of the stages of meta-analysis. The case illustrations also reveal a level of flexibility in the meta-analytic process. For example, literature searches (part of the data collection phase) may reveal that we have too few studies to answer the initial questions specified as part of the problem formulation. Problem formulation is likely to be iterative and can become more clearly focused over the course of the review. Or as the analyst progresses through the stages, the meta-analysis can be broadened. For example, statistical analyses may reveal additional hypotheses that could be assessed by returning to the original articles and extracting information on new variables of interest.

The flexibility of the method and the logical, roughly sequential ordering of the tasks should not instill a false sense of security. As our case illustrations show and as Olkin has stated (quoted in Mann 1990), doing a meta-analysis is easy, but doing one well is hard. The difficulty is due, in part, to the increasing complexity of the questions that meta-analysts are beginning to ask. Many of the early meta-analyses focused on simply describing—in quantitative terms—the results of prior studies. Having established that interventions produce robust effects that are nontrivial leads to such questions as what types of interventions work best and what causal mechanisms are at work? These questions move beyond description to explanation.

The basic data for meta-analysis is information contained in or e-duced from prior studies. Relying on this "observational data" has its strengths and weaknesses, as our case illustrations point out. This chapter examines some of the common problems confronted by meta-analysts, as well as their solutions. To organize our observations, we discuss key problems and constraints for each stage of the meta-analytic process. The reader will see that—as with any process—clear delineations of stages are almost impossible. For example, it is difficult to distinguish crisply aspects of data gathering from data evaluation. The borders are fuzzier than portrayed in this discussion.

Problem Formulation in Meta-Analysis

Scientific endeavors start with the formulation of the research problem, which can be simple and bivariate or complex and multivariate. In general, problem formulation in meta-analysis is not much different from that in primary research. Questions need to be clearly posed, and key constructs need to be distinguished from irrelevant ones.

In formulating research problems to be solved by meta-analysis, however, investigators face some constraints or pressures that differ at least in degree from those present in primary research. Chief among these constraints is the fact that answers to questions depend on the state of the literature. Cooper (1989) says, "Primary researchers are limited only by their imaginations, but research reviewers must study topics that already appear in the literature" (p. 19). A well-formulated set of questions to be answered by meta-analysis is of little use if the problem has not been investigated.

Less obvious is the constraint that for some frequently researched topics the depth, breadth, and technical adequacy of existing knowledge can fall short of that required to answer the meta-analytic questions. The problem formulation stage should anticipate these issues, devising strategies for data collection that account for features of the existing studies which influence whether the questions can be answered well, or at all.

Problem Formulation and Objectives

The rationale for conducting meta-analyses can vary from one review to the next. Each meta-analysis might contain several rationales or objectives. For example, Devine's analysis of the effects of psychoeducational interventions had two interrelated objectives: (1) to ascertain whether recent studies continued to support the positive results of prior syntheses and (2) to probe several different theoretical models in order further to develop treatment theories and practical strategies. Answering the first objective was relatively straightforward, representing a replication of prior work. Her second objective involved charting some new territory for meta-analysis. In formulating this aspect of the synthesis, she needed to derive explicit hypotheses so as to test the adequacy of each theory. Added to the original coding protocol, then, were theory-relevant variables. This set of a priori classifications provided the basis for selecting relevant studies.

Similarly, Becker formulated a sequence of goals for her meta-analysis. Starting with theoretical models, she first attempted to find what is known about factors that influence science achievement. Consistent with expectations, the simpler and more basic relationships had been studied more thoroughly and offered a platform for further work.

Becker's next objective involved an examination of models. Her data were useful in identifying the available support for some aspects of various models and for helping us to compare the merits of the models. Her data were also useful in constructing new models. Becker's analysis included additional variables not considered in the original models she relied upon in formulating her analysis.

In Devine's and Becker's work we see some similarities and some differences. Becker based her search for relevant studies on a set of theoretical models. Devine, on the other hand, first demonstrated that the overall findings still applied to more recent studies and then attempted to find explanations for these results. Both then used their data to inform further development of theories and treatment. Both analysts also found considerable gaps in the literature, curtailing complete assessments of their original meta-analytic questions.

Complex Versus Simple Questions

Meta-analysis has been especially effective in dealing with sharply posed questions: Does the drug streptokinase reduce the probability for further myocardial infarctions? The drug is well defined, the disease can be given an operational definition that few experts will decline to accept, and the outcomes—further heart attacks and deaths—are firm.

The current works represent more complex kinds of problems. For example, Becker's analysis focuses on an examination of *processes* that may engage young people in science and mathematics, especially girls and women. As a starting point, she modified an existing theory from the literature and attempted to verify it from the existing empirical research. Because different authors have different theories and because processes involve connections and steps, the meta-analytic enterprise becomes complicated. More is left to the investigators' ingenuity than in the simpler, sharply focused approach.

This ingenuity can take a number of forms. The form that is most clearly demonstrated in the case examples is the specification of variables that are likely to help explain differences among study results. Shadish, for example, imposes an "implementation theory" as a means of trying to explain the heterogeneity that he observed in study results. This was not a matter of simply recording information from research reports. Rather it involved a level of conceptualization that had to be superimposed upon the available studies. Similar types of ingenuity are seen in Lipsey's characterization of treatment conditions and methodological features of studies.

Although it is far more appealing to have simple answers to relatively simple questions, studying social processes through meta-analysis will probably require complex coding schemes that can account for substantial diversity in the methods, subjects, treatments, and measures used across primary studies. Whereas initial meta-analyses were predominately focused on simple questions about average effects, we expect that they will become more multivariate.

The problem becomes even more complex when causal models, like Becker's, are at issue. It is not unusual, when investigating a social problem, to focus on multiple constructs with causal flows that can be bidirectional. Causality can then operate in many ways between and among constructs. For example, enjoying an activity may contribute to improving one's performance, while better performance increases one's enjoyment. Thus, it is important to use whatever devices are available to reduce the number of eligible models. For example, a child's school performance ordinarily does not change the extent of a mother's formal education, but the extent of her education may contribute to the child's performance. In this example, we have eliminated one causal direction.

Such eliminations make a greater contribution than we might think at first. Each pair of constructs in such a model could allow causal influence in no direction, or in each of two directions, or in both directions simultaneously, making four possibilities in all. With 10 constructs, there are 45 pairs of points and so there are 1.2×10^{27} possible directional models. Pinning down the situation for one pair of constructs divides the possible number of models by four. Thus, one needs to be parsimonious about the number of constructs in a model. Unless the theory provides strong information about causal structure, we need to cut down the total number of models that are in competition.

Operationalization of Constructs

In meta-analysis, the problem of operational definitions does not go away simply because the reviewer uses results from primary literature. We still have to decide what sorts of measures to use. In many social problems, a single outcome is not satisfactory to represent the product of the process. For example, in Becker's study, positive end products could include (1) a good attitude toward science, (2) taking additional courses in science, (3) taking employment in a science-based field, or even (4) becoming a scientist. Society needs to know about all of these outcomes, including the final two. However, in the population as a whole the proportion of scientists is modest, and so, unfortunately, the outcome of choosing science as a vocation cannot be easily studied in young people. Thus, as a practical matter, positive attitudes and additional courses are likely to be the kinds of outcomes studied.

Knowledge Gaps

Typically, primary investigators have carried out substantial numbers of original studies in a general field. If we ask whether busing to achieve racial integration in schools also improves academic performance, we will find many studies that seem to deal with this topic. The problem for explanatory meta-analysis occurs when questions become more specific. Desegregation studies can be sorted on many variables, such as method of investigation, grade level studied, academic fields evaluated, and method of assessment. When the studies are sorted into what seem to be relatively homogeneous groups, most groups have few studies. The reason is simple: In a completely crossed system of categorical variables, the number of cells is the product of the number of categories in the variables. For instance, with two sexes, six grades, three methods of study, and three methods of assessment, 108 cells are created.

When too few studies are available to combine, the synthesizer may wish to relax the conditions for combining. If the outcome measurements in the studies are not comparable even though the methods and questions seem much the same, reviewers then must decide whether some important questions have to be set aside until suitable primary studies are carried out. Easing the comparability rule may make it possible to pool several cells that were originally separated. The sort of data that are available will help determine how the evidence from the studies can be combined, which in turn helps determine how specifically a question can be answered.

Meta-Analysis Can Tell Us What We Don't Know

While most meta-analyses emphasize an effort to summarize existing findings in a systematic way, it is possible that the most important finding from an analysis will be a finding that the existing data *do not answer* a policy question in a definitive way. For example, in his meta-analysis of research on well-baby care, William R. Shadish noted that millions of parents in America routinely take their young children to a pediatrician for "well-baby" care. He asked what the evidence shows about the effectiveness of such care on measurable outcomes for children's health. He found little evidence either for or against the value of such care, despite the billions of dollars and many hours of doctors' time spent on it. He concludes his meta-analysis: "The methodological quality of the 38 studies is mediocre on the whole, and extremely poor in some cases. Given this, the lack of attention to assessing the plausibility of rival hypotheses is unfortunate. . . . On the whole, then, the credibility of these findings is not high. . . . However, it is quite possible to begin to accumulate more useful evidence."

Source:

From Shadish, W. R., Jr.
   1982   A review and critique of controlled studies of the effectiveness of preventive child health care. *Health Policy Quarterly* 2:24–52.

In the absence of previous studies, meta-analysis cannot add information except to report on the need for more research. For policy purposes, the identification of gaps in knowledge can be especially helpful. For example, an evaluation synthesis conducted by the U.S. General Accounting Office (1986) sought to determine if a proposed piece of legislation to deal with teenage pregnancy would work if passed and funded. In refining the questions to be addressed, GAO characterized the conceptual model underlying the proposal and proceeded to synthesize the program evaluation literature on the strength of associations among proposed program elements (e.g., Is there evidence that comprehensive services yield increased health outcomes?). As with Becker's example, the conceptual model served as the basis for the literature review. The GAO synthesis revealed some suggestive evidence on paths within the model and also revealed those aspects of the model where relevant information was not available. GAO cautioned against implementing a full-scale program based on so many untested linkages

> Meta-analyses are ideal at identifying "research holes" that must be filled before a comprehensive synthesis can be performed.

in the model, arguing that well-evaluated demonstration projects would be more appropriate given the lack of relevant, credible data in the existing literature. A similar conclusion could be reached based on the gaps found by Becker in the science and gender literature.

Data Collection and Literature Search

Critics of meta-analysis sometimes point out that the outcome of any review depends heavily on which studies are used in the analysis and which are excluded. After all, in any substantive area where conflicting results exist, an advocate can steer a meta-analysis toward the conclusion sought simply by choosing the subset of studies that reach the favored conclusion. In this section, we explore several challenges for gathering studies and deciding which studies should go into a meta-analysis.

Organizing a Search

There are at least three specific steps essential to organizing a systematic search for studies for a meta-analysis.

The *first step* is to use computerized databases, accessing the databases by choosing keywords for your meta-analysis. Each of the case studies in this volume carried out such a search.

Becker, in her search for articles on science achievement and gender differences, accessed three sources: the *Educational Resources Information Center* (ERIC) database, the *Psychological Abstracts* (PSYC) database, and the *Dissertation Abstracts* (DISS) database. These three sources led to Becker's identifying about three-fourths of the articles she ultimately used in her meta-analysis.

Shadish examined six computer-based data sets. He completed a computer search of *Psychological Abstracts, Dissertation Abstracts International*, the *Social Science Citation Index*, the *National Center for Family Research*, the *National Clearinghouse for Mental Health*, and *Mental Health Abstracts*.

Devine faced the challenge of searching through both the psychological literature and the medical research literature. She used three computerized databases: *Dissertation Abstracts, Psychological Abstracts,* and *Medlars.*

Finally, Lipsey scanned the largest set of databases for his meta-analysis of delinquency programs. He felt that since delinquency spans many fields, it would be too narrow to search only in psychology or criminal justice. Therefore, he included 24 databases in his extensive search.

While a search of computerized databases is a crucial first step, it is not the final step. In the medical area, for example, searches based on computers and databases find roughly half the published articles available in English.

A good *second step* is to examine the lists of references at the end of key research reports. Such cross-checking will often turn up additional relevant information not initially located by keyword searching strategies because the meta-analyst used keywords that the authors of the original article did not use. If several articles are identified in this way, a constructive next step is to see if the large computerized database actually has these additional articles in its list and, if so, what keywords these articles use. Finding a new keyword may lead to additional articles.

A *third step* is to contact colleagues and fellow scholars around the country. Each of the four authors of the case studies in this book asked many colleagues for suggestions about key articles and research reports. For example, in an extraordinarily meticulous effort, Devine actually contacted 138 graduate programs accredited by the National League for Nursing. Studies that appeared relevant from their title or abstract were then obtained.

The goal that drives all three of these approaches is that of inclusiveness. There is some risk, in the real world, that the computerized database searches will turn up so many articles that the meta-analysis becomes unwieldy. For example, when searching for studies of delinquency treatment, Lipsey reports in his case study that he found more than 8,000 citations that, in principle, were potentially useful for his meta-analysis. So a meta-analyst must be prepared to deal with an enormous set of potential studies.

Many think that it is essential that every study be found. More important is that the methods of finding studies avoid special biases. If it were essential that every study be found, then the absence of the very next study in the literature would invalidate the meta-analysis. Of course, the next study might be so authoritative as to dominate the information in a meta-analysis, or it might reveal a key variable that was not present in earlier studies and thus invalidate all of them, but such major

> ### Sampling in Time
>
> One might suppose that if our MEDLARS approach were perfect and produced all the papers we would have a census rather than a sample of the papers. To adopt this model would be to misunderstand our purpose. We think of a process producing these research studies through time, and we think of our sample—even if it were a census— as a sample in time from the process. Thus, our inference would still be to the general process, even if we did have all appropriate papers from a time period.
>
> *Source:*
>
> Gilbert, J. P.; B. McPeek; and F. Mosteller
> 1977 Progress in surgery and anesthesia: benefits and risks of innovative surgery. In J. P. Bunker, B. A. Barnes, and F. Mosteller, eds., *Costs, Risks, and Benefits of Surgery.* New York: Oxford University Press, p. 127.

breakthroughs are uncommon. Except for keeping abreast of developments, we have no way to protect ourselves against scientific advances, nor do we wish to.

Narrowing a Broad Search

Of the 8,000 candidate studies that Lipsey identified, he used 443, or only about 5 percent. Of the several hundred that Devine identified, she used 171. Shadish's six computer searches turned up 2,539 articles that met the criterion of his keywords, but he used only 163. Becker identified 522 documents, but used only 39. Thus, the cases illustrate how the meta-analyst needs to implement a thorough search but will generally use only a small proportion of the studies identified and examined. How does the meta-analyst make the "cuts"?

Choosing which among a large list of potential candidates to include is one of the most crucial decisions to be made in organizing a meta-analysis. *Therefore, the meta-analyst must tell the audience specifically what was done.* The audience may or may not agree with every decision— indeed it won't—but at least it will know the decision rules. The authors of the four case studies have done an exemplary job here. For example, Becker, in her methods section, gives a step-by-step description of how she made decisions to include or exclude different kinds of studies.

Studies may be excluded if the treatment is not precisely defined or if the treatment is different from the standard or model treatment that is being included. For example, in Lipsey's collection of treatments for juvenile delinquency, he might have chosen to exclude any treatment that was not typical or usable for policy purposes, such as electric shock treatments. Again, whether or not it is appropriate to use such treatments in the real world is not the point here; the point is that the meta-analyst must make decisions about inclusion and then present them clearly in the report.

Unpublished Studies

All four cases in this book illustrate the importance of including unpublished studies, which may be doctoral dissertations or master's theses. In some cases, the data from the dissertation or thesis may have later appeared as a published study. Shadish, addressing this possibility directly, reports that "when we located a study in both dissertation and published form, we coded the former on the assumption that dissertations report more complete results and thus yield more accurate estimates of population effect sizes."

There are at least two good reasons why it is so important to track down these unpublished documents. First, a dissertation may be high-quality science, even if the author chose not to submit it to a journal or other source of publication. Second, obtaining such documents will provide some concrete evidence of whether or not the "file-drawer problem" is in fact a problem. Rosenthal (1979) identified the file-drawer problem as a situation in which most of the published studies report significantly larger effects of a treatment or program than unpublished studies. The reason is that many researchers hesitate to publish results of a study in which no significant treatment effects turn up. Indeed, Greenwald (1975), in an empirical survey of researchers and editors of scientific journals, asked respondents to tell what they do if their research turns up a statistically significant finding versus what they do if their research turns up a nonsignificant result. He found strong empirical evidence that suggests the possibility of publication bias. Respondents to his survey were eight times more likely to submit the results of their work if they turned up statistically significant effects than if they did not. Greenwald concludes that it is crucial, for getting a full picture of the range of findings on any research problem, to consider unpublished documents.

All four meta-analyses presented in this volume searched for and used unpublished studies. Devine made a substantial effort to track down unpublished documents. As a result, only 50 of her final group of 171 studies were published in journals. Indeed, the majority (99) were master's theses. Becker's meta-analysis had several unpublished studies among the sources that were ultimately used. Lipsey found that only 45.1 percent of the sources in his meta-analysis were published as journal articles, book chapters, or actual books. Shadish found that, in summarizing the results of studies with behavioral outcomes, the published studies have more than twice the effect size (an average of 1.09) of the unpublished studies (an average of .53). His meta-analysis illustrates in an especially graphic way the importance of tracking down unpublished work.

Primarily Descriptive Studies

Meta-analysts have learned from experience—which is illustrated in the four cases in this book—the importance of the verbal descriptions in studies that accompany quantitative data. In early work on meta-analysis, some authors (e.g., Light and Smith 1971) argued that its biggest strength is that it provides somewhat more "objective" information than narrative research reviews. They argued that meta-analysis provides a more rigorous effort to pull together results from many studies into a coherent, useful picture. Therefore, studies that emphasize narrative descriptions of a treatment or setting, and certainly narrative descriptions of an outcome, should not be included in a meta-analysis.

In one sense this still remains true. A study with *no quantitative measurement at all* cannot be included in a meta-analysis. It would be, for example, impossible to estimate any effect size. The four cases in this book illustrate how many studies are discarded after they are identified by a computer search simply because they do not have concrete quantitative data. But the qualitative information that accompanies the quantitative information is crucial. Meta-analysts should pay close attention to the descriptive information that accompanies data because it provides important insights about whatever program or process is being studied.

The Shadish meta-analysis illustrates how much we can learn from the narrative that accompanies any quantitative data. First, a central finding from Shadish is that "experimenter allegiance" is a crucial predictor of success. But how would we know this? We know it only from the text that surrounds that actual data. By identifying this variable, coding it, and including it in his analysis (Table 5.6), Shadish found that, on average, when experimenter allegiance to a treatment for marital therapy is high, the therapy has a strong, positive effect. When the experimenter's allegiance to a treatment is low, the effect size on average drops nearly to zero. The bottom line from Table 5.6 is that we get a much fuller understanding of *when* marital therapy works, and when it works best, from the narrative, nonquantitative descriptions in studies.

How Far Back in Time a Search Should Go

Since most meta-analyses involve examining a treatment to see how effective it is compared with some alternative, the meta-analyst must make a decision as to whether the treatment is the same in 1991 as it was in 1971. If yes, then the meta-analyst should include as many studies as can be found regardless of when they were carried out. If no, then the meta-analyst must decide on a "start" date for the search for studies and then inform the audience of the reasons for that decision. In general, as technology moves forward over time, some procedures (e.g., open heart surgery) really are different, substantively, in 1991 than they were in 1971. Other treatments, perhaps psychotherapy for marital problems, may not be so very different in 1991 from what they were in 1971. This is a somewhat arbitrary decision, but the point is that decision, whatever it is, should be presented to the audience in the write-up.

Our four cases show how different meta-analysts handle this challenge in practice. Shadish decided that marital psychotherapy has not changed so much in recent years that older studies should be ignored; so he included studies from a long range of time. Devine included published and unpublished sources from 1961 to 1988. In Table 3.2, she provides a summary of how effect sizes look for different time periods over the last 30 years and shows that there is no significant change, over time, in the value of psychoeducational interventions. Devine's finding tells us that we can, with reasonable confidence, assume a steady "effect size" over time for psychoeducational interventions for surgery. Future research might provide a basis for challenging this finding, but the constancy of effect sizes over 30 years certainly suggests that new research is unlikely to alter this relationship.

Data Evaluation

Meta-analysis is not simply a mechanical exercise. Research findings are not simply transcribed onto coding sheets, entered into a database,

On Good Technology Assessment

Authors of technology assessments often minimize the effect of bias and inadvertent errors by following steps similar to those expected of any scientific study. They make available the data used for the analysis, either including it in the report or furnishing it on request. For example, assessments that employ a meta-analysis of randomized, controlled trials often display in detail the data and the criteria for the inclusion of the data sources. In addition, the investigators describe clearly and prominently the assumptions used in the analysis and perform a sensitivity analysis to show the consequences of altering uncertain assumptions. Studies that follow these steps allow knowledgeable readers to reject or confirm the conclusions in a way rarely feasible with laboratory research.

Source:

Fuchs, V. R., and A. M. Garber
  1990   The new technology assessment. *New England Journal of Medicine*
      323:673–677.

and subjected to statistical evaluation. Rather, several forms of data evaluation must be undertaken prior to statistical analysis.[1] These entail judgments and decisions on the relevance and technical adequacy of information from primary studies. Further, to ensure the integrity of the coding process, specialized methods have been adapted from other arenas (e.g., behavioral observation techniques) to minimize human error and biases that may be interjected by these judgments. Our cases illustrate commonly accepted practices in data evaluation and also point out unique procedures that may be used.

Deciding Which Studies Should Remain in the Meta-Analysis

If the meta-analyst were interested in the effects of taking aspirin on adult women's body temperature, it might be easy to decide to include all the studies identified. This would be especially true if every study

1. used the same treatment (e.g., the same 365 mg aspirin tablet), regardless of where the study was carried out;

---

[1] Some degree of data evaluation occurs throughout most phases of meta-analysis (see Cordray 1990b). This section focuses on procedures and issues associated with a priori decision rules that have been built into the overall analysis strategy.

When Shall We Mount the 51st Study?

The meta-analytic process of cleaning up and making sense of research literatures not only reveals the cumulative knowledge that is there, but also provides clearer directions about what the remaining research needs are. That is, we also learn what kinds of primary research studies are needed next. However, some have raised the concern that meta-analysis may be killing the motivation and incentive to conduct primary research studies. Meta-analysis has clearly shown that no single primary study can ever resolve an issue or answer a question. Research findings are inherently probabilistic (Taveggia 1974), and, therefore, the results of any single study could have occurred by chance. Only meta-analytic integration of findings across studies can control chance and other artifacts and provide a foundation for conclusions. And yet meta-analysis is not possible unless the needed primary studies are conducted. In new research areas, this potential problem is not of much concern. The first study conducted on a question contains 100% of the available research information, the second contains roughly 50%, and so on. Thus, the early studies in any area have a certain status. But the 50th study contains only about 2% of the available information, and the 100th, about 1%. Will we have difficulty motivating researchers to conduct the 50th or 100th study? The answer will depend on the future reward system in the behavioral and social sciences.

Source:

Hunter, J. E., and F. L. Schmidt
  1990 *Methods of Meta-Analysis: Correcting Error and Bias in Research Findings.* Newbury Park, CA: Sage, pp. 38–39.

2. used exactly the same outcome measure (change in women's body temperature two hours after taking the tablet);
3. used the same population (adult women); and
4. used the same "good" research design, where women with a fever were randomly assigned to two groups, one group receiving aspirin and the other group receiving a placebo tablet.

How often do the population of studies approach this ideal situation? Almost never. Treatments are rarely so standardized across different studies. Indeed, sometimes even treatments with the same name are more variable across studies than the standardized aspirin tablet. For example, in Lipsey's meta-analysis of delinquency treatment programs, over 400 studies include a variety of different treatments and different goals. Similarly, Shadish's meta-analysis includes some studies in which a treatment is simply labeled "marital therapy," some studies in which marital therapy is used to ameliorate a specific individual's problem, such as agoraphobia, and some studies in which the presenting problems are sexual in nature. Both Lipsey and Shadish chose to deal with these differences by being explicit about their decisions to include or exclude certain studies.

The point here is not that there is a "right answer" as to which studies should be included or excluded, but that each meta-analyst should state clearly at the outset the criteria for excluding some of the studies identified, and then explain why the criteria are appropriate for the specific meta-analysis.

Minimizing Error and Bias in Judgments

Meta-analysis has been criticized for (1) camouflaging a "garbage in–garbage out" process by using fancy statistical techniques, (2) failing to take into account differential quality of evidence in primary studies, (3) "crowding out wisdom" by routinizing coding schemes and relegating the task of coding studies to research assistants, and (4) allowing "bad science" (i.e., poor studies) to drive out good science by weight of numbers (see Wachter 1988). There is some validity to these concerns (see Sacks et al. 1987). However, our case illustrations show that these criticisms can be overcome by careful consideration of the relevance and technical merits of individual studies or data sets. Indeed, the application of carefully constructed data evaluation protocols shows how wisdom can be properly applied so that meta-analytic studies rely most heavily on the best available scientific evidence in answering questions. Although competent judgment is essential for high-quality meta-analytic research, safeguards against error and bias resulting from human judgment must be installed as part of the synthesis protocol (Cordray 1990a, 1990b). Bias can enter at various stages of the evaluation process. Cooper (1989) has reported on several empirical studies suggesting that judgments of technical quality are influenced by the analysts' predispositions. For example, studies show that ratings of technical quality can be colored by knowledge of research results. Further, studies of the peer review system show considerable lack of agreement among experts on the merits of manuscripts submitted for publication. And even if the meta-analyst is aware of his or her predispositions and guards against them, the data coding task is tedious, leaving room for errors due to fatigue and bordom. Our case illustrations demonstrate several procedures designed to minimize the influence of these errors and biases.

Synthesis Protocol

In principle, data evaluation tasks for meta-analysis parallel tactics that should be employed in any form of research. Few primary studies are launched and successfully executed without a protocol for data collection, evaluation, and analysis; the same is true for meta-analysis. Although this is obviously good scientific practice, Sacks et al. (1987) found that fewer than 20 percent of the meta-analyses they reviewed on health-related topics presented information on the use of a data collection protocol.

This protocol is not merely a coding sheet. In addition to detailing how contextual, methodological, and participant data are to be extracted and coded, the protocol should specify procedures for determining whether errors and bias in primary studies are sufficient to warrant exclusion or differential weighting prior to aggregation. Although these topics are treated in the next section on statistical analysis, planning for these analyses must be part of the coding scheme.

Judgments and Dimensions of Study Relevance

As discussed in the previous section, the raw output of the search process is likely to yield many studies that do not have direct bearing on the meta-analysis questions under consideration. This makes sense inasmuch as the goal of the bibliographic search process is to assure comprehensiveness. However, comprehensiveness per se is not the only goal of meta-analysis. The data evaluation phase serves as a means of keeping track of which studies should be retained for analysis and which should be set aside as irrelevant. There are no absolute standards for these judgments. A narrowly constrained question can limit the pool of prior studies or findings. Similarly, broad questions or loose criteria open the floodgates, allowing many studies to be included. There are clear trade-offs associated with each strategy. An advantage of using restrictive criteria, of course, is that it minimizes the number of studies that must be coded. On the other hand, questions of completeness and selection bias can arise. If criteria are used to exclude studies from further consideration—as opposed to serving as a basis of categorizing evidence—there is little opportunity for statistically examining the consequences of such exclusion rules.

Given the conditional nature of judgments of relevance, it is not possible to provide a set of general guidelines for deciding whether a particular study should be retained in any meta-analysis. However, prior meta-analyses and our case illustrations provide several dimensions worth considering.

One particularly useful set of criteria is provided by Bryant and Wortman (1984). Using Cook and Campbell's typology of "threats to validity" (1979), they argue that judgments of relevance can be framed in terms of construct and external validity. Here, construct validity refers to the correspondence between operationalization of treatments or outcomes with the conceptual variables implied by the meta-analytic questions.

Devine's decision rules regarding treatment constructs are consistent with the Bryant and Wortman framework. That is, in an effort to conduct a "strong test" of the effects of psychoeducational care, she selected studies in which the treatment group was known to differ in level of care, but excluded those below a minimum level. Excluded were studies of patients scheduled for diagnostic tests or therapeutic abortions and studies involving nonsurgical interventions (e.g., effects of medications). The theory-probing nature of Devine's synthesis also established constraints on her choice of study outcomes. Of particular relevance to available theories were outcomes associated with psychological stress, pain, recovery, and other theory-relevant outcomes (e.g., indices of treatment implementation); other outcomes (blood loss) were excluded from further consideration.

In Lipsey's synthesis, construct validity played a less dominant role in screening studies. Given the broad nature of Lipsey's research questions, his definitions provided wide latitude with respect to treatment and outcome constructs. Unlike Bryant and Wortman's decision rule that requires the exclusion of studies judged low in construct validity, Lipsey's scheme made provisions for coding the degree to which a study represented a coherent treatment package.

The second dimension of the Bryant-Wortman framework—external validity or generalizability—concerns the extent to which settings, populations, or time periods covered by particular studies correspond to the questions being investigated in the synthesis.

In Lipsey's review, studies were judged eligible if the target population were juveniles, concretely defined as aged 6 to 21. Using concepts of external validity similar to Bryant and Wortman, policy considerations dictated that studies conducted outside the United States and earlier than 1950 be excluded from the synthesis.

Becker's analysis of evidence underlying models of gender differences in science achievement represents an unusual application of meta-analytic techniques. She attempted to assess relationships among theory-relevant paths depicted in two models. Studies were deemed relevant if they contained information (zero-order correlations) on at least one pair of variables prescribed in the models, for males and females separately. Constructs in each model were defined broadly allowing specific operationalizations to be categorized as relevant.

Becker's model-driven synthesis not only provides a basis for including relevant studies, but also tells us where there are gaps in the literature. (Devine's use of theories serves a similar function.) That is, paths in the model serve as the road map for compiling available relevant information. When relevant paths have not been studied, there will be gaps. Knowledge of such gaps provides a useful basis for planning future research.

Judgments of Technical Adequacy

At least two positions have emerged on how differential technical quality should be handled in meta-analysis. Some analysts argue that all studies should be classified according to their level of technical merit and included in the meta-analysis. By virtue of the individual ratings, it is possible to examine, as part of the statistical analysis, whether methodological transgressions are systematically related to effect sizes. Other analysts argue that methodologically inferior studies should be excluded entirely. But how do analysts answer these questions: What constitutes adequate criteria for judging technical merit? Is there a consensus on these criteria? Are the criteria equally applicable or should they be tailored to the specific types of information demanded of the meta-analysis? Studies of the lack of correspondence among reviewers for scientific and professional journals (see Cooper 1989) cast doubt on whether such judgments can be meaningfully rendered.

The debates on whether all studies should be included as part of the database or only those that are technically adequate have been reviewed elsewhere (e.g., Cooper 1989; Wachter 1988). For the most part, the consensus appears to favor a compromise of the two extremes. This position is seen in the case illustrations. Lipsey's synthesis of delinquency studies used a three-step evaluation procedure. First, he excluded from consideration studies using poorly controlled designs (i.e., studies with no control group and post-test only comparisons with no information on group equivalence). For the remaining studies, he coded over two dozen methodological features underlying each outcome. That is, he excluded the very worst studies and through elaborate coding accounted for variations in the adequacy of design, measurement, and statistical properties of the remaining studies. Prior to performing his statistical analyses, his final data evaluation step Winsorized studies with extreme effect sizes or sample sizes.

Devine used a similar strategy. As a first step, her selection criteria limited studies to those (1) involving treatment and control conditions, (2) in which treatment and control subjects were obtained from the same setting, and (3) with at least four subjects in each treatment group. As a second step, the remaining studies were reviewed and coded according to methodological features of the studies. This resulted in the additional exclusion of about 10 percent of the outcomes due to further discovery of inadequacies in the study design, substantial preexisting differences among groups, and outcomes with heterogeneous variances across treatment and control conditions, among other inconsistencies. Final analyses retained all remaining studies and tested the influence of methodological deviations across studies.

Shadish used, in general, the presence of random assignment as the basis for inclusion in his study on the effects of family therapy orientations.

STUDIES WITH MANY DIFFERENT KINDS OF RESEARCH DESIGNS. For the internal validity of any study to be strong, most analysts believe that random assignment of people to treatments, or treatment to sites, is crucial. Randomization allows us to make causal inferences from treatments to outcomes, and any meta-analysis on the effectiveness of a treatment should therefore lean heavily on randomized studies wherever possible (see Hoaglin et al. 1982; Boruch and Riecken 1974).

There is vigorous debate among investigators, however, as to whether only randomized studies should be included. Some meta-analyses will, because of their goals and the pool of available studies, necessarily include nonrandomized studies. While acknowledging the importance of good research design as a prerequisite for including a study, our cases illustrate the variability in the decision-making process since they use different criteria for inclusion. At one extreme, Shadish uses only studies with randomization, with the exception of a few that in his judgment are important and have no detectable selection bias. Devine uses nearly as limited a criterion for research design; and of her 171 studies that yield a usable outcome measure, 70 percent (120 studies) use random assignment of people to treatment condition. Devine includes 51 studies without randomization, but then subdivides these 51 into three categories—high, medium, and low quality. This allows her, and readers of her work, to examine whether research design is important as a predictor of substantive outcomes.

The other two cases use broader criteria for inclusion. Lipsey looks for studies that use randomization, but he is willing to include some that have a reasonable description of an effort to make different groups comparable. Therefore, Lipsey includes some studies that use matching of groups, and others with pre-treatment and post-treatment measures. Altogether, 44 percent of Lipsey's studies used randomization.

Becker's examination of forces affecting gender differences in school science achievement takes a dramatically different approach to choosing studies. Since Becker is not doing a meta-analysis of a treatment or program, the outcome measure in Becker's work is not the standardized mean difference that the other cases all use. Rather, Becker uses correlation coefficients as her outcome measure. Therefore, Becker's meta-analysis uses 446 zero-order correlation coefficients, culled from the 32 different studies that she found examining men's versus women's achievement in science. A meta-analysis of relationships in which there is no active intervention, and therefore no controllable "treatment," requires a different criterion for including studies than one in which the impact of a particular treatment is being assessed. Randomization in such studies of relationships is not the crucial issue.

RATINGS OF TECHNICAL MERIT.   Some progress in developing measures of quality has been made (see, e.g., Chalmers et al. 1981; Urkowitz and Laessig 1985). Cooper (1989) delineates three approaches for categorizing research methods along technical quality dimensions. His first approach entails making judgments about threats to validity that exist in each study. Bryant and Wortman (1984) epitomize this approach. For assessing technical quality of comparative studies, Cook and Campbell's (1979) threats to internal validity and statistical conclusion validity are most relevant. For example, studies are rated as high, moderate, or low on internal validity based on consideration of assignment rules, attrition, pre-treatment equivalence, and so on. Previous meta-analyses or meta-evaluations using this type of framework have generally shown that studies of lower methodological quality can (but not always) produce biases in aggregate effects. Cooper details several arguments against this strategy, citing limitations due to a lack of consensus on which threats to count and how threats should be combined. Some of these problems are quite daunting, while others have been resolved (see Boruch and Gomez 1977).

Cooper's second approach is more descriptive. Here the analyst is required to describe the objective design features of each study, as reported by the primary researcher. That is, rather than making judgments about quality, design features are coded and the influence of variations across studies is assessed empirically. As noted for the "threats to validity" approach, prior meta-analyses have demonstrated that variations in study features have been associated with biased effects.

Cooper's third strategy involves a mixture of the "threats to validity" and "objective features" approaches. This is nicely illustrated and expanded upon in our case examples.

Lipsey's assessment of the methodological features of delinquency studies is one of the most comprehensive to date. It entails a mixture of description and evaluation of each study. Whereas the descriptive aspect follows Cooper's notions, the evaluative ratings depart from a standard "threats to validity" approach and avoid some of the problems raised above. That is, rather than categorizing studies as high or low in internal validity, Lipsey's format confronts the implications of threats to validity directly by rating the magnitude of the problem created by methodological transgressions. For example, the absence of random or known assignment to conditions generally results in some degree of pre-treatment nonequivalence. To assess this, Lipsey's coding scheme required an explicit rating of the overall similarity of treatment and control conditions, supported by researcher-generated evidence and the calculation of a pre-test effect size. Recall that Devine used the latter as a basis for excluding studies from her database.

Other direct ratings used in the cases include judgments of the representativeness of sampling (as in Becker's coding scheme), the overlap of a measure with the content of treatment, blinding in collection of outcome data, and statistical power. Coding schemes also included a mixture of judgments (i.e., subjectivity of measures, level of internal validity) and descriptive coding of study features.

Discussions of how to derive quality judgments are usually based on schemes rooted in notions about threats to valid causal inference (or their implications). Becker's meta-analytic questions required examination of correlational data. Therefore, her ratings of technical adequacy involved examining factors that affect the precision and accuracy of correlational indices. Hoaglin et al. (1982) provide checklists for a variety of methodologies (e.g., surveys, simulations, expert opinion) that can be drawn upon for structuring judgment protocols involving other forms of data.

Quality Control: Evaluating Coding Decisions in Meta-Analysis

Data evaluation in meta-analysis hinges on numerous judgments rendered by coders. Although the data collection protocol provides a means of standardizing these judgments, it is necessary to assess whether the decision rules and coding conventions are followed. Several quality control procedures have been developed to assess the integrity of these processes.

TRAINING OF CODERS. The synthesis protocol serves as the template for judgments about the quality of data found in primary studies and for subsequent decisions on whether data are included or excluded from the meta-analysis. Some aspects of coding are self-evident. Others require expert judgment or, at the very least, sufficiently clear decision rules to allow nonexperts to consistently apply the scheme.

Coder training was conducted in each of our case studies. This is not a particularly mysterious process. For example, Becker used a preliminary version of the coding protocol as a pilot-test and training device. This form of supervised administration not only served as a basis for training coders but also resulted in modifications to the coding scheme. Similarly, Lipsey noted the iterative feature of the training-coding scheme development process.

CODER AGREEMENT. Several tactics have been used to assure that coding is done reliably within and across coders. As above, there is nothing particularly esoteric about this aspect of the process. Experience suggests that many of the disagreements between coders arise because a coder filling out the protocol misses something in the primary report. Thus, it may be worthwhile to have more than one coder read each report, even though this is an expensive procedure. Coders should be asked to identify where the material reported was found, making adjudication between coders a fairly easy job.

Practices do vary across meta-analyses, however. Becker had each study screened and coded by at least two individuals. She found that agreement rates, calculated for each variable, ranged from 56 to 100 percent, despite training. Devine and Shadish used several indicators of agreement (e.g., Cohen's Kappa, percentage agreement), yielding a similar range across types of variables.

BLINDING OF CODERS. Ratings of methodological elegance, quality, or simple descriptive classifications of study characteristics can be influenced by knowledge of the study results. To combat this potential bias, intentional or not, it has been suggested that results and study characteristics should be coded separately (Sacks et al. 1987). Both Devine's and Becker's procedures employed some form of blinding. Becker used "differential photocopying" as a means of keeping knowledge of the results independent of the coding of methodological characteristics.

Devine's procedures were more informal, but the spirit of avoiding unintentional bias was clearly present.

Deficient Reporting: Confidence Coding and Other Tactics

Meta-analytic procedures have been used to aggregate primary research that dates back several decades or longer. Since much of the early research literature could not foresee that researchers at a later time would try to statistically aggregate results across studies or probe causal theories (some of which were not developed at the time a study was published), it is not surprising that Lipsey, Becker, and Devine found numerous gaps in reports. Further, as Orwin and Cordray (1985) demonstrate, data that are reported are not always presented clearly enough for meta-analytic purposes.

There are several solutions to these reporting problems. (1) External sources can be used to obtain information about instrumentation that was not reported in the primary study. Becker, for example, relied on published sources to ascertain the reliability and validity of tests reported in studies that she and her team reviewed. (2) The primary investigator can be contacted to obtain additional data or clarification of procedures. As the historical frame for the synthesis reaches back further in time, this strategy is likely to be limited. Increasing use of archives and other repositories for storing microdata and documentation can fix this problem to a certain extent (see Cordray, Pion, and Boruch 1990). (3) Deficiencies in reporting can be accounted for as part of the coding process. That is, by explicitly recording the recorder's confidence in the codes assigned to extracted data, it is possible to distinguish data that are poorly reported or difficult to code (e.g., it is not clear if random assignment was undertaken) from data that are well reported.

The coding schemes of Becker, Lipsey, and Shadish made provisions for recording confidence ratings. Shadish's application is instructive in that it shows how confidence ratings can be incorporated into the overall analytic plan. His average confidence ratings were quite high, suggesting that data are not as poorly disclosed in family therapy studies as in other areas (Orwin and Cordray 1985).

Data Analysis and Interpretation

The purpose of data analysis in research synthesis is to organize the information about studies and their findings to reveal patterns. In many respects, data analysis for a meta-analytic study is similar to that for a primary study, but they differ in two important ways that present challenges to the meta-analyst. First, meta-analysts depend on primary studies as their basic units of information, which forces them to combine evidence over studies with dissimilar designs and endpoints. Second, meta-analysts are always observational researchers in that they have no direct control over the design of primary studies, which poses the challenge of disentangling possible associations between study attributes and outcomes to uncover valid treatment effects or relationships. (Primary researchers face similar problems, but to a lesser degree.)

This section will sketch analysis and interpretation issues that arise because of these differences. This focus is not intended to diminish the importance of more standard statistical summaries. These are well documented in the work of Hoaglin et al. (1985), Light and Pillemer (1984), and their references. Hedges and Olkin (1985) present and explain more advanced techniques with specific reference to meta-analysis.

Discovery of Meaningful Patterns

The simplest statistical methods used in this volume organize information so that patterns are visible. The studies examined by Lipsey typically obtained rather small effects, many of which are not statistically significant. Yet, the histogram of effects and the tabular summary

Exploring, Selecting, and Peeking

In the game of bridge, they say "a peek is worth two finesses" because without knowing which direction to take a finesse, it has only a 50–50 chance of success. Accidentally seeing an opponent's cards can often assure success. In a similar way, exploratory data analysis by discovering accidental correlations can drive probability levels to extremes if the data analyses cannot take proper account of the extent of the explorations, and usually they cannot. For example, suppose that we have nine uncorrelated factors unrelated to a tenth which we want to explain. If each of the nine is correlated with the tenth, then 40 percent of the time the correlation coefficient of at least one of the nine factors will be significant beyond the 5 explored, and we have no way to allow for the exploration except to make a new study on fresh data. Exploring, while good practice, does not lead to trustworthy p values.

showing means and confidence intervals demonstrate that the preponderance of effects is positive. Across all 238 effects, a global pattern emerges, made visible through a numerical summary.

The effect magnitudes reported by Devine also varied substantially. The average effects and measures of variability reported by Devine provide an indication that the typical effect obtained in studies of this type is of moderate positive size against a background of substantial variation.

More sophisticated statistical summaries are provided by Lipsey and Becker, who use statistical adjustments to compute the relation between one set of study characteristics and an index of effect while controlling for the influence of other study characteristics. For example, in the analysis reported in Table 4.7, Lipsey rank orders the effects of various types of treatments while controlling for the effects of study methodology and study context. Becker uses a similar modeling strategy in her approach to estimate correlation matrices for males and females while controlling for various study characteristics that might influence those correlations.

Becker's analysis also demonstrates an application of a statistical method that is even more unusual in research synthesis: the use of statistical methods to combine information from many studies to estimate effects (in this case, path coefficients) that may not have been estimated within any single study. By synthesizing evidence about the correlation matrix of variables of interest, she computed estimates of correlations in her path model, even though no single study actually measured all of the variables involved.

Although the discovery of patterns is important, meta-analysis need not be purely exploratory. As our case illustrations show, meta-analysis is also used to test hypotheses to confirm the existence of patterns of results. A few meta-analyses will be entirely confirmatory, seeking to test hypotheses that are well articulated in advance of the study. In meta-analysis as in all other statistical work there is a tension between exploration and confirmation.

Conceptualization of Between-Studies Variation

A key consideration in the analysis and interpretation of evidence from a review is how to think about between-studies variation in effects. The conception of between-studies variation determines the details of the analyses to be done, the computation of the uncertainty of the combined results, and their interpretation. Both Lipsey and Shadish have investigated these influences. At least two potential sources contribute

Between-Experiment Biases in the Physical Sciences

We have argued that random effects models are desirable in research synthesis because the differences among the results of experiments are frequently greater than would be expected given the sampling uncertainty of the experimental results. Such larger-than-expected differences may persist even after controlling for between-experiment differences in study design, sampling plan, and study context. Such differences arise not only in the social and medical sciences but are found in the physical sciences as well.

Studies that estimate the value of physical constants (such as the mass and charge of the electron, Planck's constant, the fine structure constant, or Avogadro's number) provide a case study for illustrating how well physical experiments agree. Periodically physicists derive "recommended values" by reviewing all relevant experimental evidence or by conducting single high-accuracy experiments. Taylor, Parker, and Langenberg (1969) examined the changes over time in the recommended values of the five constants mentioned above. The graph illustrates their findings. Note that the differences between the recommended values from each redetermination (replication) are typically several standard errors of measurement. For example, the difference between the 1963 and 1969 recommended values of all five of the constants was three to five standard errors of the 1963 value. Such differences are highly statistically significant, reflecting between-experiment differences much larger than would be expected due to sampling uncertainty.

Such larger than expected between-experiment variations that are greater than would be expected given the within-experiment sampling uncertainty are not restricted to experiments that measure fundamental physical constants. They are also found in physical chemistry, astronomy, and certain areas of biology such as x-ray crystalography (see Hedges, 1987).

Reference

Taylor, B. N.; Parker, W. H.; and Langenberg, D. N.
   1969 *Fundamental Constants and Quantum Electrodynamics.* New York: Academic Press.

to variation in the empirical evidence examined in research reviews. One is a consequence of differences among subjects within a particular research study, and the other is a consequence of differences between studies. The variation between subjects or units within an individual

Figure 7.1 **Recommended Values of Five Fundamental Physical Constants Between 1952 and 1969 with Associated 68 Percent (one standard error) Confidence Intervals (values are expressed as deviations from 1969 values in parts per million)**

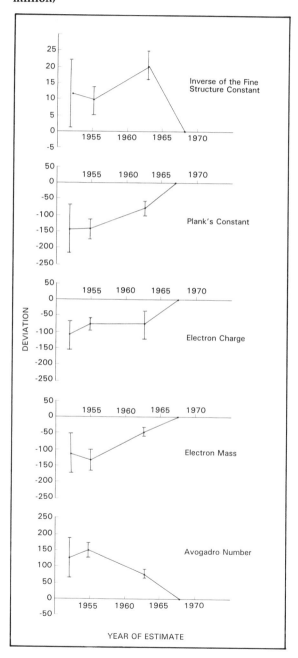

study (within-study variation) is at least partially the result of chance (sampling) processes. One could also conceive of differences between studies arising as a result of a chance or sampling process. For example, it might be useful to conceive of a universe of conditions of treatment implementation, each leading to a somewhat different treatment effect. Then, the particular collection of studies obtained is a sample from that universe, and between-studies variation provides information about the variation of the treatment implementations. This model is particularly attractive when considering (1) studies that are quite heterogeneous, (2) treatments that are ill-specified, and/or (3) effects that are complex and multi-determined.

This random-effects model may seem less attractive when studies are relatively homogeneous, treatments are relatively precise, and the mechanisms by which treatments produce effects are well understood. If this is the case, some researchers treat between-studies variation as a consequence of known (or at least knowable) characteristics of studies, such as treatment variety, duration, or intensity. For known characteristics, a regression model can be developed to explain part of the between-studies variation. The explained variation is not considered random but a consequence of a relatively small number of knowable and controllable factors (fixed effects). Tests of consistency of effects or fixed-effects model specification can suggest whether the observed between-studies variation in effects is consistent with a specified fixed-effects model.

Such explanatory regressions must be recognized as exploratory analyses. Because they are chosen from many possible variables, often by data dredging, the notion that they are part of a fixed-effects model may well be regarded skeptically.

The model of between-studies variation drives both the type of analysis and the range of generalization that are appropriate. When between-studies variation is treated as fixed, the only source of variation treated as nonsystematic is the within-study sampling variation and the analysis may be constructed accordingly. When between-studies variation is also treated as at least partially random, the statistical techniques must incorporate two sources of random variation. This affects the computation of the precision of estimates, usually decreasing precision to reflect the additional uncertainty arising from the sampling of studies. To the extent that the studies are conceived as a sample from a putative universe of studies, generalizations may be drawn about that universe of studies, including those that are unlike those actually observed. If the studies are conceived as a fixed universe, the generalizations are sharper but apply only to studies like those observed.

To clarify the issue of models for between-studies differences, imagine several studies, each extremely large, so that there is essentially no within-study variation on summary statistics. That is, if the study were repeated, the summary would not change. Inevitably, in such a situation we would still expect between-studies differences in outcome resulting from differences in design, types of subjects, measurement methods, and analytic methods. Some of the heterogeneity can be "explained" by, for example, covariance adjustment of age distributions, but some variation will be inexplicable using the available information.

Analyses based on the random-effects model take into account this unexplained variation. Failure to do so (when between-studies differences actually are random) can result in underestimates of variability of average effects and underestimates of the strength of relationship between study characteristics and outcomes. Statistical and substantive discussion of these consequences, preventions, and cures appears in Colton et al. (1987), Hedges and Olkin (1985), and Louis (1990).

Because we usually do not have a complete list of studies all of which would always be included, the fixed-effects model is often suspect.

The Random Effects Model

According to this formulation there is no single true or population effect of the "treatment" across studies. Rather, there is a *distribution* of true effects; each treatment implementation (site) has its own unique true effect. This leads naturally to the consideration of an average true effect of the treatment as an index of overall efficacy. However, this average true effect will not be very meaningful without some measure of the variation in the true effect of the treatment. For example, it is quite possible for the average true effect to be greater than zero, whereas the true effect of the treatment is negative in nearly half the implementations. The problem of estimating the variability in the true effects is further complicated by the fact that the true effect in any treatment site (or study) is never known. We must estimate that true effect from sample data, and that estimate will itself be subject to sampling fluctuations.

Source:

Hedges, L. V., and I. Olkin
   1985 *Statistical Methods for Meta-Analysis.* Orlando, FL: Academic Press,
       pp. 190–191.

Furthermore, in most of science, we do *not* have a list of studies from which we draw a random sample, which would correspond to the random-effects model.

Instead we usually have a set of studies that has been generated by some process that we can only partly describe. If we apply our random-effects model, it applies to the unknown process that chooses these studies. If the unknown sampled population is not the one that interests us, then additional variance beyond that of the random-effects model applies. This latter situation is not at all special to meta-analysis, but is a common feature of most field sciences such as sociology, biology, astronomy, and engineering.

To see how the random-effects approach operates, consider an artificial example with data similar to the length of stay (LOS) information in Devine, but where all studies use a randomized treatment versus control design and the same number of patients. Each study provides an estimate of the reduction in LOS and its associated standard error, where the standard errors are all equal. Table 7.1 presents the artificial data with summary statistics approximating those in the "days difference" row of Table 3.5 in Devine. The mean LOS reduction is 1.5 days, and a fixed-effects analysis (assuming no between-studies variation) computes a standard error for the men of .373 [$= \text{sqrt} (1.25/9)$]. An analysis dealing directly with the LOS reductions computes a sample

Table 7.1   Artificial Estimated Length of Stay (LOS) Reduction from Nine Studies

Study	LOS Reduction	Within-Study Variation
1	−0.25	1.25
2	1.50	1.25
3	2.40	1.25
4	4.00	1.25
5	−0.60	1.25
6	2.00	1.25
7	3.40	1.25
8	0.20	1.25
9	0.85	1.25
Mean	1.50	
Sample Variance	2.56	1.25

Note: Within-study variation is the square of the within-study standard error.

> To obtain appropriate estimates and standard errors, components of variance must be mapped and taken into account in the analysis.

variance of 2.56 and therefore a standard error for the mean of .533 [= sqrt (2.56/9)]. Between-studies variation produces this 43 percent increase in standard error (thus a 43 percent increase in the length of a confidence interval). The larger value more accurately represents the true variation in inferences that extend beyond the nine studies in the meta-analysis.

We can estimate the between-studies variation by computing the difference between the overall variation and that "explained" by the within-study variation, obtaining 1.31 [= 2.56 − 1.25]. Therefore 51 percent of the total variation is between studies. If each study were increased in size to drive the within-study variation to zero, we would still expect to see a sample variance of 1.31 and a standard error of the mean of .382 [= sqrt (1.31/9)]. Alternatively, the fixed-effects analysis would report a standard error of zero in this hypothetical example. The use of similar, but more complicated, variance components analyses using the program BMDP5V (Dixon 1988) is illustrated by Shadish. There is no mathematical "proof" that including the between-studies variation is always the best analysis, and controversy surrounds the question (Colton et al 1987). We recommend that variance components be identified and their influence incorporated in analyses whenever it is feasible.

When investigations regularly use the same variables for control, their use for variance components seems justified. If such variables come about from exploratory searches such as stepwise regression, their reduction of variance may be interesting but less compelling. In any case, unless our studies are sampled from a population of interest to us, we should add a grain of salt even to the inference based on the random-effects model.

Use of Tests of Heterogeneity

A statistically significant test for heterogeneity implies that between-studies variation is reliably bigger than zero. Hence the test can focus attention on finding covariates to explain the unexplained variation. Between-studies variation can exert an important influence on standard errors of combined estimates even when heterogeneity tests are not statistically significant. For example, a test for heterogeneity producing a chi-square statistic value of 12 on eight degrees of freedom is not statistically significant, but there may be a covariate (with one degree of freedom) that explains much of the chi-square value. Thus, the finding of no omnibus variation does not imply lack of significant variation associated with particular covariates. This is analogous to the finding that the overall F test for a factor in ANOVA may be nonsignificant, but contrasts among factor levels may be significant.

The studies in this volume exhibit several choices of models for between-studies variation. Lipsey most clearly adheres to a random-effects conceptualization. His collection of studies exhibits marked heterogeneity in treatment types, treatment implementation, study design, context, and outcome. Lipsey's analytic approach is that of modeling the between-studies variation in effects in as parsimonious a fashion as possible, given the substantial irreducible variation in effects. He emphasizes that the effects of treatments he examined are multiply determined by method, treatment, and context variables; and that estimated treatment effects, even after controlling for method and context, "include instances of varying efficacy ranging above and below the category mean and they overlap considerably for those many treatments with multiple elements, for example, school-based behavioral contracting program." Indeed, he cautions against the search for a single fixed treatment that would be "a 'magic bullet,' a specific treatment concept or program alleged to be a superior approach to delinquency."

Devine's use of combinations of effect estimates in probing the links of the model for the effects of information and skills teaching represents a successful example of the fixed-effects approach to modeling. The studies of each link were relatively few in number and were reasonably homogeneous. They tended to yield effects that were about as consistent as could be expected given the within-study sampling variation. Devine used both tests of consistency of results across studies (Q statistics) and between-studies variance component estimates to probe the degree of variation across studies whose results were combined in her analyses.

Becker also examined a somewhat more homogeneous collection of studies than did Lipsey. Her modeling of the between-studies variation in the correlation matrices as a function of study characteristics is also an example of a fixed-effects modeling strategy. Her use of the model specification statistic as a guide to model adequacy is a generalization of the strategy employed by Devine. Becker's search for an adequately specified fixed-effects model for the correlation matrices relies on the concept that the correlation matrix (and hence the path coefficients implied by it) is a function of a few fixed study characteristics.

Shadish used both fixed- and random-effects procedures in his modeling of the effects of marital and family therapy. His analysis illustrates both the differences in the results yielded by the two methods and the conceptual difficulties in completely justifying either approach. His work demonstrates that a random-effects approach mitigates the influence of large studies by evening out the weights given to studies in a weighted regression. It also increases the standard error for estimated population means. In exchange, the review is allowed an inference broadened from "these studies" to "all similar studies."

Making Studies Comparable and Combinable

As we have mentioned, studies can be noncomparable for a wide variety of reasons including differences in design, types of subjects, measurement procedures, and analytic methods. We should consider adjusting for between-studies differences (such as age distributions) before combining evidence. We discuss briefly the principal adjustments and standardizations.

DESIGN DIFFERENCES.   Important aspects of design include the type of intervention, subject attributes, context, basic design of the study (random assignment, matched control, static group), length of follow-up time, precision of the estimated study results, precision of the explanatory variables, and sample size. Weighted analyses deal directly with sample size, but the other aspects are more troublesome.

Studies can differ in the distribution of important covariates such as gender, age, or, more specifically in our cases, previous police record or disease status. Adjustments for differences in distribution of these covariates can make studies more comparable. The meta-analyst frequently encounters a collection of studies in which some have performed, for example, age adjustments and others have not. Between-studies variation can be reduced by adjusting all studies. Procedures need to be developed that approximate such adjustments even when explicit adjustments are not possible, but until they are available more ad hoc approaches are necessary. Some analysts insist that the raw data from each study be obtained so that such adjustments and other computations requiring such detail can be made.

Differing precisions of explanatory variables can introduce heterogeneity between studies through the attenuation effect on regression coefficients. Since in many applied contexts the attenuation effect can be as high as 50 percent, it can be important to de-attenuate coefficients before performing a meta-analysis (MacMahon et al. 1990). Sometimes the uncertainty in a variable can be deduced from the reported measurement techniques, and sometimes it is associated with coding ambiguities. This latter can be documented by having coders report their level of confidence for important items (Orwin and Cordray 1985). Confidence levels can be used to adjust regression slopes or included as an explanatory variable. Both Lipsey and Shadish take this latter approach.

SENSITIVITY ANALYSIS.   Weightings and adjustments, clear and insightful reporting of possible biases, and sensitivity studies for key assumptions and methods will produce a credible meta-analysis. Sensitivity analyses are especially important, since many assumptions cannot be verified empirically and no single analysis is indisputably superior. Sensitivity studies can be quite basic or complicated and multivariate.

Devine makes effective use of the basic approach to sensitivity analysis. She analyzed her data on the effects of psychosocial interventions by several methods. Each produced qualitatively, and quantitatively similar results, increasing confidence in the sturdiness of conclusions.

How Many Outliers?

There is considerable evidence that real data contain occasional observations that do not fit simple models well. The early developers of statistical methodology certainly believed that the exclusion of a certain amount of data from statistical analyses, solely on the basis of deviant values, was a good practice. Legendre, who is credited with the invention of the important statistical idea of least squares, recommended (in 1805) the use of his method after rejecting all observations whose errors "are found to be such that one judges them to be too large to be admissible" (Stigler 1973). Edgeworth (1887), another important contributor to the foundations of data analysis, reached the same conclusion: "The Method of Least Squares is seen to be our best course when we have thrown overboard a certain portion of our data—a sort of sacrifice which has been often made by those who sail upon the stormy seas of Probability (p. 269)."

Source:

Hedges, L. V., and Olkin, I.
   1985   *Statistical Methods for Meta-Analysis.* Orlando: Academic Press, pp. 249–250.

Similarly, Shadish explored the sensitivity of the results of his regression analyses to changes both in methods of computing study effect sizes and in methods of imputing missing effect size values. His finding that Winsorizing study effect sizes identified as outliers did not alter the relationships estimated (but improved the fit of the model) increased confidence in the robustness of those relationships.

Shadish takes another instructive approach to design differences by using a linked analysis to combining evidence over studies that compare different treatments. For example, a comparison of treatments A and C can be made using studies that compare A and B and studies that compare B and C. This is done by adding the A/B contrast and the B/C contrast. Unlike a main-effects model, this approach uses only within-study contrasts to build the indirect comparison. Such "chain-of-mail" approaches should be considered more frequently.

Covariates: Study-Level and Subject-Level

Increasing statistical power and precision by combining evidence over several primary studies is a major goal and virtue of meta-analysis. It is the ability to study a rich variety of covariate relations, however, that gives meta-analysis its true policy relevance. These relations may occur at the study level and can, therefore, be investigated only through meta-analysis. Alternatively, they may be available in individual primary studies. Investigations of the relation of outcome to study quality, publication date, type of publication, and funding source are possible only in a meta-analysis. Investigations of relations between experimental and subject factors are available to the primary researcher, but take on a greatly expanded form in meta-analysis.

This potential power must be used carefully, however, for the meta-analyst has not had control over the design of individual studies. Investigators may choose to study interventions that they know well and use subjects that are considered good candidates for the intervention. Therefore, it will be impossible to eliminate completely the possibility that apparent associations are the product of idiosyncracies. But, as in all observational studies, adjustment and discussion of possible biases can produce the best estimate of effects and a range of credible values. This creativity, however, requires the use of additional procedures to assure that we have not merely taken advantage of chance associations. Although none of our case studies reserved (through sample splitting) a fraction of the studies in their sample for purposes of *cross-validating* their models, such a tactic is simple to use. Of course, to be viable, cross-validation requires a large number of studies relative to the number of explanatory variables (say 20 to 1) used in the model.

Missing Data

The prevalence of missing data is most obvious when quantitative methods are used in research reviews, but missing information compromises the interpretability of all research syntheses. Sophisticated multivariate analyses examining the joint behavior of several variables are particularly vulnerable to problems of missing data, since computations of joint behavior ordinarily require complete data on all of the variables involved.

Information on details of treatment, context, and methodology were frequently missing in Lipsey's study. At least one variable in some critical clusters of variables was missing for over 75 percent of the studies. In part this may have been a consequence of Lipsey's extensive coding scheme, which was, however, essential, given his program of explaining variability in effects via coded study characteristics. The problem of substantial amounts of missing data also arose in Becker's study. Elements of the correlation matrix either were not computed or were not reported.

The challenge of carrying out sophisticated analyses in the presence of substantial amounts of missing data is a problem that has received considerable attention in applied statistics. Methods such as multiple imputation (Rubin 1987) and model-based estimation based on the EM (Estimation/Maximization) algorithm (Little and Rubin 1987) would seem to have considerable promise for meta-analysis, particularly as it moves toward more complicated multivariate data-analytic strategies.

Publication Bias

Authors of our case studies carefully document their extensive and exhausting searches for the published and unpublished literature. Yet, what they find may still not be representative of all studies performed. Several researchers (Devine and Cook 1986; Light and Pillemer 1984; Begg and Berlin 1988) have documented that generally the published literature reports more strongly significant findings than does the unpublished literature; published small studies generally report larger estimated effects than do published large studies. These features are likely the consequence of publication bias—the tendency of authors to submit and journals to accept statistically significant findings. Although this turns science on its head by declaring a question interesting because of its answer rather than an answer interesting because of the question, it is a current reality.

Many proposals have been made to deal with publication bias. These

range from the file-drawer method (estimating the number of unpublished studies with an overall effect that sums to zero that would be
needed to reduce the finding to nonsignificance; Rosenthal 1979) to
quite technical adjustments based on a model for the filtration process
that selects studies for publication (e.g., Iyengar and Greenhouse 1988).

Empirical studies of the magnitude of the bias in different disciplines
and for different study designs would help pin down the appropriate
adjustments. A meta-analysis should deal with this bias by some method,
but no simple method is likely to "solve" the problem in all situations.
Recent proposals and implementations of registries, databases, and
publication agreements are targeted at preventing publication bias.

In today's editorial processes, few articles in well-referred journals
are published as originally submitted. Consequently findings from unpublished work will not necessarily be comparable to findings from
published articles. The editorial process tends to reduce what is published. Some studies in the medical area suggest that failure to publish
results comes not so much from rejection by journals as from lack of
submission by authors.

Analyzing Studies with Multiple Outcomes

Social science is increasingly turning to studies with multivariate research designs. Such studies present difficulties to research reviewers,
who often treat multiple outcome variables by ignoring the multivariate
structure of the data: They address one variable at a time. Several of
the studies in this collection used that approach.

The study by Becker was the exception. Becker provided a multivariate treatment of the correlation matrices produced by her collection of
studies. By treating the correlation matrix as a vector of stochastically
dependent outcomes, she combined the evidence across studies using
generalized least squares methods that took into account the (large
sample) dependence structure of the information from different studies. Such methods should be used more widely in research syntheses,
in preference to alternatives such as discarding information or ignoring
dependence (see Raudenbush, Becker, and Kalaian 1988).

Making Results Meaningful

Whatever strategy is used to produce comparable outcome measures,
one should usually attempt to translate findings back to substantive
units. For example, an average effect size of .95 indicates a 75 percent
probability that a randomly selected response in the treatment group

> If at all possible, report the results of a meta-analysis in meaningful units. For example, reporting a typical reduction in length of stay in days will be far more meaningful and policy-relevant than reporting an effect size.

exceeds a randomly selected response in the control group, but this quantification may have very little substantive meaning. Mapping back to length of stay, recidivism rates, or achievement percentiles will produce more policy-relevant summaries.

Meta-analysis faces all of the challenges present in a primary study. Added are the complications of synthesizing information from what may be an extremely heterogeneous collection of primary studies that may or may not be representative of the population of studies. Data collection, evaluation, analysis, and reporting must take these complications into account. Clear documentation of the research questions, procedures used by the meta-analyst, assumptions, approximations, and methods, coupled with sensitivity analyses, are vital components of a valid and persuasive meta-analysis.

8

What Have We Learned About Explanatory Meta-Analysis?

In this final chapter we examine how the eight explanatory tasks set out in Chapter 2 were handled in the cases described in this volume. To reiterate, these tasks require identifying (1) those mediating processes that causally link one construct to another, particularly a cause and an effect; (2) those components of a treatment responsible for influencing a particular outcome; (3) those components of an outcome that have been impacted by a causal agent; (4) those person, setting, and time variables that moderate a descriptive causal relationship; (5) the treatment classes that influence an outcome; (6) the theoretical integrity of treatment or outcome; and (7) the differential consequences of different dosage levels. Since these research goals are only meaningful when the phenomenon-to-be-explained is real, it is also important (8) to construct an argument that the phenomenon-to-be-explained is not spurious.

In discussing these tasks we also reflect a little on what the four meta-analysts might have done differently and on some issues that future researchers might consider as they use meta-analysis in the service of scientific explanation. We organize the discussion around the three models of explanation introduced in Chapter 2: the manipulability model and the scientific model in either the form that seeks to account for the variance in effect sizes or the form that seeks to provide a comprehensive and true description of mediating processes. We also briefly assess the implication of these four explanatory meta-analyses for how meta-analysis might be improved and how it relates to the formation of public policy.

A Rare Example of a Meta-Analysis
Explicitly Testing a Theory of Causal Mediation:
Harris and Rosenthal (1985)

Rosenthal (1973) has proposed four constructs to explain why adults' expectations of high performance enhance performance in others. Written in a way that applies to teachers and students (rather than, say, experimenters and rats that are also part of the database), the four explanatory constructs are: "*Climate* refers to the warmer socio-emotional climate that teachers tend to create for high-expectancy students, a warmth that can be communicated both verbally and non-verbally. The *feedback* factor refers to teachers' tendency to give more differentiated feedback to their special high-expectancy students. . . . The *input* factor refers to the tendency for teachers to attempt to teach more material and more difficult material to high-expectancy students. . . . The *output* factor refers to teachers' tendency to give their special students greater opportunities for responding" (Harris and Rosenthal, p. 365).

Harris and Rosenthal found that they could construct effect sizes for 31 behavioral variables that seemed to measure all or part of one of these four explanatory constructs. They then used each of these individual measures (let us call them B measures) to explore whether they were related *both* to an expectancy manipulation (A) *and* a performance outcome (C). This is their criterion of mediation, and it required 31 separate meta-analyses on their part.

Let us first consider tests of the A-B links. In 10 of the 31 mediating behaviors the number of studies with both a manipulation and mediator exceeded 13; for 9 of the mediating behaviors the number of tests was between 7 and 12; and for 11 it was between 4 and 6. In testing the B-C link between a potential mediator and effect, for one mediating behavior the sample size of the studies exceeded 13; for 2 it was between 7 and 12; for 5 it was between 4 and 6; while for 16 it was under 4. The mode was one study per B-C link!

Harris and Rosenthal sought to get around this sample size problem at the individual variable level by classifying each variable into one of their four superordinate explanatory categories: classroom climate, performance feedback, enhanced input, and more possibilities for output. They then estimated average effect sizes for each category based on a total of 135 relevant studies. This is a number that few meta-analyses of micro-mediating processes can hope to match, given how many expectancy studies have been conducted over the last 40 years. Harris and Rosenthal estimate this to be in the thousands. However, these 135 studies are a small percentage of all the studies conducted, and this should prompt us to ask: Why are there relatively so few studies with mediating variables, given the volume of studies in the area over its atypically long and active history?

A second generic difficulty in meta-analytic studies of mediating processes is the validity of the causal model specified. Rosenthal's explanatory theory of expectancy effects is not the only one possible. Raudenbush (1984) has advanced a dissonance theory explanation. Braun's model (1976) includes an explanatory variable not in Rosenthal's model—that is, teacher expectancies influence what children expect from themselves. The same is true of Brophy and Good's model (1970), which postulates that teacher expectancies alter a child's self-concept and motivation to do well in school. The most we can conclude from the meta-analysis of Harris and Rosenthal is that the relationships they tested are not inconsistent with Rosenthal's model, though they do not rule out alternative models which are plausible enough to be already in the published literature.

The work Harris and Rosenthal present does not even rule out all alternative models within their system of 31 individual behaviors and four behavioral classes. For example, it is possible to argue that the four behavioral classes they examined are not unique causes but are instead temporally linked. One example of this is a model where an induced expectancy first changes climate, then leads to more opportunities for responding, and then leads to greater demands being made of high-expectancy students. Many other combinations of time links are also possible and, as Harris and Rosenthal themselves point out, their four variables may be reciprocally related to each other and to performance changes. When coarse-grained knowledge of temporal relationships emanates from the studies being meta-analyzed, fine-grained probes of temporal relationships are not possible.

One of the advantages of all causal modeling is parameter estimation—specifying the strength of a presumed causal link. Although they did not provide a point estimate of the link between feedback and performance, Harris and Rosenthal concluded that it was less strong than the links between performance and each of the three other classes of mediating variable examined. But even this modest comparative conclusion needs further probing. If the feedback variable were measured consistently less validly than the other measures of mediators, this alone would lead to it correlating less highly with performance. Moreover, if positive and negative feedback operate differently, as Harris and Rosenthal suggest, then combining them into an overall feedback category would obscure the possibility that positive feedback mediates expectancy effects differently from negative feedback. Misspecifying either the validities (implicitly assumed in Harris and Rosenthal to be equal across measures) or the underlying feedback process could lead to the pattern of differential correlation that the authors used to draw their particular substantive conclusions.

Model specification and data availability are likely to be chronic difficulties in all meta-analyses that aspire to test mediating processes, as they indeed are in most other forms of empirical research.

Theories of Explanation

Explanation via Specification of Contingencies

Perhaps the most widely used theory of explanation in meta-analysis to date has characterized explanation as defining the set of contingencies on which the effect depends. Meta-analyses have frequently made use of this strategy by examining whether effects vary across studies with differing setting, subject, or temporal characteristics. In fact, the earliest aspiration of meta-analysts was the search for robust main effects that would permit both simple generalization and correspondingly simple explanations. This theory of explanation is relatively easy to implement in meta-analysis and is sometimes successful in generating persuasive evidence of generalizability. For example, Devine demonstrated that psychoeducational care led to substantial reductions in length of postsurgical hospital stay and to reductions in other undersired outcomes. Setting and person characteristics were not strongly related to treatment effects when each was considered in isolation, and, in some cases, the results were so consistent across studies that the hypothesis of no variation in effect sizes could not be ruled out.

While this theory of explanation is the most widely used in meta-analysis, it is limited. It directly incorporates only one of the tasks of explanation discussed earlier (task 4). It can be extended, however, to incorporate analyses of the effects of treatment dosage (task 7) and the relative efficacy of various classes of treatment (task 5). By examining the relative effects of studies deemed to be more or less vulnerable to threats to internal validity, the method can also be extended to address the critical task of determining whether the phenomenon to be explained is an artifact (task 8).

This theory of explanation does not, however, incorporate what may be the most important explanatory tasks: identifying the mediating variables in the causal process (task 1); identifying the causal components of the treatment (task 2); or identifying the components of the outcome on which the treatment has a causal effect (task 3). It is not well suited for elaborating a theory of linked mediating processes that demonstrates why the treatment has an effect or for generating predictions about the effect that would be expected in some new situation (task 6). These attributes of explanations are necessary for the most intellectually satisfying and practically useful explanations.

Whatever the limitations of meta-analysis for explanation, it can be made most effective if it attempts as many of the tasks of explanation as feasible. For instance, the task of establishing that the observed treatment effects are not artifacts is an explanatory task that takes logical precedence over all others. Thus the partitioning of studies according to their vulnerability to various sources of bias is an important explanatory task in meta-analysis. In addition, the examination of the relative efficacy of different classes of treatment and various dosages is almost always desirable and is usually quite feasible.

Although it is usually more difficult, it may be possible to categorize studies into different groups according to the components of the treatment that are present and the type of outcomes that are measured in each study. By comparing the effects among different groups of studies, it may be possible to identify which components of the treatment affect which outcomes (tasks 2 and 3). An example of an analysis of this sort was carried out by Giaconia and Hedges (1982), who coded studies according to theoretical dimensions of a rather diffuse treatment (open-education programs) and then demonstrated that the theoretically relevant dimensions of treatment were strongly associated with positive effects on the particular outcome constructs that theory suggested should be affected by them.

Explanation via Accounting for Variance in Effects

A second widely used model of explanation in meta-analysis is that of systematically accounting for between-studies variance in the treatment effects. This model differs from the specification of contingencies for treatment efficacy in that it is inherently multivariate in its outlook. While it is conceivable to attempt to identify contingencies for treatment efficacy one at a time, accounting for variance in treatment effects necessarily involves the use of several explanatory variables at the same time.

Although it may be less obvious, this model of explanation also underlies meta-analyses that use the strategy of homogeneity testing and its generalization, model-fit statistics, as the primary analytic tool. The goal in either analytic strategy is to explain the between-studies variance in effect sizes; they differ only in how they assess the adequacy of explanation of the variance in effects.

Conventional regression analysis concentrates on the square of the multiple correlation coefficient (R^2) as a quantitative index of variance accounted for. In conventional regression analyses the only limit on the potential size of the R^2 is imposed by the reliability of the variables; with highly reliable variables R^2 values near 1 are possible, in principle.

In contrast there is a limit to the *systematic* between-studies variance that can be accounted for in meta-analysis. Nontrivial variables measured at the study level cannot account for the between-studies variance that is attributable to within-study sampling error. Consequently the maximum possible squared correlation between study characteristics and sample effect sizes is determined by the relative size of the systematic variance and the sampling error variance.

If the (unbiased) sample effect size, d $(=\delta+e)$, is composed of a systematic part δ, and an unsystematic part e, then the expected value $E[S_d^2]$ of the variance of d is given by

$$E[S_d^2] = \sigma_\delta^2 + \sigma_e^2,$$

where σ_δ^2 is the systematic variance (variance in effect size parameters) and σ_e^2 is of the sampling error variance across all studies. Since only systematic variance can be accounted for by systematic between-studies differences, the largest possible proportion of variance accounted for is given by

$$\rho^2 = \frac{\sigma_\delta^2}{\sigma_\delta^2 + \sigma_e^2}$$

Consequently when σ_δ^2, the between-studies variance component of the effects is large, the R^2 can be large, but when it is small in comparison with sampling error the R^2 can *never* be large. One way to interpret this result is by analogy to classical measurement theory. Within-study sampling error is analogous to the error of measurement of the population effect size. The maximum possible squared correlation ρ^2 is analogous to a reliability coefficient—the reliability of the (typical) sample effect size as a measure of population effect size.

One implication of this mathematical result is that collections of studies in which treatment effects do not exhibit much real variability cannot yield large R^2 values. Because the sampling error variance is largely a function of (the inverse of the) sample size, collections of studies with small sample sizes will be particularly prone to small R^2 values. Hence when models of explanation based on variance accounted for are used, interpretations should incorporate the idea that the obtained proportion of accounted for variance should be compared to the maximum possible R^2, and not to 1, as is often the case in conventional regression analysis. Since the maximum possible R^2 is frequently much less than 1, multivariate models may explain much more of the *explainable* variance than might be immediately apparent under the conventional interpretation of R^2.

Analyses using homogeneity statistics to characterize the explained variance implicitly compare the variance accounted for by the between-studies model to the maximum *explainable* variance. In fact, the tests of homogeneity or model specification that the four chapter authors routinely used are tests that the model explains all of the explainable between-studies variance. The shortcoming of these analyses is that they do not provide a quantitative index of the proportion of systematic variance that is explained.

How well does the variance-accounted-for model of explanation accomplish the explanatory tasks described earlier, which are about attributing variability in effect sizes to particular variables or classes of variables? The first issue, of course, is to make sure that the effect sizes under analysis are not themselves artifacts. In the model being discussed here, this is tested by entering methodological attributes of studies into the multivariate analysis first and by assuming that the attributes studied provide a complete model of between-studies differences. Given these assumptions, the variance-accounted-for model of explanation seems well suited to identifying efficacious components of treatments (task 2), the causally affected component of effects (task 3), the person, setting, and time variables that condition an effect (task 4), the classes of more effective treatments (task 5), the fidelity of treatments (task 6), and the impact of variations in dosage levels (task 7).

The success of the multivariate strategy in these regards is ably demonstrated in the chapters by Lipsey and Shadish. They demonstrate that purely methodological variables accounted for a substantial portion of the between-studies variance in effect sizes, and they account for this variance before going on to explore more substantive explanatory constructs. The addition of variables distinguishing treatment, subject, and setting variations further increased the accounted-for variance and permit us to specify which factors are more or less strongly related to effect sizes. Moreover, the R^2 values found by Lipsey and Shadish were not only large in absolute terms, but also they approached the maximum that could be accounted for. Within the framework of a predictive model of explanation, this suggests that their models are comprehensive and fully predictive.

Identifying the specific treatment, person, and setting variables responsible for effect size variance helps specify some of the conditions under which a treatment is most and least effective, and careful study of the chapters by Lipsey and Shadish should alert readers to the procedures required to carry out a state-of-the-art meta-analysis of sources of variability in effect sizes.

However, the model of explanation based on accounting for variance is less well suited to probing the mediating processes leading to treatment effects (task 1). With all causal modeling there is the possibility that the assumptions of the multivariate model are wrong or poorly tested. Even without considering mediating processes, models may still be seriously incorrect if important causal variables are omitted, if the form of the model is incorrectly specified (with respect to, say, causal orderings, multiple interacting causes, or selection artifacts), or if the stochastic part of the model is incorrect. The variance-accounted-for model does not necessarily begin with a model of a causal process in time whose truth value is assumed (or hopefully tested). Instead, prediction is the central objective, leading to a higher likelihood that the causal model will be poorly specified from the start. When Shadish wanted to test an explicitly mediational model he had to turn away from the multiple regression format he had used in the rest of his chapter, and Becker did the same. The issue is to estimate parameters for time-bound links between constructs that are presumed to be causal rather than to predict the variance accounted for by a variable or a class of variables like methodological characteristics of studies.

Explanation via Explication of Mediating Processes

The third model of explanation is that of explicating the network of mediating variables that lead to treatment effects. The most difficult to apply in meta-analysis, this model has only rarely been used. Most of the difficulties stem from a lack of available data on the mediating variables posited in the theories under study. In spite of longstanding concerns about the necessity for probing causal mechanisms, relatively few studies seem to investigate mediating processes extensively.

Why are there so relatively few studies with mediating variables? One possibility is that process measures are derived from theory and that substantive theories keep changing as they are improved or abandoned and as new theoretical orientations emerge. New theories invoke new constructs that, by definition, are not likely to have been measured in past work. Moreover, the new theories imply a negative judgment on old theories and explanatory constructs, reducing the need felt to measure these constructs.

A second reason for the low availability of micro-mediational data may be that most of the studies entering into meta-analyses are experiments. Experimenters prefer to achieve explanation through the choice of theory-relevant independent and dependent variables and sometimes through constructing contingency hypotheses involving a very small number of moderator variables. We believe that the radical experimentalist's preference is slowly changing, but even today few experimenters are prepared to devote the same level of resources to the careful measurement of process as to the careful manipulation of an independent variable or the careful measurement of some outcome.

A third reason for the low availability is that mediating variables are documented less systematically in research reports than molar cause and effect constructs. This holds, we believe, not only in journals and books, but also in unpublished reports. When process variables are assessed, it is usual to measure many of them and the quality of measurement is likely to be highly variable. Moreover, the time pressure during data analysis is so real that some process measures do not get analyzed or are analyzed in only perfunctory fashion. Even if process data are of high quality and are extensively analyzed, they sometimes lead to considerable conceptual complication, and there are, unfortunately, analysts who prefer to leave them out of the report in order to be able to tell a simpler story. Also, editors urge researchers to be succinct rather than comprehensive. For all these reasons, the meta-analyst interested in a particular mediating construct will have many fewer studies to analyze compared with the meta-analysts interested in describing a causal connection.

A fourth possibility is that mediating variables are often of a quite different type than the outcome variables and often difficult to measure. Their measurement may require research skills from a different discipline than those required to measure the outcome variables. Consider the psychoeducational treatments examined by Devine. Her theory posited mediating variables of the kind most typically measured by social psychologists. The measurement of these social variables requires different skills and training than the measurement of the outcomes of patient recovery, wound healing, and so on.

Yet, the examination of mediating processes via meta-analysis is clearly possible, as demonstrated by the work of Becker and to a lesser extent by Shadish and Devine. Becker extracted and synthesized a surprising amount of data about the processes mediating the development of science-achievement behaviors. Devine attempted the same kind of aggregation of effects on a smaller scale, but was less successful in uncovering data on the mediating variables she sought.

While both Becker and Devine aggregated within-study relationships, Shadish sought to explore the processes that mediate treatment effects by examining between-studies relationships. While the aggregation of within-study relationships is generally preferable, the estimation of relationships from between-studies analyses may sometimes be the only feasible strategy when the relationships are not measured within studies. It remains to be seen whether this strategy will prove generally useful in meta-analysis.

Implications for Methodology in Meta-Analysis

The use of meta-analysis for explanatory purposes poses serious challenges for both conceptual and statistical aspects of methodology. On the qualitative side, the use of meta-analysis or any other quantitative research strategy for explanation requires far greater explicitness of theory than is needed for using theories in descriptive research. To support explanatory meta-analysis, primary researchers must be more explicit in specifying their theory, in defining the constructs relevant to that theory, and in reporting results. Meta-analysts must also be more explicit about the theories examined in their meta-analyses, the constructs associated with them, and the analyses used to examine theoretical predictions. Devine and Becker provide excellent examples of the explicit specification of theories of mediating processes, but such explicitness is not yet common in meta-analysis.

The use of meta-analysis for explanatory purposes poses greater, but not qualitatively different, demands on statistical methodology than does the use of meta-analysis for purely descriptive purposes. In fact, several problems that plague descriptive meta-analysis become more important in explanatory meta-analysis.

Missing Data

The problem of how to handle missing data occurs in every meta-analysis, but it becomes more important in explanatory meta-analyses. Many of the variables that are the most important for explanation (e.g., mediating variables or codings of treatment or outcome components) are among the ones that are most frequently missing. Because key aspects of the explanation may hinge on the relationships between variables that may have many missing values, the treatment of missing data can have a profound effect on the validity and credibility of the explanation.

Although sophisticated methods have been developed for conducting statistical analyses when missing data are present, these have not been widely used in meta-analyses. Greater use of model-based methods for estimation with missing data (see Little and Rubin 1986) is a desirable direction for future meta-analyses to pursue. There are, however, three major obstacles to the use of these methods. First, analyses based on these methods do not currently yield easily computed standard errors that can be used in significance testing. Consequently, meta-analysts desiring significance tests might not find the methods entirely satisfactory whatever their other merits. Second, software to carry out these analyses is not readily available (at least to social scientists). Third, relatively few social scientists are familiar with these methods as they may be applied to primary analysis. Even fewer have sophisticated enough statistical training to adapt them to the meta-analytic context.

Statistical methods for handling missing data via so-called multiple imputation (see Rubin 1987) are also promising possibilities for meta-analysis. In some ways methods based on multiple imputation may be more readily applicable to meta-analysis than the methods described above. Methods based on multiple imputation do yield standard errors that can be used in significance tests. Although they require somewhat less elaborate software than do model-based methods, specialized software is still necessary. The greatest barrier to the use of multiple imputation in meta-analysis may be that it has a rather sophisticated statistical rationale—one that has not been entirely convincing even to professional statisticians.

Modeling Dependence Among Effects

The use of meta-analyses for explanation requires greater specificity of coding of both mediating and outcome variables. Greater specificity of coding and the use of analyses designed to exploit this specificity naturally leads to situations where statistical analyses must take into account the nonindependence of several variables coded from the same study, be they outcomes or moderators. In descriptive meta-analyses specific treatment of these dependencies is often not crucial. For example, if effect size estimates on several different versions of the same outcome construct can be calculated in one study, a descriptive meta-analysis would lose little by using the average of these estimates as the value for that study. In an explanatory meta-analysis, it may be conceptually important to preserve the effects for different dimensions of the same broad outcome construct. Hence, the statistical methods used in explanatory meta-analysis must explicitly take dependencies into account.

Methods do exist for treating dependent data in meta-analysis (see Hedges and Olkin 1985, chap. 10; or Raudenbush, Becker, and Kalaian 1988). In fact, Becker explicitly modeled the dependencies among correlations estimated from the same studies in her analysis in this volume. The use of multivariate methods, however, poses problems similar to those that occur when sophisticated methods for handling missing data are used. The analyses require a greater level of statistical sophistication than is typical among social researchers. These methods also pose another problem: Software for computing these analyses is not readily available. Finally, the multivariate methods require data on within-study intercorrelations among variables that are often poorly reported. Thus, using more sophisticated analyses poses greater demands for data and may exacerbate the problem of missing data.

Understanding Between-Studies Variation

The cases in this volume implicitly illustrate the need for greater understanding of the nature of between-studies variation. Many questions that arise in meta-analyses can be answered either by aggregating estimates of within-study relationships or by computing between-studies relationships. Cooper (1989) has distinguished the evidence generated by these two strategies, calling them "study-generated evidence"

Causation May Be Timebound

In the years between 1932 and 1980 the size and geographic distribution of grocery stores changed from small neighborhood enterprises to enormous buildings in malls. The concept of neighborhood cannot be the same at the two periods if one is to include consumer behavior as part of the notion of a neighborhood.

The treatment of schoolchildren by their teachers, parents, and peers changes from generation to generation, to say nothing of differences in the time spent watching movies and television or listening to the radio. Thus, we are not comparing the same sorts of people when we compare 15-year-olds in 1980 with those in the good old depression days of 1932. Even the proportion of children attending school at age 15 has changed. Since the population has changed in so many ways, even those differences in the cognitive performance of 15-year-olds then and now that appear large would leave us wondering about the causes.

Source:

Hoaglin, D. C.; R. J. Light; B. McPeek; F. Mosteller; and M. A. Stoto
    1982 *Data for Decisions*. Cambridge, MA: Abt Books, pp. 68–69.

and "review-generated evidence," respectively. While all critics would probably agree that conclusions based on study-generated evidence are subject to fewer threats to their validity, the ability of meta-analyses to *create* such evidence is limited. Studies frequently do not carry out the contrasts that may be desirable to the reviewer. In situations where study-generated evidence is not available, review-generated evidence can often be developed by computing relationships between studies with different characteristics while controlling for other confounding variables.

For example, Becker developed her models of processes based on within-study relationships. This entailed considerable effort to locate and extract data on within-study relationships (correlations) among variables. Devine tried, but was generally unable, to locate data on within-study relationships. Shadish examined between-studies relationships (review-generated evidence) to investigate the kinds of questions that Becker investigated and Devine tried to investigate via within-study relationships.

Meta-Analysis and Public Policy

The studies in this volume describe and analyze cause and effect: How well does patient education promote recovery from surgery? How well do juvenile delinquency programs prevent recidivism? How do teacher expectancies influence student performance? How well do family therapies work and under what conditions? If a meta-analysis reveals that the causal relationship under consideration holds in all—or nearly all—of the contexts examined, it helps policymakers who want to know which programs to develop and to fund and which policy and program guidelines to issue. Policymakers particularly need to know what works "generally"—namely, at a wide variety of program sites and with a wide range of human populations. The local world in which policies and programs are actually implemented is highly variable. Practitioners have much more discretion than central planners and regulators would care to admit. Therefore, knowledge about causal factors that are manipulable, that people are willing to implement locally, and that are robust across many settings of application can be critical to a social program's overall success. Widely replicated causal relationships also increase the chances that the local stakeholders will find some circumstances in the total data set that resemble those for which they have responsibility and to which they want to generalize.

Knowing *why* a treatment works is also important for policymakers.

Such explanatory knowledge enhances the transfer of effective treatments to new settings where they might never before have been examined. For example, once we know the mechanisms through which teacher expectancies impinge on student performance, we can make sure that the mediating processes located by the explanatory theory are known in schools throughout a nation serving quite different kinds of children (or even adults).

While social science knowledge is sometimes used to formulate policy or improve service delivery, it is probably most often used for more general purposes that Weiss (1987) has labeled "enlightenment." These include giving members of the policy-shaping community a new definition of a social issue, or a new sense of its importance. They also include providing a deeper understanding of the implementability of an intervention, a fresh sense of the tractability of the social problem a program is meant to address, or novel insight into the difficulties of implementing certain classes of services. Research rarely provides information that would have an immediate effect on a pending decision. Rather it provides background knowledge that changes understanding and might influence later deliberations, not only about a policy or program that has been evaluated but about others as well. In such a policy-making system, what role can meta-analysis play that is more effective than the role of individual studies? And what special role can meta-analysis play if it is more explanatory than descriptive in its major focus?

Causal Contingencies and Robust Main Effects

One model of research utilization posits that it is particularly valuable to gain knowledge about whether a treatment is broadly robust in its effectiveness. A prime advantage of meta-analysis is that it can provide assurance to policymakers that the manipulations and measures studied are what they are supposed to be. Meta-analysis probes whether the same treatment-outcome relationship emerges despite differences in how individual researchers define or operationalize their variables. A conclusion is strengthened if its effects are robust despite heterogeneity in definitions and other sources of irrelevant variance, including how measures are made and treatments are implemented.

The outcome measures that meta-analysts typically use are also quite important to policymakers. In education, for example, academic achievement would be central to most stakeholders, especially the science achievement that Becker studied; in juvenile delinquency programs, the recidivism that interested Lipsey is of general concern (in a

Findings for a Nutrition Program

The Senate Committee on Agriculture, Nutrition and Forestry invited the U.S. General Accounting Office to do a meta-analysis of findings about the effectiveness of the Women, Infants, and Children (WIC) nutrition program. In 1991, this program cost over $1 billion per year and served 30 percent of all children under age 5 in America. The General Accounting Office's meta-analysis found that for some outcomes, such as reducing the proportion of low birthweight babies, the WIC nutrition supplements were clearly helpful. For other outcomes, such as the mental health and rate of mental retardation of children of WIC-eligible mothers, it found no evidence that the program offered any benefits. This finding led the U.S. Department of Agriculture, which sponsors the WIC program, to commission a new data-gathering effort, so that evidence could be gathered about the effectiveness of WIC for children's mental development.

The GAO report concludes: "The major benefit of the synthesis is that, beyond the literature review, it analyzes the quality of each evaluation finding in terms of the evidence supporting it and yields refined information about what is known on a particular topic at a particular time. General knowledge is strengthened by the findings of several soundly designed and well-executed evaluations when they are consistent, even though they may have used different methods. No matter how high its quality, a single evaluation can rarely do this."

Source:

GAO-PEMD 84-4
  1984 *WIC Evaluations Provide Some Favorable but No Conclusive Evidence on the Effects Expected for the Special Supplemental Program for Women, Infants, and Children.* (See especially pp. 9 and 10.)

way that, say, satisfaction with the detention experience would not be); in psychotherapy research, the behavioral outcomes that Shadish stressed are almost universally considered important; in Devine's work, the length of hospital stay captures general attention, primarily because of its financial implications.

The criterion issue is important, not only for reasons of policy relevance but also because meta-analysis emphasizes effect sizes rather than statistical significance. Average, standardized effect sizes are not meaningful to most persons, so it is desirable to translate them into more easily understood metrics. For Devine, the task was simple. Her single most important outcome could be expressed as days-in-hospital. She had no need to standardize the measure. But the other chapter authors were not so lucky. To help readers they had to translate their average effect sizes into several different measures. Lipsey presented results in terms of the percentage of persons with lesser recidivism due to attending a juvenile delinquency program compared with controls. Since magnitude estimates depend on the reliability of measures, Lipsey also corrected his average effect size for unreliability in the recidivism measure. This doubled the estimated effect, suggesting that the assumptions he built into the reliability correction need special scrutiny because of the large correction they brought about.

To members of the central decision-shaping community, an average effect size, no matter how well expressed, would have little relevance if the causal connection could not be generalized to many types of programs, settings, or persons. In this context, consider the results from Devine's meta-analysis. She was able to show that the link between patient education and recovery from surgery was constant in causal direction whatever the measure of recovery used, whatever the time period studied (over 30 years), whatever the type of hospital studied (private or public), whatever the type of surgery involved (orthopedic, thoracic, gastroid, etc.), whatever the type of person involved, and whoever delivered the patient education (nurses, physicians, clergy, or others).

Could hospital administrators, nursing managers, insurance agencies, politicians, and federal and state officials use Devine's information if they wanted to? The robust findings and the absence of obvious null or negative effects argue that a similar effect is likely to emerge, even in unique and still unstudied populations and times. If central decision-makers look at the range of stratification variables examined by Devine, they will conclude that the patient education effect holds comprehensively.

Consider next the person responsible for a particular hospital in the private sector in the southern United States. She can examine Devine's database by the criteria that interest her and conclude that, at least

> In a meta-analysis documenting heterogeneity over units can be as important as reporting central tendency. The heterogeneity invites explanation by finding situations where an intervention works and where it does not. Such information can be key for policy decisions and the design of subsequent studies.

when these variables are examined separately, the effect does not disappear. Someone interested in furthering the profession of nursing can examine the data from the perspective of their interests and conclude that their protégés are capable of producing the desired result. No single study could meet such a variety of needs; and even the meta-analysis could not have met them if Devine had not included so many analyses, for which a large set of studies is a precondition.

Had Devine examined statistical interactions she might have found that some of her stratification variables interacted with patient education to determine variability in the size of effects. Probing such variability is an important scholarly concern and was paramount in the work of Shadish and Lipsey. But central policy actors are not always able to do much with causal contingency variables. A medical insurance company could theoretically mandate that patient education will be reimbursed in some types of hospitals but not others; but we doubt very much that this policy could be defended to hospital administrators, journalists, and the general public. It does not matter much, therefore, if the size of effect varies so long as the causal sign is the same. Even sign reversals, if few in number, may be ignored at the central decision-making level, but they should not be ignored locally. If negative causal signs are relatively frequent they should at a minimum raise a red flag of caution, even if the overall average effect size is positive and of a policy-relevant magnitude. Contingent policy is often called for in this circumstance, if it is politically feasible.

Devine's meta-analysis consisted nearly exclusively of interventions conducted by researchers rather than practicing nurses. Indeed, average effect sizes approached zero only in those circumstances in which staff nurses provided the treatment. Do we have, therefore, a patient education phenomenon that cannot be realized in hospital practice because practicing nurses, for whatever reasons, cannot implement the treatment well? To fill this gap Devine and her colleagues conducted a primary study (Devine et al. 1988) and found the same effect when practicing nurses delivered the patient education. This highlights an incidental advantage of individual meta-analyses. They can identify gaps in the knowledge base, hence justifying the next study and creating the next stage in what is, in essence, a partly systematic *program* of research in which the studies in the meta-analysis constitute an earlier stage.

One inferential problem with meta-analyses concerns sources of bias that run through all the studies examined, as when researchers deliver all the treatments. But this aside, meta-analysis has already demonstrated its potential to help central, and perhaps even peripheral, policy and program personnel in deciding what is likely to be effective over many different populations, settings, and time periods and what is likely to be effective in particular circumscribed contexts of obvious relevance to some actors in the policy world.

Variations in Dosage Levels and Treatment Types

Meta-analysis has great potential to uncover two types of knowledge of great value to the policy-shaping community. One concerns the functional form of the treatment-effect relationship and the other concerns the types of treatment responsible for desirable outcomes. Devine, Lipsey, and Shadish all address these issues.

To examine how much of the treatment is required for a given level of effect, Lipsey tested a model in which dosage was examined after the effects of several method factors had been removed from the effect sizes. To the same end, Devine probed which combination of information, social support, and skills training was needed for reliable effects to be obtained. However, neither of them constructed a simple graph relating a meaningfully scaled independent variable to outcomes. If done responsibly, this is likely to be particularly useful.

Since there are many unique project managers and service providers, it is always useful for policy if it can be shown that different treatments have roughly comparable effects. Then one can move toward a smorgasbord model of research utilization where several treatments are defined as effective, and local officials are left free to implement whichever they choose as most appropriate to their local circumstances. All three of the chapter authors who explored causal generalization also probed the effects of different classes of treatment. Devine's work addressed the consequences of variation in the number of components of patient education. Shadish probed the effect that therapist allegiance had within each type of marital and family therapy, seeking to identify whether one theoretical allegiance was more effective than another. Lipsey also examined whether different types of juvenile delinquency treatment have different effects, cautioning us lest we take too seriously the descriptive labels that primary invesitgators used to classify their treatments. The smorgasboard approach to policy-relevant meta-analysis obviously presumes some validity to the labels used to describe treatment types.

Questions about treatment types may seem most germane to the search for the variant that produces the largest effects. But from a policy perspective, wherever there is considerable local discretion, knowing the several treatment options available, the levels of resources each requires, and the size of effect each achieves is also important. If more than one treatment realization is demonstrably effective, local actors are provided with a chance to choose among options, selecting those that best fit their budgets, the size of effect they need to achieve, and other aspects of the settings for which they are responsible. This often increases local acceptability of at least one of the options and can avoid the situation where local officials react negatively to central policymakers trying to promote just one alternative and seeming to force it on state and local personnel. There is obvious utility to learning about more than one effective treatment type, which is why some theories of program evaluation are dedicated to this end.

This is not to deny the rationalist's search for the one best strategy of treatment, whether conceived as the largest average effect size or the most advantageous cost-benefit ratio. (The latter is particularly assumption-riddled, of course.) It is merely to reiterate that in some real-world contexts of application it is not easy to implement the one best strategy and that trying to operate this way can alienate local practitioners and so backfire.

The Mediational Model

Explanatory meta-analyses that identify the mediating processes through which effects occur are especially helpful to policymakers. So is the identification of components of treatments and outcomes that have causal relationships. Such knowledge often results in explanations that help streamline treatment or broaden their applicability. For example, if marital therapy were found to be effective because it makes spouses confront their relationship problems more honestly, it might be possible to develop vehicles other than therapy for eliciting honest discussion within families. Causal explanatory knowledge also makes it easier for local personnel to determine how they want to achieve a desired outcome. While they must reproduce the effective mechanisms they may be able to set them in motion with interventions that are locally acceptable and different from those used elsewhere. They may even be less expensive.

Becker's work exemplifies the potential utility of knowledge of causal processes. Such knowledge has four major advantages. It promises effects that are larger, more regularly predicted, and applicable across a broader range of contexts; also, effects that can be produced in novel contexts where they may never have been studied before (Cronbach 1982). Thus, if the socialization practices of teachers are the components most strongly related to gender differences in science achievement, then we should look there to improve girls' performance in science. There may be less payoff, for example, in attempting to influence student liking for science. Or if it turns out, as in Shadish's causal model, that differences in behavioral outcome arise because therapies are implemented with different degrees of standardization in universities as opposed to therapists' offices, then policymakers in mental health might devote their resources to better implementing the therapies we already have rather than to developing new ones. (Lipsey implies something similar with his finding that researchers produce larger effect sizes than practitioners.) All policy actors stand to gain from knowledge of the principles that bring about valued ends.

The discussion thus far has had a decidedly instrumental flavor, implying that policy actors look to research for help in making decisions. But descriptive research on knowledge utilization suggests that many policy decisions are "slipped into" rather than "rationally" made and that most social science is used more for enlightenment than for decision-making. A key question thus becomes: How might meta-analysis with an explanatory flavor contribute to social science being used more often and more fruitfully for enlightenment purposes? One possibility is that conducting a meta-analysis will often force into the light of day important research and policy questions about which little is known. This then helps set the agenda for future data collection efforts. Of particular importance here is the likelihood that meta-analyses will more and more highlight the absence of knowledge about mediating processes, helping create a climate that favors collecting such data. Another possibility is that a meta-analysis comparing the efficacy of different treatment classes might identify those that are singularly ineffective, even if it does not identify those that are clearly more effective than others. In the last analysis, though, enlightenment-based usage is difficult to predict, for the enlightenment notion suggests that empirical results generated at one time and place will eventually be evaluated within a different cognitive framework at a different time or in a different place. What confuses one generation may enlighten another.

There can be no doubt of the desirability to the policy (and scientific) world of full knowledge of causal mediating processes. The real issue is the likelihood of meta-analysis delivering such knowledge, given how difficult it is to achieve even with the most carefully considered individual study. There is no doubt that, in theory, meta-analysis can deliver. If the specification of the causal models were independently known to be true; if measures of all the relevant constructs were included in the primary studies to be synthesized; and if, therefore, within-study estimates of relationships could be aggregated in the meta-analysis, then valid and precise estimates of causal parameters could be achieved. But these conditions do not always hold. In reality we need to test abbreviated models whose correct specification is not independently known, where crucial measures may not always be available, and sometimes we will need to examine between-studies estimates given the paucity of within-study ones. Useful knowledge can be gained, even in these real-world conditions, as Becker's work on mediating processes shows. But it is not complete knowledge. The challenge facing meta-analysis is to develop better techniques for assessing causal mediation, given what a good job can often be done in predicting the variability in effect sizes and exploring many of the individual treatment, setting, person, and time factors that condition descriptive causal relationships.

The Promise of Explanatory Meta-Analysis

It may seem that much of what we have learned about the use of meta-analysis for explanatory purposes is concerned with the difficulties of the enterprise. While all explanation is fraught with difficulties, we are optimistic about the prospects for the use of meta-analysis in explanation. The case studies in this volume have shown that meta-analysis can be effectively used to generate useful knowledge about all or part of the eight explanatory tasks we explicated in Chapter 2.

It seems well suited to exploring contingencies affecting treatment efficacy, particularly those relevant to attributes of person, settings, and the time interval over which a phenomenon has been studied. We are also impressed that thoughtful meta-analytic work can led to the explanation of very substantial proportions of the between-studies variance in effects, leading particularly to assessments of treatment dosage and fidelity issues, as well as to comparisons between different treatment classes. But other potential moderator variables can also be explored within the same framework that depends on predicting variability in effect sizes.

Analyses of mediating variables may be stymied by the quality of substantive theory and the failure of the original researcher to report information on mediating processes. But sophisticated analyses of some links in simpler causal models will often be possible, as illustrated in the work of Becker, Devine, and Shadish. We believe that the cases examined in this volume demonstrate both the feasibility and desirability of using meta-analysis in scientific explanation. They also suggest the need for more technical and conceptual work on fostering an even greater explanatory emphasis in future meta-analytic practice.

References

Alwin, D. F., and R. C. Tessler
  1985 Causal models, unobserved variables, and experimental data. In H. M. Blalock, ed., *Causal Models in Panel and Experimental Designs.* New York: Aldine.

Anderson, G. J.
  1969a Effects of classroom social climate on individual learning. Unpublished doctoral dissertation, Harvard University.

Andrews, D. A.; I. Zinger; R. D. Hoge; J. Bonta; P. Gendreau; and F. T. Cullen
  1990 Does correctional treatment work? A clinically relevant and psychologically informed meta-analysis. *Criminology* 28:369–404.

Archuletta, V.; O. B. Plummer; and K. D. Hopkins
  1977 *A Demonstration Model for Patient Education: A Model for the Project "Training Nurses to Improve Patient Education."* Boulder: Western State Commission for Higher Education.

Astin, H. S.
  1968 Career development of girls during the high school years. *Journal of Counseling Psychology* 15:536–540.

Averill, J. R.
  1973 Personal control over aversive stimuli and its relationship to stress. *Psychological Bulletin* 80:286–303.

Bailar, J. C., III, and F. Mosteller, eds.
  1986 *Medical Uses of Statistics.* Waltham, MA: New England Journal of Medicine Books.

Baker, D. R.
1981   The differences among science and humanities males and females. Paper presented at the annual meeting of the National Association for Research in Science Teaching, Ellenville, NY, April. (ED 204 143)

Bandura, A.
1977   *Social Learning Theory*. Englewood Cliffs, NJ: Prentice-Hall.

Becker, B. J.
1986   Influence again: An examination of reviews and studies of gender differences in social influence. In J. S. Hyde and M. C. Linn, eds. *The Psychology of Gender: Progress Through Meta-analysis*. Baltimore: Johns Hopkins University Press.
1989a   Gender and science achievement: A reanalysis of studies from two meta-analyses. *Journal of Research in Science Teaching* 26:141–169.
1989b   Model-driven meta-analysis: Possibilities and limitations. Paper presented at the annual meeting of the American Educational Research Association, San Francisco, March.

Beecher, H. K.
1955   The powerful placebo. *Journal of the American Medical Association* 159:1602–1606.

Begg, C. B., and J. A. Berlin
1988   Publication bias: a problem in interpreting medical data (with discussion). *Journal of the Royal Statistical Society 151*(Series A):419–463.

Bem, D. J.
1972   Self-perception theory. In L. Berkowitz, ed., *Advances in Experimental Social Psychology*. New York: Academic Press.

Bendixen, H.; L. Egbert; J. Hedley-White; M. Laver; and H. Pontoppindan
1965   *Respiratory Care*. St. Louis: Mosby.

Bentler, P. M.
1989a   Comparative fit indices in structural models. *Psychological Bulletin* 107:238–246.
1989b   *EQS: A Structural Equations Program Manual*. Los Angeles: BMDP Statistical Software.

Berman, J. S.; R. C. Miller; and P. J. Massman
1985   Cognitive therapy versus systematic desensitization: Is one treatment superior? *Psychological Bulletin* 97:451–461.

Bernard, M. E.
1979   Does sex role behavior influence the way teachers evaluate students? *Journal of Educational Psychology* 71:553–562.

Bernstein, S., and S. Small
1951   Psychodynamic factors in surgery. *Journal of Mount Sinai Hospital* 17:938–958.

Bhaskar, R.
1975   *A Realist Theory of Science*. Leeds, England: Leeds.

Bird, B.
1955   Psychological aspects of preoperative and postoperative care. *American Journal of Nursing* 55:685–687.

Birge, R. T.
1932   The calculation of errors by the method of least squares. *Physical Review* 40:207–227.

Blau, P. M., and O. D. Duncan
1967   *The American Occupational Structure.* New York: Wiley.

Bogner, I., and H. Zielenbach-Coenen
1984   On maintaining change in behavioral marital therapy. In K. Hahlweg and N. S. Jacobson, eds., *Marital Interaction: Analysis and Modification.* New York: Guilford Press.

Bollen, K. A.
1989a   Overall fit in covariance structure models: Two types of sample size effects. *Psychological Bulletin* 107:256–259.
1989b   *Structural Equations with Latent Variables.* New York: Wiley.

Bornstein, R. F.
1989   Exposure and affect: Overview and meta-analysis of research, 1968–1987. *Psychological Bulletin* 106:265–289.

Boruch, R. F., and H. Gomez
1977   Sensitivity bias, and theory in impact evaluation. *Professional Psychology* 10:411–434.

Boruch, R. F. and H. W. Riechen, eds.
1974   *Experimental Testing of Public Policy: The Proceedings of the 1974 Social Science Research Council Conference on Social Experiments.* Boulder, CO: Westview Press.

Bowers, T. G., and G. A. Clum
1988   Relative contribution of specific and nonspecific treatment effects: Meta-analysis of placebo-controlled behavior therapy research. *Psychological Bulletin* 103:315–323.

Braun, H. I.
1988   Empirical Bayes methods: A tool for exploratory analysis. In R. D. Bock, ed., *Multilevel Analysis of Educational Data.* San Diego: Academic Press.

Brewer, M. B., and M. W. Blum
1979   Sex-role androgyny and patterns of causal attribution for academic achievement. *Sex Roles* 5:783–796.

Bridgeman, D. J.; S. Oliver; and R. D. Simpson
1985   Relationships of attitude toward science and family environment. Paper presented at the annual meeting of the National Association for Research in Science Teaching, French Lick Springs, IN, April. (ED 255 388)

Bridgeman, P. W.
  1927   *The Logic of Modern Physics.* New York: Macmillan.

Bridgham, R. G.
  1969   Classification, seriation, and the learning of electrostatics. *Journal of Research in Science Teaching* 6:118–127.

Bruer, J. T.
  1983   Women in science: Lack of full participation. *Science* 221:1339.
  1984   Women in science: Toward equitable participation. *Science, Technology & Human Values* 9:3–7.

Bryant, F., and P. M. Wortman
  1984   Methodological issues in meta-analysis of quasi-experiments. *New Directions in Program Evaluation* 24:25–42.

Bryk, A. S., and S. W. Raudenbush
  1988   Heterogeneity of variance in experimental studies: A challenge to conventional interpretations. *Psychological Bulletin* 104:396–404.

Campbell, D. T.
  1957   Factors relevant to the validity of experiments in social settings. *Psychological Bulletin* 54:297–312.

———, and J. C. Stanley
  1966   *Experimental and Quasiexperimental Designs for Research.* Chicago: Rand McNally.

Cannon, R. K., Jr., and R. D. Simpson
  1985   Relationships among attitude, motivation, and achievement of ability grouped seventh-grade life science students. *Science Education* 69:121–138.

Chalmers, T. C.; H. Smith, Jr.; B. Blackburn; B. Silverman; B. Schroeder; D. Reitman; and A. Ambroz
  1981   A methodology for assessing the quality of randomized control trials. *Controlled Clinical Trials* 2:31–49.

Cline, V. B.; J. M. Richards, Jr.; and W. E. Needham
  1963   Creativity tests and achievement in high school science. *Journal of Applied Psychology* 47:184–189.

Cloninger, C. R.; D. C. Rao; J. Rice; T. Reich; and N. E. Morton
  1983   A defense of path analysis in genetic epidemiology. *American Journal of Human Genetics* 35:733–756.

Cochran, W. G.
  1937   Problems arising in the analysis of a series of similar experiments. *Journal of the Royal Statistical Society (Supplement)* 4:102–118.

Cohen, F., and R. S. Lazarus
  1973   Active coping processes, coping dispositions, and recovery from surgery. *Psychosomatic Medicine* 35:375–389.

Cohen, J.
1960   A coefficient of agreement for nominal scales. *Educational and Psychological Measurement* 1:37–46.
1969   *Statistical Power Analysis for the Behavioral Sciences.* New York: Academic Press.
1988   *Statistical Power Analysis for the Behavioral Sciences,* 2nd ed. Hillsdale, NJ: Lawrence Erlbaum.

————, and P. Cohen
1983   *Applied Multiple Regression/Correlation Analysis for the Behavioral Sciences,* 2nd ed. Hillsdale, NJ: Erlbaum.

Cohen, M. P.
1979   Scientific interest and verbal problem solving: Are they related? *School Science and Mathematics* 79:404–408.

Collingwood, R. G.
1940   *An Essay on Metaphysics.* Oxford, England: Clarendon Press.

Colton, T.; L. S. Freedman; and A. L. Johnson, eds.
1987   Proceedings of the workshop on methodological issues in overviews of randomized clinical trials, May 1986. *Statistics in Medicine Volume 6.*

Cook, T. D.
1991   Meta-analysis: Its potential for causal description and causal explanation within program evaluation. In Albrecht, Günter and Otto, Hans-Uwe, eds., *Social Prevention and the Social Sciences: Theoretical Controversies, Research Problems, and Evaluation Strategies.* Berlin-New York: Walter de Gruyter.

————; A. Anson; and S. Walchli
In press Evaluating adolescent health programs: The challenge for the next decade. In S. G. Millstein, A. C. Petersen, and E. O. Nightingale, eds., *Adolescent Health Promotion.* Washington, DC: Carnegie Council on Adolescent Development.

Cook, T. D., and D. T. Campbell
1979   *Quasi-Experimentation: Design and Analysis Issues for Field Settings.* Boston: Houghton-Mifflin.

Cook, T. D.; L. C. Leviton; and W. R. Shadish
1985   Program evaluation. In G. Lindzey and E. Aronson, eds., *The Handbook of Social Psychology,* 3rd ed. New York: Random House.

Cooper, H. M.
1982   Scientific guidelines for conducting integrative research reviews. *Review of Educational Research* 52:291–302.
1984   *The Integrative Research Review: A Systematic Approach.* Beverly Hills, CA: Sage.
1986   On the social psychology of using research reviews: The case of de-

segregation and black achievement. In R. Feldman, ed., *The Social Psychology of Education*. Cambridge, England: Cambridge University Press.

1987   Literature searching strategies of integrative research reviewers: A first survey. *Knowledge* 8:372–383.

1988   Organizing knowledge synthesis: A taxonomy of literature reviews. *Knowledge in Society* 1:104–126.

1989   *Integrating Research: A Guide for Literature Reviews*, 2nd ed. Newbury Park, CA: Sage.

Corah, N. L., and J. Boffa
1970   Perceived control, self-observation and response to aversive stimulation. *Journal of Personality and Social Psychology* 16:1–4.

Cordray, D. S.
1986   The future of meta-analysis: An assessment from the policy perspective. Paper presented at the Committee on National Statistics workshop on "The Future of Meta-Analysis," Hedgesville, WV, October.

1990a   Meta-analysis: An assessment from the policy perspective. In K. Wachter and M. Straf, eds., *The Future of Meta-Analysis*. New York: Russell Sage Foundation.

1990b   Strengthening causal interpretations of nonexperimental data: The role of meta-analysis. In L. Sechrest; E. Perrin; and J. Bunker, eds., *Research Methodology: Strengthening Causal Interpretations of Nonexperimental Data*. Washington, DC: U.S. Department of Health and Human Services Public Health Service, Agency for Health Care Policy and Research, 151–172.

————; G. M. Pion; and R. F. Boruch
1990   Data sharing: With whom, when and how much? Paper presented at the Public Health Service Workshop on Scientific Integrity, Washington, DC.

Crandall, V. J.; W. Katovsky; and W. W. Preston
1960   A conceptual formulation for some research on children's achievement development. *Child Development* 31:787–797.

Cronbach, L. J.
1980   *Towards Reform of Program Evaluations: Aims, Methods, and Institutional Arrangements*. San Francisco: Jossey-Bass.

1982   *Designing Evaluations of Educational and Social Programs*. San Francisco: Jossey-Bass.

————, and R. E. Snow
1981   *Aptitudes and Instructional Methods: A Handbook for Research on Aptitude-Treatment Interactions*, 2nd ed. New York: Irvington.

Cullen, F. T., and K. E. Gilbert
1982   *Reaffirming Rehabilitation*. Cincinnati: Anderson.

Deaux, K.
1976   Sex: A perspective on the attribution process. In J. Harvey, W. Ickes, and R. Kidd, eds., *New Directions in Attributional Research*, vol. 1. Hillsdale, NJ: Erlbaum.

DeBoer, G. E.
1984   A study of gender effects in the science and mathematics course-taking behavior of students who graduated from college in the late 1970s. *Journal of Research in Science Teaching* 21:95–103.

Devine, E. C.
1984   Effects of psychoeducational interventions: A meta-analytic analysis of studies with surgical patients. *Dissertation Abstracts International* 44:3356B. (University Microfilms no. DEQ 84-04400)
1990   Effects of psychoeducational care: A meta-analysis of the effects of variability in type of treatment. Unpublished manuscript.

———, and T. D. Cook
1983   A meta-analytic review of psychoeducational interventions on length of postsurgical hospital stay. *Nursing Research* 32:267–274.
1986   Clinical and cost relevant effects of psychoeducational interventions with surgical patients: A meta-analysis. *Research in Nursing and Health* 9:89–105.

———; F. W. O'Connor; T. D. Cook; V. A. Wenk; and T. F. Curtin
1988   Clinical and financial effects of psychoeducational care provided by staff nurses to adult surgical patients in the post-DRG environment. *American Journal of Public Health*, 78:1293–1297.

Dixon, W. J., ed.
1988   *BMDP Statistical Software Manual*, vol. 2. Los Angeles: UCLA Press.

Dripps, R. D., and M. V. Demming
1946   Postoperative atelectasis and pneumonia. *Annals of Surgery* 124:94–109.

Dunteman, G. H.; J. Wisenbaker; and M. E. Taylor
1979   *Race and Sex Differences in College Science Program Participation*. Research Triangle Park, NC: Research Triangle Institute. (ED 199 034)

Dush, D. M.; M. L. Hirt; and H. E. Schroeder
1989   Self-statement modification in the treatment of child behavior disorders: A meta-analysis. *Psychological Bulletin* 106:97–106.

Dziuban, C. D., and E. C. Shirkey
1974   When is a correlation matrix appropriate for factor analysis? Some decision rules. *Psychological Bulletin* 81:358–361.

Eccles, J.; T. F. Adler; R. Futterman; S. B. Goff; C. M. Kaczala; J. L. Meece; and C. Midgley
1983   Expectations, values, and academic behaviors. In J. T. Spence, ed., *Perspectives on Achievement and Achievement Motivation*. San Francisco: Freeman.

Elman, R.
1951   *Surgical Care: A Practice Physiologic Guide.* New York: Appleton-Century-Crofts.

Erb, T. O., and W. S. Smith
1984   Validation of the attitude towards women in science scale for early adolescents. *Journal of Research in Science Teaching* 21:392–397.

Ethington, C. A., and L. M. Wolfle
1988   Women's selection of quantitative undergraduate fields of study: Direct and indirect influences. *American Educational Research Journal* 25:157–175.

Feingold, A.
1988   Matching for attractiveness in romantic partners and same-sex friends: A meta-analysis and theoretical critique. *Psychological Bulletin* 104:226–235.

Feldman, K. A.
1971   Using the work of others: Some observations on reviewing and integrating. *Sociology of Education* 44:86–102.

Festinger, L., and M. Carlsmith
1959   Cognitive consequences of forced compliance. *Journal of Abnormal and Social Psychology* 58:203, 210.

Fisher, R. A.
1935   *The Design of Experiments.* London: Oliver & Boyd.

Fiske, S. T.; D. A. Kenny; and S. E. Taylor
1982   Structural models for the mediation of salience effects on attribution. *Journal of Experimental Social Psychology* 18:105–127.

Fleiss, J. L.
1981   *Statistical Methods for Rates and Proportions.* New York: Wiley.

Foreman, M. D.
1982   Effects of relevant sensory information on recovery from carotid artery surgery. Unpublished master's thesis. Medical College of Ohio.

Fox, J.
1987   Statistical models for nonexperimental data: A comment on Freedman. *Journal of Educational Statistics* 12:161–165.

Freedman, D. A.
1987   A rejoinder on models, metaphors, and fables. *Journal of Educational Statistics* 12:206–223.

Frieze, I. H.; B. E. Whitley, Jr.; B. H. Hanusa; and M. C. McHugh
1982   Assessing the theoretical models for sex differences in causal attributions for success and failure. *Sex Roles* 8:333–343.

Garfield, S. L.
1980   *Psychotherapy: An Eclectic Approach.* New York: Wiley.

Garrett, C. J.
1984   Meta-analysis of the effects of institutional and community residential treatment on adjudicated delinquents. Unpublished doctoral dissertation, University of Colorado.
1985   Effects of residential treatment on adjudicated delinquents: A meta-analysis. *Journal of Research in Crime and Delinquency* 22:287–308.

Geiselman, R. E.; J. A. Woodward; and J. Beatty
1982   Individual differences in verbal memory performance: A test of information-processing models. *Journal of Experimental Psychology: General* 111:109–134.

Gendreau, P., and B. Ross
1979   Effective correctional treatment: Bibliotherapy for cynics. *Crime and Delinquency* 25:463–489.

Giaconia, R. M., and L. V. Hedges
1982   Identifying features of effective open education. *Review of Educational Research* 52:579–602.

Gilmartin, K. J.; D. M. McLaughlin; L. L. Wise; and R. J. Rossi
1976   *Development of Scientific Careers: The High School Years*. Palo Alto, CA: American Institutes for Research in the Behavioral Sciences. (ED 129 607)

Glass, G. V.; B. McGaw; and M. L. Smith
1981   *Meta-Analysis in Social Research*. Beverly Hills, CA: Sage.

Glymour, C., and R. Scheines
1986   Causal modelling with the TETRAD program. *Syntheses* 68:37–63.

Glymour, C.; C. Sprites; and R. Scheines
1987   *Discovering Causal Structure: Artificial Intelligence, Philosophy of Science, and Statistical Modelling*. Orlando, FL: Academic Press.

Gottfredson, M. R.
1979   Treatment destruction techniques. *Journal of Research in Crime and Delinquency* 16:39–54.

Gottschalk, R.; W. S. Davidson; L. K. Gensheimer; and J. Mayer
1987   Community-based interventions. In H. C. Quay, ed., *Handbook of Juvenile Delinquency*. New York: Wiley.

Greenberg, D. F.
1977   The correctional effects of corrections: A survey of evaluations. In D. F. Greenberg, ed., *Corrections and Punishment*. Newbury Park, CA: Sage.

Greenwald, A. G.
1975   Consequences of prejudice against the null hypothesis. *Psychological Bulletin* 82(1):1–20.

Grieb, A., and J. Easley
1984   A primary school impediment to mathematical equity: Case studies

in rule-dependent socialization. In M. W. Steinkamp and M. L. Maehr, eds., *Advances in Motivation and Achievement: Women in Science*, vol. 2. Greenwich, CT: JAI Press.

Grobman, H.
1965   Identifying the "slow learner" in BSCS high school biology. *Journal of Research in Science Teaching* 3:3–11.

Gurman, A. S., and D. P. Kniskern
1978   Research on marital and family therapy: Progress, perspective, and prospect. In S. L. Garfield and A. E. Bergin, eds., *Handbook of Psychotherapy and Behavior Change: An Empirical Analysis*, 2nd ed. New York: Wiley.

Gurman, A. S., and W. M. Pinsof
1986   Research on marital and family therapies. In S. L. Garfield and A. E. Bergin, eds., *Handbook of Psychotherapy and Behavior Change: An Empirical Analysis*, 3rd ed. New York: Wiley.

Guzzo, R. A.; S. E. Jackson; and R. A. Katzell
1987   Meta-analysis analysis. In B. M. Staw and L. L. Cummings, eds., *Research in Organizational Behavior*, vol. 9. Greenwich, CT: JAI Press.

Hahlweg, K., and H. J. Markman
1988   Effectiveness of behavioral marital therapy: Empirical studies of behavioral techniques in preventing and alleviating marital distress. *Journal of Consulting and Clinical Psychology* 56:440–447.

Hahlweg, K.; L. Schindler; D. Revenstorf; and J. C. Brengelmann
1984   The Munich marital therapy study. In K. Hahlweg and N. S. Jacobson, eds., *Marital Interaction: Analysis and Modification*. New York: Guilford Press.

Haladyna, T.; R. Olsen; and J. Shaughnessy
1982   Relations of student, teacher, and learning environment variables to attitudes toward science. *Science Education* 66:671–687.

Handley, H. M., and L. W. Morse
1984   Two-year study relating adolescents' self-concept and gender role perceptions to achievement and attitudes toward science. *Journal of Research in Science Teaching* 21:599–607.

Harris, M. J., and R. Rosenthal
1985   Mediation of interpersonal expectancy effects: 31 meta-analyses. *Psychological Bulletin* 97:363–386.

Heckmann, J. J.; V. J. Hotz; and M. Dabos
1987   Do we need experimental data to evaluate the impact of manpower training on earnings? *Evaluation Review* 11:395–427.

Hedges, L. V.
1981   Distribution theory for Glass's estimator of effect size and related estimators. *Journal of Educational Statistics* 6:107–128.

1982   Estimation of effect size from a series of independent experiments. *Psychological Bulletin* 92:490–499.

1982   Fitting continuous models to effect size data. *Journal of Educational Statistics* 7:245–270.

1984   Advances in statistical methods for meta-analysis. *New Directions for Program Evaluation* 24:25–42.

1986   Issues in meta-analysis. *Review of Research in Education* 13:353–398.

1987   How hard is hard science, how soft is soft science? The empirical cumulativeness of research. *American Psychologist* 42:443–455.

1988   The meta-analysis of test validity studies: Some new approaches. In H. Wainer and H. Braun, eds., *Test Validity*. Hillsdale, NJ: Erlbaum.

————, and I. Olkin
1985   *Statistical Methods for Meta-Analysis.* New York: Academic Press.

Hoaglin, D. C.; R. J. Light; B. McPeek; F. Mosteller; and M. A. Stoto
1982   *Data for Decisions: Information Strategies for Policy Makers.* Cambridge, MA.: Abt Books.

Hoaglin, D. C.; F. Mosteller; and J. W. Tukey
1983   *Understanding Robust and Exploratory Data Analysis.* New York: Wiley.
1985   Eds. *Exploring Data Tables, Trends, and Shapes.* New York: Wiley.

Hsu, L. M.
1989   Random sampling, randomization, and equivalence of contrasted groups in psychotherapy outcome research. *Journal of Consulting and Clinical Psychology* 57:131–137.

Hyde, J. S.
1981   How large are cognitive gender differences? A meta-analysis using ω and d. *American Psychologist* 36:892–901.

————, and M. C. Linn
1988   Gender differences in verbal ability: A meta-analysis. *Psychological Bulletin* 104:53–69.

Ignatz, M.
1982   Sex differences in predictive ability of tests of Structure-of-Intellect factors relative to a criterion examination of high school physics achievement. *Educational and Psychological Measurement* 42:353–360.

1985   In E. Chelimsky, ed., *Program Evaluation: Patterns and Directions.* Washington, DC: American Society for Public Administration.

Iyengar, S., and J. B. Greenhouse
1988   Selection models and the file drawer problem. *Statistical Science* 3:109–135.

Jackson, G.
1980   Methods for integrative reviews. *Review of Educational Research* 50:438–460.

Jacobs, J. E., and A. Wigfield
  1989   Sex equity in mathematics and science education: Research-policy links. *Educational Psychology Review* 1:39–56.

Jacobson, N. S.
  1977   Problem-solving and contingency in the treatment of marital discord. *Journal of Consulting and Clinical Psychology* 45:92–100.

———; W. C. Follette; D. Revenstorf; D. H. Baucom; K. Hahlweg; and G. Margolin
  1984   Variability in outcome and clinical significance of behavioral marital therapy: A reanalysis of outcome data. *Journal of Consulting and Clinical Psychology* 52:497–504.

Jacobson, N. S., and A. S. Gurman, eds.
  1986   *Clinical Handbook of Marital Therapy.* New York: Guilford Press.

Janis, I. L.
  1958   *Psychological Stress.* New York: Wiley.

Jennrich, R. I., and P. Sampson
  1988   General mixed model analysis of variance. In W. J. Dixon et al., eds., *BMDP Statistical Software Manual,* vol. 2. Berkeley: University of California Press.

Jensen, J. A.
  1966   An analysis by class size and sex of orthogonalized interest and aptitude predictors in relation to high school chemistry achievement criteria. Unpublished doctoral dissertation, University of Rochester.

Johnson, B. T., and A. H. Eagley
  1989   Effects of involvement on persuasion: A meta-analysis. *Psychological Bulletin* 106:290–314.

Johnson, J. E.; S. S. Fuller; M. P. Endress; and V. H. Rice
  1978a   Altering patient's response to surgery: An extension and replication. *Research in Nursing and Health* 1:111–121.
  1978b   Sensory information, instruction in a coping strategy, and recovery from surgery. *Research in Nursing and Health* 1:4–17.

Johnson, J. E.; H. Leventhal; and J. Dabbs
  1971   Contribution of emotional and instrumental response processes in adaptation to surgery. *Journal of Personality and Social Psychology* 20:55–64.

Joreskog, K. G., and D. Sorbom
  1986   *PRELIS: A Program for Multivariate Data Screening and Data Summarization.* Mooresville, IN: Scientific Software.
  1988   *LISREL 7: A Guide to the Program and Applications.* Chicago: SPSS, Inc.

Kaminski, D. M., and E. Erickson
  1979   The magnitude of sex role influence on entry into science careers. Paper presented at the meeting of the American Sociological Association, Boston. (ED 184 855)

Karlin, S.; E. C. Cameron; and R. Chakraborty
  1983   Path analysis in genetic epidemiology: A critique. *American Journal of Human Genetics* 35:695–732.

Kaufman, P.
  1985   Meta-analysis of juvenile delinquency prevention programs. Unpublished master's thesis, Claremont Graduate School.

Kavrell, S. M., and A. C. Peterson
  1984   Patterns of achievement in early adolescence. In M. W. Steinkamp and M. L. Maehr, eds., *Advances in Motivation and Achievement: Women in Science*, vol. 2. Greenwich, CT: JAI Press.

Keeves, J. P.
  1975   The home, the school, and achievement in mathematics and science. *Science Education* 59:439–460.

Kelly, A., and H. Weinreich-Haste
  1979   Science is for girls? *Women's Studies International Quarterly* 2:275–293.

Kelly, A.; J. Whyte; and B. Smail
  1984   *Girls into Science and Technology: Final Report.* Manchester, England: GIST, Department of Sociology, University of Manchester. (ED 250 203)

Kenny, D. A., and C. M. Judd
  1986   Consequences of violating the independence assumption in analysis of variance. *Psychological Bulletin* 99:422–431.

Kopeikin, H. S.; M. J. Goldstein; and V. Marshall
  1983   Stages and impact of crisis-oriented family therapy in the aftercare of acute schizophrenia. In W. R. McFarlane, ed., *Family Therapy in Schizophrenia.* New York: Guilford Press.

Kreinberg, N.
  1982   The math and science education of women and minorities: The Californian perspective. Paper presented at the Third Annual American Educational Research Association Northern California Regional Conference, University of California, Davis, May. (ED 218 114)

Kuhn, T. S.
  1962   *The Structure of Scientific Revolutions.* Chicago: University of Chicago Press.

LaLonde, R. J.
  1986   Evaluating the econometric evaluations of training programs with experimental data. *American Economic Review* 76:604–620.

Landman, J. T., and R. M. Dawes
  1982   Psychotherapy outcome: Smith and Glass' conclusions stand up under scrutiny. *American Psychologist* 37:504–616.

Lazarus, R. S.
  1966   *Psychological Stress and the Coping Process.* New York: McGraw-Hill.

1968   Emotions and adaptation: conceptual and empirical relations. In
        W. S. Arnold, ed., *Nebraska Symposium on Motivation*. Lincoln: University of Nebraska Press.

Leventhal, H.
1970   Findings and theory in the study of fear communications. In L. Berkowitz, ed., *Advances in Experimental Social Psychology*. New York: Academic Press.

———, and J. E. Johnson
1983   Laboratory and field experimentation: Development of a theory of self-regulation. In P. Wooldridge, et al., eds., *Behavioral Science and Nursing Theory*. St. Louis: Mosby.

Light, R. J., and D. B. Pillemer
1984   *Summing Up: The Science of Reviewing Research*. Cambridge, MA: Harvard University Press.

Light, R. J. and P. V. Smith
1971   Accumulating evidence: Procedures for resolving contradictions among different studies. *Harvard Educational Review*, 41:429–471.

Linn, M. C., and J. S. Hyde
1989   Gender, mathematics, and science. *Educational Researcher* 18:17–27.

Linn, M. C., and S. Pulos
1983   Male-female differences in predicting displaced volume: Strategy usage, aptitude relationships, and experience influences. *Journal of Educational Psychology* 75:86–96.

Lipsey, M. W.
1982   *Measurement Issues in the Evaluation of the Effects of Juvenile Delinquency Programs*. Washington, DC: National Institute of Justice, Office of Research and Evaluation Methods, National Criminal Justice Reference Service, Document no. NCJ-84968. (Project 80-IJ-CX-0036)
1983   A scheme for assessing measurement sensitivity in program evaluation and other applied research. *Psychological Bulletin* 94:152–165.

Lipton, D.; R. Martinson; and J. Wilks
1975   *The Effectiveness of Correctional Treatment: A Survey of Treatment Evaluation Studies*. New York: Praeger.

Little, R. J. A., and D. B. Rubin
1987   *Statistical Analysis with Missing Data*. New York: Wiley.

Locke, H. J., and K. M. Wallace
1959   Short-term marital adjustment and prediction tests: Their reliability and validity. *Journal of Marriage and Family Living* 21:251–255.

Loehlin, J. C.
1987   *Latent Variable Models: An Introduction to Factor, Path, and Structural Analysis Models*. Hillsdale, NJ: Erlbaum.

Louis, T. A.
1990   Assessing, accommodating, and interpreting the influences of heterogeneity. *Environmental Health Perspectives* 90:215–222.

———; H. V. Fineberg; and F. Mosteller
1985   Findings for public health from meta-analyses. *Annual Review of Public Health* 6:1–20.

Lundman, R. J.; P. T. McFarlane; and F. R. Scarpitti
1976   Delinquency prevention: A description and assessment of projects reported in the professional literature. *Crime and Delinquency* 22:297–308.

Mackie, J. L.
1974   *The Cement of the Universe: A Study of Causation*. Oxford, England: Clarendon Press.

MacMahon, S.; R. Peto; J. Cutler; R. Collins; P. Sorlie; J. Neaton; R. Abbott; J. Godwin; A. Dyer; and J. Stamler
1990   Blood pressure, stroke, and coronary heart disease: Part 1, prolonged differences in blood pressure: prospective observational studies corrected for the regression dilution bias. *Lancet* 335:765–774.

McMillen, L.
1987   Step up recruitment of women into science or risk U.S. competitive edge in field, colleges are warned: Affirmative action seen no longer as only a moral responsibility, but as a matter of national survival. *Chronicle of Higher Education* 33:9, 12.

Malcom, S.
1984   Equity and excellence: Compatible goals. An assessment of programs that facilitate increased access and achievement of females and minorities in K–12 mathematics and science education. Study done for the National Science Board Commission on Precollege Education in Mathematics, Science, and Technology, December.

Mann, C.
1990   Meta-analysis in the breech. *Science* 249:476–478.

Marjoribanks, K.
1976   School attitudes, cognitive ability, and academic achievement. *Journal of Educational Psychology* 75:86–96.

Marmor, J.
1958   The psychodynamics of realistic worry. *Psychoanalysis and the Social Sciences* 5:155–163.

Martin, S. E.; L. B. Sechrest; and R. Redner
1981   *New Directions in the Rehabilitation of Criminal Offenders*. Washington, DC: National Academy Press.

Martinson, R.
1974   What works? Questions and answers about prison reform. *Public Interest* 10:22–54.

Mathews, A.
  1978   Fear-reduction research and clinical phobias. *Psychological Bulletin* 85:390–404.

Matt, G. E.
  1989   Decision rules for selecting effect sizes in meta-analysis: A review and reanalysis of psychotherapy outcome studies. *Psychological Bulletin* 105:106–115.

Matthews, K. A.
  1988   Coronary heart disease and Type A behaviors: Update on and alternative to the Booth-Kewley and Friedman (1987) quantitative review. *Psychological Bulletin* 104:373–380.

Matyas, M. L.
  1985   Prediction of attrition among male and female college biology majors using specific attitudinal, socio-cultural, and traditional predictive variables. Unpublished doctoral dissertation, Purdue University.
  1986   Persistence in science-oriented majors: Factors related to attrition among male and female students. Paper presented at the annual meeting of the American Educational Research Association, San Francisco.
  1988   Intervention programs in mathematics and science for precollege females: Program types and characteristics. Report to the Board of Directors, Bush Foundation, Minneapolis, May.

Mazzuca, S. A.
  1982   Does patient education in chronic disease have therapeutic value? *Journal of Chronic Diseases* 35:521–529.

Meece, J. L.; J. E. Parsons; C. M. Kaczala; S. B. Goff; and R. Futterman
  1982   Sex differences in math achievement: Toward a model of academic choice. *Psychological Bulletin* 91:324–348.

Miller, P. A., and N. Eisenberg
  1988   The relation of empathy to aggressive and externalizing/antisocial behavior. *Psychological Bulletin* 103:324–344.

Miller, S. M.
  1979   Controllability and human stress: Methods, evidence, and theory. *Behavior Research and Therapy* 17:287–304.

Muthen, B., and A. Satorra
  1989   Multilevel aspects of varying parameters in structural models. In R. D. Bock, ed., *Multilevel Analysis of Educational Data*. San Diego: Academic Press.

National Research Council
  1983   *Climbing the Ladder: An Update on the Status of Doctoral Women Scientists and Engineers*. Washington, DC: National Academy Press.

Neale, D. C.; N. Gill; and W. Tismer
  1970   Relationship between attitudes toward school subjects and school achievement. *Journal of Educational Research* 63:232–237.

Nelson, G. K.
  1976   Concomitant effects of visual, motor, and verbal experiences in young children's concept development. *Journal of Educational Psychology* 68:466–473.

Neter, J.; W. Wasserman; and M. H. Kutner
  1983   *Applied Linear Regression Models.* Homewood, IL: Irwin.

Neuberg, S. L.
  1989   The goal of forming accurate impressions during social interactions: Attenuating the impact of negative expectancies. *Journal of Personality and Social Psychology* 56:374–386.

Nichols, M.
  1984   *Family Therapy: Concepts and Methods.* New York: Gardner Press.

Noone, J.
  1985   The effect of sensory information on postoperative pain and stress. Unpublished master's thesis, Adelphi University.

Nunnally, J. C.
  1978   *Psychometric Theory,* 2nd ed. New York: McGraw-Hill.

O'Connor, F. W.; E. C. Devine; T. D. Cook; V. A. Wenk; and T. R. Curtin
  1989   Enhancing surgical nurses' patient education: Development and evaluation of an intervention. Submitted for publication.

Olkin, I., and M. Siotani
  1976   Asymptotic distribution of functions of a correlation matrix. In S. Ikeda, ed., *Essays in Probability and Statistics.* Tokyo: Sinko Tsusho.

Orchen, M. D.
  1983   A treatment efficacy study comparing relaxation training, emg biofeedback, and family therapy among heavy drinkers. *Dissertation Abstracts International* 4:2565B. (University Microfilms No. 83-27825)

Ormerod, M. B.
  1975   Single sex and co-education: An analysis of pupils' science preference and choice and their attitudes to other aspects of science under these two systems. Conference at the Centre for Science Education, Chelsea College, London, England.

Orwin, R. G., and D. S. Cordray
  1985   Effects of deficient reporting on meta-analysis: A conceptual framework and reanalysis. *Psychological Bulletin* 97(1):134–147.

Palmer, T.
  1975   Martinson revisited. *Journal of Research in Crime and Delinquency* 12:133–152.

1983   The effectiveness issue today: An overview. *Federal Probation* 47: 3–10.

Parker, K. C. H.; R. K. Hanson; and J. Hunsley
1988   MMPI, Rorschach, and WAIS: A meta-analytic comparison of reliability, stability, and validity. *Psychological Bulletin* 103:367–373.

Pearson, K.
1904   Report on certain enteric fever inoculations. *British Medical Journal* 2:1243–1246.

Peng, S. S., and J. Jaffe
1979   Women who enter male-dominated fields of study in higher education. *American Educational Research Journal* 16:285–293.

Pennsylvania State Department of Education
1984   *Science Unlimited: Pennsylvania's Resource Guide for Elementary Science.* Harrisburg, PA: Pennsylvania State Department of Education.

Popper, K. R.
1959   *The Logic of Scientific Discovery.* New York: Basic Books.
1972   *Objective Knowledge: An Evolutionary Approach.* Oxford, England: Clarendon Press.

Premack, S. L., and J. E. Hunter
1988   Individual unionization decisions. *Psychological Bulletin* 103:223–234.

Raudenbush, S. W.
1984   Magnitude of teacher expectancy effects on pupil IQ as a function of the credibility of expectancy induction: A synthesis of findings from 18 experiments. *Journal of Educational Psychology* 76:85–97.
1988   Educational applications of hierarchical linear models: A review. *Journal of Educational Statistics* 13:85–116.

———; B. J. Becker; and H. Kalaian
1988   Modeling multivariate effect sizes. *Psychological Bulletin* 103:111–120.

Raudenbush, S. W., and A. S. Bryk
1985   Empirical Bayes meta-analysis. *Journal of Educational Statistics* 10:75–98.

Resnick, D. P.
1980   Minimum competency testing historically considered. *Review of Research in Education* 8:3–29.

Richter, M. L., and M. B. Seay
1987   ANOVA designs with subjects and stimuli as random effects: Applications to prototype effects on recognition memory. *Journal of Personality and Social Psychology* 53:470–480.

Roberts, M. W.; R. J. McMahon; R. Forehand; and L. Humphreys
1978   The effect of parent instruction-giving on child compliance. *Behavior Therapy* 9:793–798.

Roberts, R. J.
1965   Prediction of college performance of superior students. *National Merit Scholarship Research Reports* 1:11–19.

Robinson, L. A.; J. S. Berman; and R. A. Neimeyer
1990   Psychotherapy for the treatment of depression: A comprehensive review of controlled outcome research. *Psychological Bulletin* 108:30–49.

Rogosa, D.
1980   A critique of cross-lagged correlation. *Psychological Bulletin* 88:245–258.

Romig, D.
1978   *Justice for Our Children.* Lexington, MA: Lexington Books.

Rosenthal, R.
1973   The mediation of Pygmalion effects: A four-factor "theory." *Papua New Guinea Journal of Education* 9:1–12.
1974   *On the Social Psychology of the Self-fulfilling Prophecy: Further Evidence for Pygmalion Effects and Their Mediating Mechanisms.* New York: MSS Modular Publication, Module 53.
1978   The "file drawer problem" and tolerance for null results. *Psychological Bulletin* 86:638–641.
1984   *Meta-analytic Procedures for Social Research.* Beverly Hills, CA: Sage.
1989   Experimenter expectancy, covert communication, and meta-analytic methods. Paper presented at the annual meeting of the American Psychological Association, New Orleans, August.

————, and D. B. Rubin
1982   A simple, general purpose display of magnitude of experimental effect. *Journal of Educational Psychology* 74:166–169.

Rosnow, R. L., and R. Rosenthal
1989   Statistical procedures and the justification of knowledge in psychological science. *American Psychologist* 44:1276–1284.

Rothman, A. I.; W. W. Welch; and H. J. Walberg
1969   Physics teacher characteristics and student learning. *Journal of Research in Science Teaching* 6:59–63.

Rubin, D. B.
1987   *Multiple Imputation for Nonresponse in Surveys.* New York: Wiley.
1990   A new perspective. In K. Wachter and M. Straf, eds., *The Future of Meta-Analysis.* New York: Russell Sage Foundation.

Sacks, H. S., et al.
1987   Meta-analysis of randomized controlled trials. *New England Journal of Medicine* 316(8):450–455.

SAS Institute
1979   *SAS User's Guide 1979 Edition.* Cary, NC: SAS Institute.

Schiebinger, L.
  1987 The history and philosophy of women in science: A review essay. *Signs: Journal of Women in Culture and Society* 12:305–332.

Schluchter, M. D.
  1988 Unbalanced repeated measures models with structured covariance matrices. In W. J. Dixon et al., eds., *BMDP Statistical Software Manual*, vol. 2. Berkeley: University of California Press.

Schock, N. H.
  1973 An analysis of the relationship which exists between cognitive and affective educational objectives. *Journal of Research in Science Teaching* 10:299–315.

Science Council of Canada
  1981 *Who Turns the Wheel?* Ottawa: Science Council of Canada.

Searleman, A.; C. Porac; and S. Coren
  1989 Relationship between birth order, birth stress, and laternal preferences: A critical review. *Psychological Bulletin* 105:397–408.

Sechrest, L.; S. G. West; M. Phillips; R. Redner; and W. Yeaton
  1979 Some neglected problems in evaluation research: Strength and integrity of treatments. In L. Sechrest and Associates, eds., *Evaluation Studies Review Annual*, vol. 4. Newbury Park, CA: Sage.

Sechrest, L. B.; S. O. White; and E. D. Brown
  1979 *The Rehabilitation of Criminal Offenders: Problems and Prospects.* Washington, DC: National Academy of Sciences.

Shadish, W. R.
  1989a Critical multiplism: A research strategy and its attendant tactics. In L. Sechrest, H. Freeman, and A. Mulley, eds., *Health Services Research Methodology: A Focus on AIDS.* Rockville, MD: National Center for Health Services Research and Health Care Technology Assessment, Public Health Service, U.S. Department of Health and Human Services. DHHS Publication no. (PHS) 89-3439.

  1989b Methodological findings from a family therapy meta-analysis. American Evaluation Association, San Francisco, October.

———; M. Doherty; and L. M. Montgomery
  1989 How many studies are in the file drawer? An estimate from the family/marital psychotherapy literature. *Clinical Psychology Review* 9:589–603.

Shadish, W. R., and R. B. Sweeny
  In press Mediators and moderators in meta-analysis: There's a reason we don't let dodo birds tell us which psychotherapies should have prizes. *Journal of Consulting and Clinical Psychology.*

Shapiro, A. K., and L. A. Morris
  1978 The placebo effect in medical and psychological therapies. In S. L.

Garfield and A. E. Bergin, eds., *Handbook of Psychotherapy and Behavior Change: An Empirical Analysis*, 2nd ed. New York: Wiley.

Shapiro, A. K.; L. M. Montgomery; P. Wilson; M. R. Wilson; I. Bright; and T. M. Okwumabua
1989   Marital/family therapy effectiveness: meta-analysis of 163 randomized trials. American Psychological Association, New Orleans, August.

Siegel, S.
1956   *Nonparametric Statistics for the Behavioral Sciences.* New York: McGraw-Hill.

Sime, A. M.
1976   Relationship of preoperative fear, type of coping, and information received about surgery to recovery from surgery. *Journal of Personality and Social Psychology* 34:716–724.

Smail, B., and A. Kelly
1984b   Sex differences in science and technology among 11-year-old school-children: I—Cognitive. II–Affective. *Research in Science and Technological Education* 2:87–106.

Smith, I. R.
1966   Factors in chemistry achievement among eleventh-grade girls and boys. Unpublished doctoral dissertation, Catholic University of America.

Smith, M. L., and G. V. Glass
1977   Meta-analysis of psychotherapy outcome studies. *American Psychologist* 32:752–760.

———, and T. I. Miller
1980   *The Benefits of Psychotherapy.* Baltimore: Johns Hopkins University Press.

Steier, F.
1983   Family interaction and properties of self-organizing systems: A study of family therapy with addict families. *Dissertation Abstracts International* 44:863A. (University Microfilms no. 83-16093)

Steinkamp, M. W., and M. L. Maehr
1983   Affect, ability, and science achievement: A quantitative synthesis of correlational research. *Review of Educational Research* 53:369–396.
1984   Gender differences in motivational orientations toward achievement in school science: A quantitative synthesis. *American Educational Research Journal* 21:39–60.

Stelzl, I.
1986   Changing a causal hypothesis without changing the fit: Some rules for generating equivalent path models. *Multivariate Behavioral Research* 21:309–331.

Stigler, S. M.
1986   *The History of Statistics: The Measurement of Uncertainty before 1900.* Cambridge, MA: Harvard University Press.

Stromsdorfer, E. W.
  1987   Economic evaluation of the Comprehensive Employment and Train-
          ing Act: An overview of recent findings and advances in evaluation
          methods. *Evaluation Review* 11:387–394.

Subcommittee on Science Research and Technology, U.S. House of Represen-
tatives
  1982   *Symposium on Minorities and Women in Science and Technology.* Serial
          AA. Washington, DC: SSRT. (ED 221 378)

Swim, J.; E. Borgida; G. Maruyama; and D. G. Myers
  1989   Joan McKay versus John McKay: Do gender stereotypes bias evalua-
          tions? *Psychological Bulletin* 105:409–429.

Taveggia, T.
  1974   Resolving research controversy through empirical cumulation. *Socio-
          logical Methods and Research* 2:385–407.

Thomas, G. E.
  1981   *Choosing a College Major in the Hard and Technical Sciences and the Profes-
          sions: A Causal Explanation.* Report no. 313. Baltimore: Johns Hopkins
          University Center for the Social Organization of Schools. (ED 206
          829)

Thompson, S. C.
  1981   Will it hurt less if I can control it? A complex answer to a simple
          question. *Psychological Bulletin* 90:89–101.

U.S. General Accounting Office
  1986   *Teenage Pregnancy: 500,000 Births Yet Few Tested Programs.* Washing-
          ton, DC: U.S. General Accounting Office, Program Evaluation Divi-
          sion.

University Microfilms International
  1987   *Education: A Catalogue of Dissertations.* Ann Arbor, MI: University Mi-
          crofilms International.

Urkowitz, A. G., and R. E. Laessig
  1985   Assessing the believability of research results reported in the envi-
          ronmental health matrix.

Van Harlingen, D. L.
  1981   Cognitive factors and gender related differences as predictors of per-
          formance in an introductory level college physics course. Unpub-
          lished doctoral dissertation, Rutgers University.

Vernon, D. T., and D. A. Bigelow
  1974   Effect of information about a potentially stressful situation on re-
          sponses to stress impact. *Journal of Personality and Social Psychology*
          29:50–59.

Wachter, K. W.
  1988   Disturbed by meta-analysis. *Science* 1407–1408.

Wamboldt, F. S.; M. Z. Wamboldt; and A. S. Gurman
  1985   Marital and family therapy research: The meaning for the clinician.
          In L. L. Andreozzi, ed., *Integrating Research and Clinical Practice*, Fam-
          ily Therapy Collections, vol. 15. Rockville, MD: Aspen Systems Pub-
          lications.

Ware, N. C.; N. A. Steckler; and J. Leserman
  1985   Undergraduate women: Who chooses a science major? *Journal of Higher
          Education* 56:73–84.

Weimer, L. J.
  1985   Sex differences in achievement beliefs of bright children. Unpub-
          lished doctoral dissertation, University of Washington.

Weiss, C. H.
  1980   Knowledge creep and decision accretion. *Knowledge: Creation, Diffu-
          sion, Utilization* 1:381–404.
  1987   The circuitry of enlightenment. *Knowledge: Creation, Diffusion, Utili-
          zation* 8:274–281.

Welch, W. W.; S. J. Rakow; and L. J. Harris
  1984   Women in science: Perceptions of secondary school students. Paper
          presented at the annual meeting of the National Association for Re-
          search in Science Teaching, New Orleans, April.

White, P. A.
  1990   Ideas about causation in philosophy and psychology. *Psychological
          Bulletin* 108:3–18.

Whitehead, J. T., and S. P. Lab
  1989   A meta-analysis of juvenile correctional treatment. *Journal of Research
          in Crime and Delinquency* 26(3):276–295.

Williams, J. R., and M. Gold
  1972   From delinquent behavior to official delinquency. *Social Problems*
          20:209–229.

Wilson, G. T., and S. J. Rachman
  1983   Meta-analysis and the evaluation of psychotherapy outcome: Limi-
          tations and liabilities. *Journal of Consulting and Clinical Psychology* 51:54–
          64.

Wolf, F.
  1986   *Meta-Analysis.* Beverly Hills, CA: Sage.

Wolfer, J. A., and C. E. Davis
  1970   Assessment of surgical patient's preoperative emotional condition and
          postoperative welfare. *Nursing Research* 19:402–414.

Wood, W.; N. Rhodes; and M. Whelan
  1989   Sex differences in positive well-being: A consideration of emotional
          style and marital status. *Psychological Bulletin* 106:249–264.

Wright, S.
1983   On "Path Analysis in Genetic Epistemology: A Critique." *American Journal of Human Genetics* 35:757–768.

Wright, W. E., and M. C. Dixon
1977   Community prevention and treatment of juvenile delinquency: A review of evaluation studies. *Journal of Research in Crime and Delinquency* 14:35–67.

Yates, F., and W. G. Cochran
1938   The analysis of groups of experiments. *Journal of Agricultural Science* 28:556–580.

Index

Entries in **boldface** refer to figures and tables

A

academic achievement, 18–20, 334; and aptitude, 245–247, 265; measures, 229; and past events, 250–251, **250,** 265; and social and psychological factors, 218–220; and task value, 247, 250; and teacher expectancy studies, 322–323. *See also* science

academic choice model, 218–222, **219**

affect category, defined, 133

"aggregate" comparisons, 91

aggression, measuring, 2

agricultural experiments, 6

alcoholism, 152

Alexander's functional model, 130

Alwin, D. F., 188

ambulation after surgery, 59, 60

analysis and interpretation stage, **9,** 11–12

analysis of covariance structures (ACS) fit tests, 173n

Anderson, G. J., 217

Andrews, D. A., 86, 121, 123, 126

ANOVA, 139n, 168, 185, 314

Anson, A., 26

anxiety after surgery, 53, 54–55

Archuleta, V., 52

A Scientific Literacy Test (ASLT), 275

Astin, H. S., 225

attribution theory, 273

attrition, 91, 120, 148

Averill, J. R., 38

B

Baker, D. R., 240

Bandura, A., 38

Baucom, D. H., 138

Beatty, J., 188

Becker, B. J., 15, 19, 32, 180, 184, 210, 217, 220n, 226, 233, 272, 285–286, 287, 289, 291, 293, 299–300, 302, 303, 304–305, 314, 318, 319, 328, 329, 330, 331–332, 333, 334, 339–340, 341

Beecher, H. K., 7

Begg, C. B., 318

behavioral orientation: defined, 129–130; outcomes, compared with other orientations, 129–189

behavior category, defined, 133

Bem, D. J., 21

Bendixen, H., 37

Bentler, P. M., 173, 174, 180, 181

Berlin, J. A., 318

Berman, J. S., 131, 153, 185, 187

Bernard, M. E., 224

Bernstein, S., 64

between-studies variation, 31, 32, 52, 62, 169, 170, 231–232, 307–313, 314–315, 325–328, 329–330, 332–333; model specification tests in, 154; treatment-control comparisons, 153–154, 186

Bhaskar, R., 18

bias, 3, 17, 24, 43, 52, 125, 295, 317, 325, 337; between-experiment, in physical

367